T0350118

Supersymmetric Quantum Mechanics

An Introduction

Second Edition

Supersymmetric Quantum Mechanics

An Introduction

Second Edition

Asim Gangopadhyaya
Jeffry Mallow
Constantin Rasinariu

Loyola University Chicago, USA

World Scientific

NEW JERSEY · LONDON · SINGAPORE · BEIJING · SHANGHAI · HONG KONG · TAIPEI · CHENNAI · TOKYO

Published by

World Scientific Publishing Co. Pte. Ltd.

5 Toh Tuck Link, Singapore 596224

USA office: 27 Warren Street, Suite 401-402, Hackensack, NJ 07601

UK office: 57 Shelton Street, Covent Garden, London WC2H 9HE

Library of Congress Cataloging-in-Publication Data
Names: Gangopadhyaya, Asim, author. | Mallow, Jeffry V., author. | Rasinariu, Constantin, author.
Title: Supersymmetric quantum mechanics : an introduction / Asim Gangopadhyaya
    (Loyola University Chicago, USA), Jeffry Mallow (Loyola University Chicago, USA),
    Constantin Rasinariu (Loyola University Chicago, USA).
Description: Second edition. | Singapore ; Hackensack, NJ : World Scientific, [2017] |
    Includes bibliographical references and index.
Identifiers: LCCN 2017032702| ISBN 9789813221031 (hardcover ; alk. paper) |
    ISBN 9813221038 (hardcover ; alk. paper) | ISBN 9789813221048 (pbk. ; alk. paper) |
    ISBN 9813221046 (pbk. ; alk. paper)
Subjects: LCSH: Supersymmetry. | Quantum theory.
Classification: LCC QC174.17.S9 G36 2017 | DDC 530.12--dc23
LC record available at https://lccn.loc.gov/2017032702

British Library Cataloguing-in-Publication Data
A catalogue record for this book is available from the British Library.

Printed in Singapore

DEDICATION

AG: TO MY WIFE, ALPANA.

JVM: TO MY WIFE AND SON, ANN AND DAVID MALLOW,
AND TO THE MEMORY OF MY SISTER, BRONA
NEVILLE-NEIL.

CR: TO MY WIFE, MARIANA.

Preface

We have written this book in order to provide a single compact source for undergraduate and graduate students, as well as for professional physicists who want to understand the essentials of supersymmetric quantum mechanics. It is an outgrowth of a seminar course taught to physics and mathematics juniors and seniors at Loyola University Chicago, and of our own research over a quarter of a century. These research results appear in several chapters. They include the following:

Gangopadhyaya and Mallow found that all known $\hbar$- independent "shape invariant" potentials were solutions of a non-linear partial differential equation (NPDE). This provided a new way to determine those potentials, which we have dubbed "conventional," and to prove that the family of them was complete. Subsequently, Gangopadhyaya, Mallow and Rasinariu, joined by a new colleague, Jonathan Bougie, generalized this.

Other researchers found several new exactly solvable potentials that were extensions of some of the conventional ones. These have come to be known as extended potentials. Bougie, Gangopadhyaya, and Mallow showed that these were $\hbar$-dependent. We have changed the first edition's chapter, Generating Shape Invariant Potentials, to reflect this.

We have added a chapter on Phase Space Quantization (PSQ), an alternate path to quantum mechanics and its connection to SUSYQM studied, among others, by Rasinariu. In PSQ formalism, hamiltonians constrained by shape invariance are rendered solvable, just as in the case of canonical quantization.

The book covers a wide range of topics, some in non-traditional order. For example, we first review those parts of quantum mechanics that segue into SUSYQM. We show that some apparently unrelated potentials have almost identical spectra. We do this in order to emphasize the essential symmetry connecting different hamiltonians. One such example is that of the infinite square well and the less well-known $\operatorname{cosec}^2 x$ potential, with the startling near-identity of their eigenvalues. We also investigate the harmonic oscillator, the "proto-supersymmetric" system, as it might be designated. We defer to a later chapter, concepts such as measurements and expectation values, taught in conventional quantum mechanics courses.

We then examine more advanced topics. These include selected applications of SUSYQM, such as generating orthogonal polynomials by shape invariance; the supersymmetric version of the WKB approximation (SWKB); isospectral deformations of the conventional shape invariant potentials; the supersymmetric Quantum Hamilton-Jacobi formalism (QHJ); and deformation quantization. These topics are especially relevant both for the advanced student who is interested in gaining insight into current SUSYQM research, and for the professional physicist who is looking for a good reference source. The degree of complexity in the book slowly increases as it morphs from a text accessible to advanced undergraduates into a research monograph for professional physicists.

Each of the chapters contains problems. All are designed to give the reader hands-on experience with SUSYQM. We have provided solutions at the end.

Finally, we have compiled an extensive bibliography to help the reader orient herself/himself to the vast literature that is available. The list is by no means exhaustive, and we apologize for any omissions.

We have corrected errors on the first edition, and hopefully not introduced any more.

Acknowledgments

AG: I would like to acknowledge support from many that made this project possible. My parents who, through their massive sacrifices, instilled in me the value of learning. My mentors, Quraishi Guruji, Ratna K. Thakur, Carl M. Shakin and Uday P. Sukhatme who helped shape my academic life. My friend, Aleksandr Goltsiker, without his constant encouragement the project would not have seen light of the day. My children - for their love and their irreverence that has helped me keep a clearer perspective. Most importantly my wife for her infinite patience and love which has sustained my growth.

JVM: My wife Ann is a constant source of inspiration and support. In addition, her copy-editing of the book for language and style has greatly improved it, and her curiosity about its contents has put me through my pedagogic paces. For all of this, I am grateful.

CR: I would like to thank my mentors Miron Bur, Dan Albu, Bernhard Rothenstein, Henrik Aratyn, and Uday P. Sukhatme for their guidance. My parents Traian and Paraschiva Răşinariu for their constant encouragement to learn, and my children and wife, for their infinite love and patience.

Chicago, June 2017

Contents

Chapter 1

Introduction

Obtaining exact solutions to the Schrödinger equation (and its relativistic counterpart, the Dirac equation) has always been a focus of quantum mechanical studies.

One feature of interest was the factorization method for obtaining solutions. A hamiltonian can be rewritten as a product of two factors, usually called "raising and lowering operators." The method replaces the need to directly solve the Schrödinger equation, a second order differential equation, with the capability to solve a first order equation. Schrödinger himself noticed this, and provided a way to factorize the hamiltonian for the hydrogen atom and other potentials in 1941.[1] A decade later, Infeld and Hull generalized this to numerous other systems[2] (a set of systems now known as "shape invariant potentials"). All of these observations turned out to be hidden manifestations of an underlying symmetry, subsequently explained by supersymmetric quantum mechanics.

The next impetus came in the early 1980s, when elementary particle physicists attempting to find an underlying structure for the basic forces of nature, proposed the existence of "shadow" partner particles to the various known (or conjectured) elementary particles. Thus, to the photon is conjectured the photino; to the quark-binding gluon, the gluino; to the as-yet-unobserved graviton, the gravitino; and to the W particle, the wino (pronounced *wee-no*!). These particle partnerships and their interrelationships comprise what is called supersymmetry.

Supersymmetry with its new mathematical apparatus soon led to investigations of partnerships in other areas of physics. One was ordinary

---

[1]E. Schrödinger, "Further studies on solving eigenvalue problems", *Proc. Roy, Irish Acad.*, **46A**, 183–206 (1941).

[2]L. Infeld and T.E. Hull, "The factorization method", *Rev. Mod. Phys.* **23**, 21–68 (1951).

quantum mechanics itself. First employed as a so-called "toy model" of field theory,[3] supersymmetric quantum mechanics, based on the notion of "partner potentials" derivable from an underlying "superpotential," was born. It was soon found to have value in its own right, with application to the resolution of numerous questions in quantum mechanics and the posing of various interesting new ones.

In this chapter, we shall illustrate how certain unusual features of some well-known examples of the Schrödinger equation (and one less well-known) suggest underlying symmetries. In Chapter 2, we will work out the examples the conventional way, by direct solution of the Schrödinger equation. In later chapters, these examples will be revisited and solved more elegantly, to illustrate the ideas of supersymmetric quantum mechanics.

But first and foremost, let us replace the cumbersome "supersymmetric quantum mechanics" with the acronym SUSYQM, pronounced "Suzy-Cue-Em."

The time dependent Schrödinger equation is

$$H(\mathbf{r},t)\Psi(\mathbf{r},t) \equiv -\frac{\hbar^2}{2m}\,\nabla^2\Psi(\mathbf{r},t) + V(\mathbf{r},t)\,\Psi(\mathbf{r},t) = i\hbar\,\frac{\partial}{\partial t}\Psi(\mathbf{r},t)\,, \quad (1.1)$$

with $H(\mathbf{r},t)$ the hamiltonian of the system.

All of the problems that we shall consider will involve time-independent one-dimensional potentials: either $V(x)$ in Cartesian coordinates, or $V(r)$, a radially symmetric potential in spherical coordinates. For any time-independent potential $V(\mathbf{r})$, the time dependence and space dependence can be separated:

$$\Psi(\mathbf{r},t) = \psi(\mathbf{r})f(t)\,.$$

This yields

$$-\frac{\hbar^2}{2m}\,\frac{\nabla^2\psi(\mathbf{r})}{\psi(\mathbf{r})} + V(\mathbf{r}) = i\hbar\,\frac{df(t)/dt}{f(t)} = E\,, \quad (1.2)$$

where E is a constant, since $\mathbf{r}$ and t are independent variables.

Problem 1.1. *Obtain Eq. (1.2).*

Solving the time dependent part

$$\frac{i\hbar}{f(t)}\,\frac{df(t)}{dt} = E\,,$$

---

[3]E. Witten, "Dynamical breaking of supersymmetry", *Nucl. Phys.* B **185**, 513–554 (1981).

we obtain $f(t) = e^{-iEt/\hbar}$, while the space-dependent part gives the time-independent Schrödinger equation:

$$H(\mathbf{r})\psi(\mathbf{r}) \equiv -\frac{\hbar^2}{2m}\nabla^2\psi(\mathbf{r}) + V(\mathbf{r})\psi(\mathbf{r}) = E\psi(\mathbf{r}) . \qquad (1.3)$$

From the wave-mechanical definition of linear momentum: $\mathbf{p} \equiv -i\hbar\nabla$, the first term on the left is the quantum extension of the kinetic energy $p^2/2m$, the second term is the extension of the potential energy; thus, E must be the total energy of the system. In one dimension,

$$-\frac{\hbar^2}{2m}\frac{d^2\psi(x)}{dx^2} + V(x)\psi(x) = E\psi(x) . \qquad (1.4)$$

Let us consider the first example: the infinite square well in one dimension, $V(x) = 0$, $0 < x < L$ infinite elswhere. The energy eigenvalues are found to be

$$E_n = n^2\hbar^2/2mL^2 , \qquad n = 1, 2, 3, \dots , \qquad (1.5)$$

where m is the particle's mass and L is the width of the well.

Now let us look at another problem, less well-known: the potential $V(x) = \frac{\hbar^2}{2m}\left(2\operatorname{cosec}^2\frac{\pi x}{L} - 1\right)$, $0 < x < L$. Why we chose this particular form will become evident when we compare it to the infinite well. The energy eigenvalues turn out to be

$$E_n = [(n+2)^2 - 1]\hbar^2/2mL^2 , \qquad n = 0, 1, 2, \dots . \qquad (1.6)$$

We notice something remarkable about the energy spectrum. It is virtually identical to that of the infinite well, except for the starting value of $n = 0$ rather than 1, the shift from n^2 to $(n+2)^2$, and the constant shift -1. These distinctions are independent of the physics: we could shift the infinite well's bottom from $V = 0$ to $V = -\frac{\hbar^2}{2mL^2}$; we could just as easily start the infinite well count at $n = 0$. Then the energy spectrum for the infinite well would be given by

$$E_n = [(n+1)^2 - 1]\hbar^2/2mL^2 , \qquad n = 0, 1, 2, \dots . \qquad (1.7)$$

Thus, for some reason these two very different potentials yield virtually identical spectra. This may be a signal that there is a hidden symmetry connecting them.

Our next example is the harmonic oscillator: $V(x) = \frac{1}{2}m\omega^2 x^2$. The energy turns out to be

$$E_n = \left(n + \frac{1}{2}\right)\hbar\omega , \qquad n = 0, 1, 2, \dots . \qquad (1.8)$$

The full solution of the Schrödinger equation for the harmonic oscillator, which we shall work out in Chapter 2, is quite cumbersome. This is surprising, given how symmetric the hamiltonian is: quadratic in both momentum $p = -i\hbar\frac{d}{dx}$ and position x. It would be nice if we could factor it into two linear terms. Unfortunately, the fact that x and $\frac{d}{dx}$ do not commute seems to thwart this program. But this factorization approach is more powerful for determining the eigenenergies than is the direct solution of the Schrödinger equation. Indeed, it will prove to be the first step toward SUSYQM. We shall work it out in Section 3.4.

We have seen that the infinite well and the cosec^2 potential yield similar energy spectra. They will turn out to be each others' supersymmetric partner potentials. This, as we shall see, will provide us with a way for generating the eigenfunctions of one potential from those of the other, *provided that we know one of them already.* The harmonic oscillator is a special case, partnering with itself, and generating a single set of eigenfunctions. Its importance here resides in its being the first and simplest application of the factorization method that leads to SUSYQM.

Next we will introduce the concept of "shape invariance," which will allow us to obtain the complete energy spectrum of a single potential without knowing *a priori* the spectrum of its partner. *We will do all of this without ever directly solving the Schrödinger differential equation.*

We mentioned earlier that SUSYQM is not restricted to the Schrödinger equation. We shall examine the Dirac equation in Chapter 13.

Chapter 2

Quantum Mechanics and Clues to SUSYQM

In this chapter, we shall work out the details of the following four examples: the infinite well, the cosec^2 potential (also called Pöschl-Teller I, or trigonometric Pöschl-Teller potential), the harmonic oscillator, and the hydrogen atom. We shall do this by direct solution of the Schrödinger equation. We want to demonstrate the complexity (and tediousness) of this approach, so that it may be compared with the more elegant factorization method of supersymmetry in subsequent chapters.

Quick Facts Review

- The time-independent Schrödinger equation in one dimension is

$$-\frac{\hbar^2}{2m}\frac{d^2\psi(x)}{dx^2} + V(x)\psi(x) = E\psi(x) \ . \qquad (2.1)$$

- $|\psi(x)|^2 dx$ represents the probability of finding a particle between x and $x + dx$. If we consider only those cases where particles do not decay, the probability of finding them somewhere must be 1. I.e., the wave function is normalized so that

$$\int_{-\infty}^{\infty} |\psi(x)|^2 dx = 1 \ . \qquad (2.2)$$

- In addition, $\psi(x)$ must be continuous over its entire range. We also usually insist that $\frac{d\psi(x)}{dx}$ be continuous, in order that the term $\frac{d^2\psi(x)}{dx^2}$ in the Schrödinger equation be well-defined. This constraint is suspended when $V(x)$ has a singularity.

Now let us turn to our four examples.

Example 1: The Infinite Square Well

Let us take the well to stretch from 0 to L. Then

$$V(x) = \begin{cases} \infty , & x < 0 \\ 0 , & 0 \leq x \leq L \\ \infty , & x > L \end{cases} .$$

Outside the well, $V(x) = \infty$ requires that $\psi(x) = 0$. Inside the well, since the potential is 0,

$$-\frac{\hbar^2}{2m}\frac{d^2\psi(x)}{dx^2} = E\psi(x) . \tag{2.3}$$

The solution is $\psi(x) = A \sin kx + B \cos kx$, where $k \equiv \sqrt{2mE}/\hbar$.

The wave function must be continuous in order to insure that the probability is meaningful across a barrier; thus, ψ must vanish at the boundaries. For $x = 0$ we have $\psi(0) = 0$, which implies $B = 0$; viz., $\psi(x) = A \sin kx$. Then at the other boundary, $\psi(L) = A \sin kL = 0$. This means that $kL = n\pi$, where $n = 0, 1, 2, \ldots$. The value $n = 0$ is excluded by the normalization condition. We get $k_n = n\pi/L$, thus quantizing the energy:

$$E_n = \hbar^2 k_n^2/2m = n^2\pi^2\hbar^2/2mL^2 , \; n = 1, 2, 3 \ldots$$

The corresponding wave functions are

$$\psi_n(x) = A \sin(n\pi x/L) .$$

Since $\psi(x) = 0$ outside the well, the normalization integral need only be taken from 0 to L:

$$\int_0^L |\psi_n(x)|^2 dx = \int_0^L A^2 \sin^2(n\pi x/L)dx = 1 .$$

Thus, $A = \sqrt{2/L}$.

Summarizing

$$\psi_n(x) = \sqrt{\frac{2}{L}} \sin\left(\frac{n\pi x}{L}\right) , \tag{2.4}$$

$$E_n = \frac{n^2\pi^2\hbar^2}{2mL^2} . \tag{2.5}$$

Now let us make some useful simplifications. The first simplification is to choose physical units such that $2m = 1$ and $\hbar = 1$. (This we will continue to do throughout the book, unless we specify otherwise.) The second is to choose a length unit for the well, $L = \pi$. Then our solutions for the wave function and energy simplify:

$$\psi_n(x) = \sqrt{\frac{2}{\pi}} \, \sin nx \tag{2.6}$$

$$E_n = n^2 \,, \qquad n = 1, 2, 3, \ldots \tag{2.7}$$

The potential and first three wave functions and the corresponding eigenenergies are illustrated in Fig. 2.1.

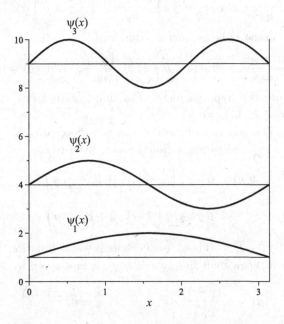

Fig. 2.1 The infinite well potential for $L = \pi$ and its first three energy levels and wave functions.

Example 2: The Trigonometric Pöschl-Teller Potential

The potential is given by

$$V(x) = 2\mathrm{cosec}^2 x - 1 \,, \quad 0 < x < \pi \,. \tag{2.8}$$

To solve the corresponding Schrödinger equation[1]

$$-\frac{d^2\psi(x)}{dx^2} + \left(2\,\text{cosec}^2 x - 1\right)\psi(x) = E\psi(x)\,, \qquad (2.9)$$

we change variables to $y = \sin^2\left(\frac{x}{2}\right)$ and $\psi(x(y)) = \frac{(-1+y)}{\sqrt{y}}\,\phi(y)$. While the domain of x is given by $(0, \pi)$, it is now mapped to $(0, 1)$; i.e., $(0 < y < 1)$. The new wave function $\phi(y)$ is related to our original wave function by

$$\phi(y) = -\frac{\sin\left(\frac{x}{2}\right)}{\cos^2\left(\frac{x}{2}\right)}\,\psi(x)\,.$$

These substitutions lead to:

$$y\,(1-y)\,\frac{d^2\phi(y)}{dy^2} + \left(-\frac{1}{2} - 2\,y\right)\frac{d\phi(y)}{dy} - \left(-\frac{3}{4} - E\right)\phi(y)\,. \qquad (2.10)$$

This equation is the hypergeometric equation[2]:

$$y\,(1-y)\,\frac{d^2\phi(y)}{dy^2} + (c - (a+b+1)\,y)\,\frac{d\phi(y)}{dy} - ab\,\phi(y) = 0\,. \qquad (2.11)$$

Its solutions are the hypergeometric functions $_2F_1(a, b, c, y)$. Comparing Eqs. (2.10) and (2.11), we identify $a = \frac{1}{2} - \sqrt{1+E}$, $b = \frac{1}{2} + \sqrt{1+E}$, and $c = -\frac{1}{2}$. The general solution of this equation, consisting of a linear combination of two independent functions, is given by

$$\phi(y) = \alpha\ _2F_1\left(\frac{1}{2} - k, \frac{1}{2} + k, -\frac{1}{2}, y\right) \qquad (2.12)$$

$$+ \beta\, y^{\frac{3}{2}}\ _2F_1\left(2 - k, 2 + k, \frac{5}{2}, y\right),$$

where $k = \sqrt{1+E}$. Thus, our original wave function $\psi(x(y)) = \frac{(-1+y)}{\sqrt{y}}\,\phi(y)$, which we shall simply call $\psi(y)$, is now given by

$$\psi(y) = \alpha\,\frac{(y-1)}{\sqrt{y}}\ _2F_1\left(\frac{1}{2} - k, \frac{1}{2} + k, -\frac{1}{2}, y\right) \qquad (2.13)$$

$$+ \beta\, y(y-1)\ _2F_1\left(2 - k, 2 + k, \frac{5}{2}, y\right),$$

---

[1]At this point, we would like to caution especially our undergraduate readers, that the traditional method of solving the Schrödinger equation, which we shall now carry out, leads to a general solution involving hypergeometric functions, which may not be familiar to you. If you do not find it very instructive at this stage, feel free to skip the details of this example at this time. Do, however, look at the final answers. In Chapter 4 we will again solve this problem using the much more accessible methods of SUSYQM. You may revisit this example at that time for a comparative study of the two methods.

[2]For a nice discussion of this, see Arfken and Weber, *Mathematical Methods for Physicists*, 5th edition, San Diego: Harcourt-Academic Press (2001).

where α and β are constants to be determined. The first term in this solution diverges at $y = 0$ (i.e., at $x = 0$), which will yield an unphysical infinite probability. So the only way this solution can be well behaved is if $\alpha = 0$. With this choice, $\psi(y)$ reduces to

$$\psi(y) = y(y - 1) \, _2F_1 \left(2 - k, 2 + k, \frac{5}{2}, y \right).$$

We wish to apply the boundary condition $\psi(x(y)) = 0$ at $y = 1$, where $V \to \infty$. But we cannot immediately do so, as the hypergeometric function becomes infinite at $y = 1$. We need to investigate the behavior of the entire function $\psi(y) = \frac{(-1+y)}{\sqrt{y}} \phi(y)$. We use an identity that extracts the essence of this singularity,

$$_2F_1 (a, b, c, y) = \frac{\Gamma(c) \Gamma(c - a - b)}{\Gamma(c - a) \Gamma(c - b)} \, _2F_1 (a, b, a + b - c + 1, 1 - y)$$

$$+ (1 - y)^{c-a-b} \frac{\Gamma(c) \Gamma(a + b - c)}{\Gamma(a) \Gamma(b)} \, _2F_1 (c - b, c - a, c - a - b + 1, 1 - y).$$

Γ denotes the Gamma function.[3] These hypergeometric functions are no longer singular when $y \to 0$ as the argument $(1 - y)$ goes to zero. The singularity is contained in the term $(1 - y)^{c-a-b}$. Thus, for $\psi(y)$, we obtain

$$\psi(y) = y(y - 1) \frac{\Gamma\left(\frac{5}{2}\right) \Gamma\left(-\frac{3}{2}\right) \, _2F_1 \left(2 - k, 2 + k, \frac{5}{2}, 1 - y\right)}{\Gamma\left(\frac{1}{2} - k\right) \Gamma\left(\frac{1}{2} + k\right)}$$

$$- \frac{y}{\sqrt{1 - y}} \frac{\Gamma\left(\frac{5}{2}\right) \Gamma\left(\frac{3}{2}\right) \, _2F_1 \left(\frac{1}{2} - k, \frac{1}{2} + k, -\frac{1}{2}, 1 - y\right)}{\Gamma(2 + k) \Gamma(2 - k)}. \tag{2.14}$$

The second term is infinite at $y = 1$ due to $1 - y$ in the denominator. The hypergeometric function is 1 when its variable argument, in this case $(1-y)$, is zero. For this wave function to be well behaved at $x = \pi$, therefore, the second term must vanish. This is possible only if

$$\Gamma(2 + k) \Gamma(2 - k) \to \infty,$$

which occurs when the arguments of $\Gamma(2 + k)$, or $\Gamma(2 - k)$ equal a negative integer or zero. Thus, we set $2 \pm k \equiv 2 \pm \sqrt{1 + E} = -n$, where $n = 0, 1, 2, \ldots$, thus quantizing the energy.[4]

Solving for E, we get:

$$E_n = (n + 2)^2 - 1 \qquad n = 0, 1, 2, \ldots. \tag{2.15}$$

---

[3] $\Gamma(x)$ behaves like $\frac{(-1)^n}{n!} \frac{1}{x+n}$ near $x = -n$, $n = 0, 1, 2, \ldots$. For positive integer n, $\Gamma(n + 1) = n!$.

[4] Careful analysis, which we gloss over, shows that the same condition that makes the second term in Eq. (2.14) vanish also truncates the first term to a polynomial of order n. Otherwise the wave functions would diverge.

If we investigate the hypergeometric function for, say, the lowest two states: $n = 0$ and 1, we find that it simplifies, yielding $\psi_0(x) \sim \sin^2 x$ and $\psi_1(x) \sim \sin x \, \sin 2x$.

Figure 2.2 shows the potential and the first two energy levels and wave functions. Note the complicated effort expended to obtain what appear

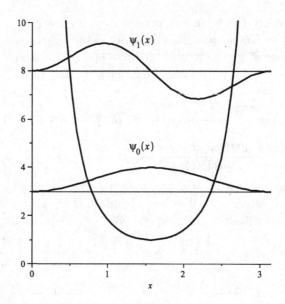

Fig. 2.2   The $2\,\mathrm{cosec}^2 x - 1$ potential and its first two energy levels and wave functions.

to be rather simple wave functions, and a very simple expression for the energies.

If we now go back to the infinite well spectrum, Eq. (2.6), and redefine $n \mapsto n + 1$, so that n starts at 0 instead of 1, and shift its bottom to $V(x) = -1$ instead of 0, neither of which affects the physics, then its energy spectrum is given by

$$E_n = (n+1)^2 - 1 \qquad n = 0, 1, 2, \ldots . \qquad (2.16)$$

This means that the infinite well and the cosec^2 potentials yield virtually identical spectra, except for the infinite well's extra state: its ground state $n = 0$, with energy E_0 now equal to 0. Why this coincidence occurs will be revealed by SUSYQM.

Example 3: The Harmonic Oscillator

The harmonic oscillator $(V(x) = \frac{1}{2}m\omega^2 x^2$ in ordinary units), is arguably the most important potential in quantum mechanics. It approximates any attractive potential in a narrow region near the minimum, as illustrated in Fig. 2.3. When we consider realistic systems, we observe that any smooth attractive potential $V(x)$ has a minimum corresponding to the equilibrium position. In this small region around the minimum, the potential can be

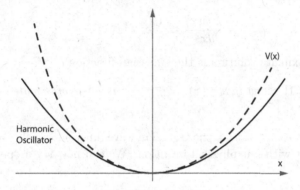

Fig. 2.3   For small x -regions near the minimum, the harmonic oscillator potential approximates any arbitrary attractive potential $V(x)$.

approximated by its Taylor series expansion

$$V(x) = V(0) + \frac{dV}{dx}\bigg|_{x=0} x + \frac{1}{2}\frac{d^2V}{dx^2}\bigg|_{x=0} x^2 + \dots , \qquad (2.17)$$

where x is the distance from the minimum. That minimum, the equilibrium position, has $dV/dx|_{x=0} = 0$. We can always set the constant term $V(0) = 0$, since this does not change the physics. Thus, the potential energy of any such system can be approximated as

$$V(x) = \frac{1}{2}k x^2 , \qquad (2.18)$$

where we define $\frac{d^2V}{dx^2}\big|_{x=0} \equiv k = m\omega^2$. This is the harmonic oscillator potential, which, in our convention ($\hbar = 1$, $2m = 1$), becomes

$$V(x) = \frac{1}{4}\omega^2 x^2 . \qquad (2.19)$$

Finding the exact solutions for the harmonic oscillator potential will give us insight into the solutions of more complicated potentials. The Schrödinger equation becomes

$$-\frac{d^2\psi(x)}{dx^2} + \frac{1}{4}\,\omega^2\,x^2\psi(x) = E\psi(x) \ . \tag{2.20}$$

Let us first analyze its asymptotic solutions at large x. When $|x| \to \infty$, the potential $V(x) \to \infty$ as well, and we expect two things: that the wave functions $\psi(x)$ will vanish and that the energy term E will be negligible with respect to the x^2 term. The latter condition means that at large $|x|$, Eq. (2.20) becomes

$$\frac{d^2\psi(x)}{dx^2} \approx \frac{1}{4}\,\omega^2\,x^2\psi(x) \ , \tag{2.21}$$

whose approximate solution is the Gaussian function $e^{-\omega x^2/4}$.

Problem 2.1. *Show that* $\psi(x) = e^{-\omega x^2/4}$ *is an approximate solution of Eq. (2.21).*

We observe that indeed, the Gaussian wave function vanishes at infinity, in agreement with our physical intuition. We can now try a "peeling-off" process:

$$\psi(x) = C\,e^{-\omega x^2/4}\,v(x) \ , \tag{2.22}$$

where C is a normalization constant, to be determined later. Thus, we have "peeled off" the Gaussian, and need only find the remaining function $v(x)$. We must keep in mind, however, that $v(x)$ cannot rise faster than $e^{\omega x^2/4}$, as $x \to \pm\infty$, otherwise $\psi(x)$ will diverge.

Substituting Eq. (2.22) into (2.20) and making a simplifying change of variables

$$\xi \equiv x\,\sqrt{\omega/2} \ , \ \ K \equiv 2\,E/\omega \ , \tag{2.23}$$

we obtain

$$\frac{d^2v(\xi)}{d\xi^2} - 2\,\xi\,\frac{dv(\xi)}{d\xi} + (K-1)\,v(\xi) = 0 \ . \tag{2.24}$$

Problem 2.2. *Do the explicit calculations to derive Eq. (2.24).*

We can solve this differential equation using the series method. We express the function $v(x)$ as a power series

$$v(\xi) = \sum_{j=0}^{\infty} v_j\,\xi^j \ , \tag{2.25}$$

where the coefficients v_j are to be chosen such that Eq. (2.24) is identically satisfied for each power of ξ. We now need to substitute Eq. (2.25) into Eq. (2.24). We obtain $dv(\xi)/d\xi = \sum_{j=0}^{\infty} j v_j \xi^{j-1}$, whence $2\xi v(\xi) = 2\sum_{j=0}^{\infty} j v_j \xi^j$. (The first term is 0, but that has no effect on the calculation.) For the second derivative, we obtain $d^2 v(\xi)/d\xi^2 = \sum_{j=0}^{\infty} j(j-1)v_j \xi^{j-2}$. We observe that the first two terms of this series are 0, so we may rewrite it starting with the third term: $\sum_{j=2}^{\infty} j(j-1)v_j \xi^{j-2}$ or, with a change of index, $\sum_{j=0}^{\infty}(j+2)(j+1)v_{j+2}\xi^j$. Now all three terms in Eq. (2.24) have the same powers of ξ:

$$\sum_{j=0}^{\infty} \left[(j+2)(j+1)v_{j+2} - 2jv_j + (K-1)v_j\right]\xi^j = 0 . \qquad (2.26)$$

This can be satisfied if and only if all the coefficients of ξ^j vanish. This leads to the following recursion relation:

$$v_{j+2} = \frac{2j+1-K}{(j+1)(j+2)} v_j . \qquad (2.27)$$

Note that this relation separates the solutions into those of odd and even parity. Given the initial coefficient v_0 we can find all even v_j; given v_1 we can calculate all odd v_j. If $v_0 = 0$ then from Eq. (2.27) we obtain that $v_2 = v_4 = \ldots = 0$. Then in the series expansion for v, Eq. (2.25), we are left with only the odd powers of ξ : $v(\xi) = v_1\xi + v_3\xi^3 + \ldots$. Similarly, if $v_1 = 0$, then $v_3 = v_5 = \ldots = 0$, and the series expansion of v will contain only even terms: $v(\xi) = v_0 + v_2\xi^2 + v_4\xi^4 + \ldots$.

It would seem that a general solution could be a mix of odd and even terms, provided we knew *a priori* the coefficients v_0 and v_1. But before we worry about this, we need to see if each of these series converges, and if so, to what. In particular, we must insure that when multiplied by the peeled-off Gaussian $e^{-\omega x^2/4}$, $v(x)$ gives a normalizable $\psi(x) = C e^{-\omega x^2/4} v(x)$.

For $j \gg 1$ and K, the recursion relation Eq. (2.27) may be approximated by

$$v_{j+2} \approx \frac{2}{j}v_j . \qquad (2.28)$$

First we need to establish that the approximate series is an upper bound of the exact series, Eq. (2.27), for if it isn't, then we would not know if the exact series gives a normalizable ψ, even if the approximate series does. We can in fact show that the approximate coefficients $\frac{2}{j}$ exceed the exact coefficients $\frac{2j+1-K}{(j+1)(j+2)}$ for every j. Since the potential well's bottom is at 0, all energies must be positive; thus, $K \equiv 2E/\omega > 0$. Consequently

$$\frac{2j+1-K}{(j+1)(j+2)} < \frac{2j+1}{(j+1)(j+2)} < \frac{2j+2}{(j+1)(j+2)} = \frac{2}{j+2} < \frac{2}{j} .$$

Thus the approximate series overestimates the exact series. So if the approximate series converges we are assured that the exact series will converge as well. The coefficients which obey the approximate recursion relation are $v_j \approx C/(j/2)!$.

Problem 2.3. *Confirm that this relation yields Eq. (2.28).*

Then the approximate series is $v(\xi) \approx C \sum_{j=0}^{\infty} \frac{1}{(j/2)!} \xi^j$. Defining $j' \equiv 2j$, $v(\xi) \approx C \sum_{j'=0}^{\infty} \frac{1}{j'!} \xi^{2j'} \approx C e^{\xi^2}$. So unfortunately the approximate series converges to $e^{\xi^2} \equiv e^{\omega x^2/2}$, which overwhelms the decaying Gaussian $e^{-\omega x^2/4}$, producing a divergent wave function at $\pm\infty$. Therefore, we need $v(\xi)$ to converge fast enough that when multiplied by the decaying Gaussian, it produces a normalizable wave function. But this approximation must be good at every ξ. In particular, it must hold for very large ξ. Now note from Eqs. (2.26) and (2.27) that the ratio of successive exact terms $v_{j+2}\xi^{j+2}/v_j \xi^j = \frac{2j+1-K}{(j+1)(j+2)}\xi^2$. Thus, as $\xi \to \infty$, the larger j terms increasingly overwhelm the smaller ones, making the large-j approximate series more and more accurate. This means that the approximate and exact series converge to each other as $\xi \to \infty$. Therefore, the exact series behaves as badly as its approximation.

The only way out of this is to make sure that the exact series will always be overwhelmed by the decaying Gaussian. We accomplish this by truncating the series at some finite index n.

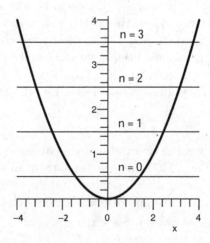

Fig. 2.4   The first four energy levels of the harmonic oscillator for $\omega = 1$.

This can be done if in Eq. (2.27) we require that at some $j_{max} \equiv n$, $v_{j_{max}+2} = 0$. This will be the case if

$$K = 2n + 1 . \tag{2.29}$$

We note now that the value of $n \equiv j_{max}$ will be odd or even, depending on whether the series is odd or even. Thus, for example, an odd n leaves the even series untruncated, converging to the prohibited $e^{\omega x^2/2}$, and conversely for even n. So the solutions for a given energy must be of either odd or even parity, with the leading term of the other parity, v_0 or v_1, set to 0.

From Eq. (2.23) we see that the constraint $K = 2n + 1$ quantizes the energy:

$$E_n = \left(n + \frac{1}{2}\right)\omega , \quad n = 0, 1, 2, \ldots \tag{2.30}$$

The energy levels are equally spaced by ω (in $\hbar = 1$ units) as illustrated in Fig. 2.4.

We now know that the solution to the Eq. (2.20) is given by the product of the Gaussian function $e^{-\omega x^2/4}$ and the solution of Eq. (2.24) when we substitute for K the value $2n + 1$. Equation (2.24) thus becomes

$$\frac{d^2 v(\xi)}{d\xi^2} - 2\xi \frac{dv(\xi)}{d\xi} + 2n\, v(\xi) = 0 , \quad n = 0, 1, 2, \ldots \tag{2.31}$$

This is the Hermite equation, whose solutions are the Hermite polynomials. The first four polynomials are

$$H_0(\xi) = 1, \; H_1(\xi) = 2\xi, \; H_2(\xi) = 4\xi^2 - 2, \; H_3(\xi) = 8\xi^3 - 12\xi . \tag{2.32}$$

Hermite polynomials can be generated directly using the Rodrigues differential formula

$$H_n(\xi) = (-1)^n e^{\xi^2} \frac{d^n}{d\xi^n}\left(e^{-\xi^2}\right) . \tag{2.33}$$

Using this, we can show by direct manipulation that they have the following properties

$$H_n' = 2nH_{n-1} , \quad H_{n+1} - 2\xi H_n + 2nH_{n-1} = 0 ,$$
$$H_n' = 2nH_{n-1} , \quad H_{n+1} = 2\xi H_n - H_n' , \tag{2.34}$$

where the prime represents the derivative with respect to ξ.

Problem 2.4. *Use the Rodrigues formula to confirm the values of the first four Hermite polynomials, H_0 to H_3, in Eq. (2.32), and to find the fifth, H_4. Test that the recursion relations Eq. (2.34) are obeyed by the polynomials in Eq. (2.32).*

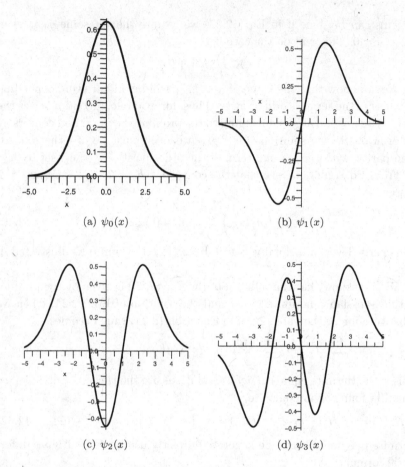

Fig. 2.5   Normalized harmonic-oscillator wave functions for $\omega = 1$ and quantum numbers $n = 0, 1, 2, 3$.

The constant C from Eq. (2.22) can be determined from the normalization condition $\int_{-\infty}^{\infty} |\psi_n|^2 \, dx = 1$. We obtain $C \equiv C_n = \left(\frac{\omega}{2}\right)^{\frac{1}{4}} [\sqrt{\pi}\, n!\, 2^n]^{-\frac{1}{2}}$.

Putting these results together, we conclude that the wave function $\psi_n(x)$ and the corresponding energy E_n for the harmonic oscillator are given by

$$\psi_n(x) = \left(\frac{\omega}{2}\right)^{\frac{1}{4}} [\sqrt{\pi}\, n!\, 2^n]^{-\frac{1}{2}} e^{-\omega x^2/4} H_n(x\sqrt{\omega/2}) \,,$$

$$E_n = \left(n + \frac{1}{2}\right) \omega \,, \quad n = 0, 1, 2, \ldots \tag{2.35}$$

The eigenfunctions $\psi_n(x)$ for $n = 0, 1, 2, 3$ are shown in Fig. 2.5.

Before we conclude this section we should make one very important observation. In Eq. (2.34) we saw that the Hermite polynomials obey certain recursion relations. These relations, in terms of the wave functions ψ_n, become:

$$\sqrt{n+1}\,\psi_{n+1} - \sqrt{2\omega}\,x\,\psi_n + \sqrt{n}\,\psi_{n-1} = 0 , \tag{2.36}$$

$$\psi'_n + \frac{1}{\sqrt{2}}\,\omega\,x\psi_n = \sqrt{n\omega}\,\psi_{n-1} , \tag{2.37}$$

$$-\psi'_n + \frac{1}{\sqrt{2}}\,\omega\,x\,\psi_n = \sqrt{(n+1)\,\omega}\,\psi_{n+1} , \tag{2.38}$$

where prime represents the derivative with respect to x. That is, some rather simple operations relate ψ_n to its neighbors ψ_{n-1} and ψ_{n+1}.

Problem 2.5. *Use the recursion relations for the Hermite polynomials, Eq. (2.34), to prove Eqs. (2.36), (2.37) and (2.38).*

The simplicity of the solution for the infinite well seemed commensurate with the amount of work expended to obtain it. This is not true for the the harmonic oscillator (as was the case for the cosec^2 problem.) The work seems quite excessive compared to the simplicity of the solutions. There is an easier way: the so-called factorization method to which we alluded in the introduction. It provided the first example of what subsequently led to supersymmetric quantum mechanics. In the next chapter, we shall work out the harmonic oscillator problem using the factorization method.

Problem 2.6. *Modify the Schrödinger equation for the harmonic oscillator, Eq. (2.20), so that $E_0 = 0$. Argue that this does not change the wave functions. (We will soon see that we always want to design potentials in SUSYQM so that the ground state energy is 0. We did this for the infinite well in Eq. (2.16)).*

Problem 2.7. *Use your results from the infinite well and the harmonic oscillator to obtain the wave functions and energies for the "half-harmonic oscillator":*

$$V(x) = \begin{cases} \infty & , \quad x < 0 \\ \frac{1}{4}\,\omega^2 x^2 & , \quad x \geq 0 \end{cases} .$$

(Hint: The well shape is that of the harmonic oscillator; the infinite wall constitutes a boundary condition on the wave functions. You can get the answer by just considering the symmetries of the problem.)

Example 4: The Hydrogen Atom

The hydrogen atom problem is also known as the Coulomb problem. The standard treatment is in spherical coordinates. The time-independent Schrödinger equation in three dimensions for the Coulomb potential[5] is

$$H(\mathbf{r})\psi(\mathbf{r}) \equiv -\nabla^2\psi(\mathbf{r}) - \frac{e^2}{|\mathbf{r}|}\psi(\mathbf{r}) = E\psi(\mathbf{r}) \tag{2.39}$$

or,

$$-\frac{1}{r^2}\frac{\partial}{\partial r}\left(r^2\frac{\partial\psi}{\partial r}\right) - \frac{1}{r^2\sin\theta}\frac{\partial}{\partial\theta}\left(\sin\theta\frac{\partial\psi}{\partial\theta}\right) - \frac{1}{r^2\sin^2\theta}\left(\frac{\partial^2\psi}{\partial\phi^2}\right) - \frac{e^2}{r}\psi = E\psi. \tag{2.40}$$

Since the Coulomb potential is independent of angle, we can write the angular dependence as a single function $Y(\theta,\phi)$, and the wave function as $\psi(r,\theta,\phi) = R(r)Y(\theta,\phi)$, where only $R(r)$ depends on the potential. The Schrödinger equation becomes

$$Y(\theta,\phi)\frac{1}{r^2}\frac{d}{dr}\left(r^2\frac{dR(r)}{dr}\right) + R(r)\frac{1}{r^2\sin\theta}\frac{\partial}{\partial\theta}\left(\sin\theta\frac{\partial Y(\theta,\phi)}{\partial\theta}\right) \tag{2.41}$$

$$+ R(r)\frac{1}{r^2\sin^2\theta}\left(\frac{\partial^2 Y(\theta,\phi)}{\partial\phi^2}\right) = \left(E + \frac{e^2}{r}\right)R(r)Y(\theta,\phi).$$

Multiplying by $-r^2/R(r)Y(\theta,\phi)$ separates the radial and angular parts:

$$\left[\frac{1}{R(r)}\frac{d}{dr}\left(r^2\frac{dR(r)}{dr}\right) + r^2\left(\frac{e^2}{r} + E\right)\right] \tag{2.42}$$

$$+ \frac{1}{Y(\theta,\phi)}\left[\frac{1}{\sin\theta}\frac{\partial}{\partial\theta}\left(\sin\theta\frac{\partial Y(\theta,\phi)}{\partial\theta}\right) + \frac{1}{\sin^2\theta}\left(\frac{\partial^2 Y(\theta,\phi)}{\partial\phi^2}\right)\right] = 0.$$

Each of the terms must be constant. Presciently calling that constant $l(l+1)$, we obtain

$$\frac{1}{R(r)}\frac{d}{dr}\left(r^2\frac{dR(r)}{dr}\right) + r^2\left(\frac{e^2}{r} + E\right) = l(l+1); \tag{2.43}$$

$$\frac{1}{Y(\theta,\phi)}\frac{1}{\sin\theta}\left[\frac{\partial}{\partial\theta}\left(\sin\theta\frac{\partial Y(\theta,\phi)}{\partial\theta}\right) + \frac{1}{\sin\theta}\left(\frac{\partial^2 Y(\theta,\phi)}{\partial\phi^2}\right)\right] = -l(l+1). \tag{2.44}$$

The angular solutions, which are the same for all radially dependent V are the spherical harmonics $Y_l^m(\theta,\phi)$. We will not concern ourselves with them here, except to note that they require that l be a non-negative integer. Note that the radial equation alone determines the energy.

---

[5]In our unit system, $4\pi\epsilon_0$ is equal to 1.

With the substitution $u(r) = R(r)r$ the radial equation reduces to

$$-\frac{d^2u}{dr^2} + \left[\frac{-e^2}{r} + \frac{l(l+1)}{r^2}\right] u = Eu . \tag{2.45}$$

Problem 2.8. *Check that this substitution yields Eq. (2.45) from Eq. (2.43).*

Note that the effective potential is composed of two terms: the Coulomb attraction $-e^2/r$, plus a centrifugal repulsion $l(l+1)/r^2$.

Since $u(r)$ in Eq. (2.45), due to normalizabilty, must vanish as $r \to \infty$, E must be negative. We define $\kappa \equiv \sqrt{-E}$, and divide Eq. (2.45) by $-\kappa^2$. Separating the derivative term on the left side gives

$$\frac{1}{\kappa^2}\frac{d^2u}{dr^2} = \left[1 - \frac{e^2}{\kappa^2 r} + \frac{l(l+1)}{\kappa^2 r^2}\right] u . \tag{2.46}$$

We now make two additional simplifying substitutions: $\rho \equiv \kappa r$ and $\rho_0 \equiv e^2/\kappa$. This yields

$$\frac{d^2u}{d\rho^2} = \left[1 - \frac{\rho_0}{\rho} + \frac{l(l+1)}{\rho^2}\right] u . \tag{2.47}$$

In a peel-off procedure analogous to that for the harmonic oscillator, we examine the approximate behavior of the equation at the asymptotes $\rho \to 0$ and $\rho \to \infty$. As $\rho \to \infty$, the first term dominates, giving

$$\frac{d^2u}{d\rho^2} \approx u . \tag{2.48}$$

The solution is $u \sim Ae^{-\rho} + Be^{\rho}$. We set $B = 0$; otherwise, u would diverge[6] as $\rho \to \infty$. So $u \sim Ae^{-\rho}$ at the large distance limit.

Now we consider the approximate behavior of u near the origin ($\rho \to 0$). In this case the dominant term is the centrifugal one: $\frac{l(l+1)}{\rho^2}$, for $l > 0$. (We shall consider the special case of $l = 0$ shortly.) In that case

$$\frac{d^2u}{d\rho^2} \approx \frac{l(l+1)}{\rho^2} u . \tag{2.49}$$

The solution is $u \sim C\rho^{l+1} + D\rho^{-l}$.

Problem 2.9. *Derive the above solution of Eq. (2.49).*

---

[6]Note that $\int u^2 dr = \int R^2 r^2 dr$ and $\int |\psi|^2 d^3r = \int Y_l^m(\theta, \phi)^2 \sin^2\theta d\theta d\phi \int R^2 r^2 dr$. So if u diverges, then $\int |\psi|^2 d^3r$ diverges, which would yield an infinite probability.

We must set $D = 0$ in order that u not diverge at 0. So $u \sim C\rho^{l+1}$.

We are now ready for the peel-off: $u \sim \rho^{l+1}e^{-\rho}v(\rho)$. We substitute this into Eq. (2.47) and obtain, after some work, the following equation for v:

$$\rho\frac{d^2v}{d\rho^2} + 2(l+1-\rho)\frac{dv}{d\rho} + [\rho_0 - 2(l+1)]v = 0 . \qquad (2.50)$$

Problem 2.10. *Work out the details, obtaining this equation by direct substitution of $u \sim \rho^{l+1}e^{-\rho}v(\rho)$ into Eq. (2.47).*

As before we try a series expansion for v:

$$v = \sum_{j=0}^{\infty} v_j\rho^j .$$

Then $dv/d\rho = \sum_{j=0}^{\infty} jv_j\rho^{j-1}$, which, since the leading term is 0, is the same as $\sum_{j=0}^{\infty}(j+1)v_{j+1}\rho^j$.

The second derivative is

$$\frac{d^2v}{d\rho^2} = \sum_{j=0}^{\infty} j(j+1)v_{j+1}\rho^{j-1} .$$

Then Eq. (2.50) becomes

$$\rho\frac{d^2v}{d\rho^2} + 2(l+1-\rho)\frac{dv}{d\rho} + [\rho_0 - 2(l+1)]v$$

$$= \sum_{j=0}^{\infty} \{j(j+1)v_{j+1} + (2l+1)(j+1)v_{j+1} - 2jv_j$$

$$+ [\rho_0 - 2(l+1)]\,v_j\}\,\rho^j = 0 . \qquad (2.51)$$

This can be satisfied if and only if all the coefficients of ρ^j vanish. This leads to the following recursion relation between the coefficients v_j:

$$v_{j+1} = \frac{2(j+l+1) - \rho_0}{(j+1)(j+2l+1)}\,v_j . \qquad (2.52)$$

Now we need to examine the behavior of this recursion relation. We need to assure ourselves that $v(\rho)$ does not overwhelm ρ^{l+1} near 0, nor $e^{-\rho}$ near ∞, otherwise u would diverge.

The $\rho = 0$ limit is $u \sim \rho^{l+1}\sum_{j=0}^{\infty} v_j\rho^j$. Since l and j are both positive, there is no possibility of a singularity for small ρ.

The $\rho = \infty$ limit *is* a problem. For large ρ, large values of j might produce a strongly divergent v, one that might dominate the decaying exponential $e^{-\rho}$, analogous to what we saw for the harmonic oscillator. We therefore examine the behavior of the recursion relation for large values of j. For $j \gg l$ Eq. (2.52) becomes

$$v_{j+1} \approx \frac{2}{j} v_j . \tag{2.53}$$

Problem 2.11. *Confirm that this relation yields Eq. (2.53).*

Problem 2.12. *In analogy with what we did for the harmonic oscillator, show that the large-j approximate series coefficients, Eq. (2.53), overestimate the exact series coefficients, Eq. (2.52).*

Substituting Eq. (2.53) into Eq. (2.51) we obtain

$$v \approx v_0 + \rho \sum_{j=1}^{\infty} \frac{2^{j-1}}{(j-1)!} v_0 \rho^{j-1} = v_0 \left[1 + \rho \sum_{k=0}^{\infty} \frac{(2\rho)^k}{(k)!} \right] , \tag{2.54}$$

where $k \equiv j - 1$. At large ρ,

$$v \approx v_0 \rho \, e^{2\rho} .$$

Then the function $u \sim \rho^{l+1} e^{-\rho} v(\rho) \to v_0 \rho^{l+2} e^{\rho}$, which diverges as $\rho \to \infty$. The only way to eliminate this problem is to truncate the exact series at some j_{max}, such that $v_{j_{max}+1} = 0$. From Eq. (2.52) this can be accomplished by setting $2(j_{max} + l + 1) - \rho_0 = 0$. We define the "principal quantum number" $n \equiv j_{max} + l + 1$. (Note that this requires $l \leq n - 1$.) Then the truncation requirement is $\rho_0 = 2n$. This gives us the quantized energy levels. From $\rho_0 \equiv e^2/\kappa$ and $\kappa \equiv \sqrt{-E}$, the energies are

$$E_n = -\kappa_n^2 = -\frac{e^4}{4n^2} . \tag{2.55}$$

(This equals $-13.6 \, eV/n^2$ in atomic units.)

Problem 2.13. *Work out the details for Eq. (2.55).*

What about $l = 0$? In this case, Eq. (2.47) becomes

$$\frac{d^2 u}{d\rho^2} = \left[1 - \frac{\rho_0}{\rho} \right] u . \tag{2.56}$$

In the $\rho \to 0$ limit

$$\frac{d^2 u}{d\rho^2} \approx -\frac{\rho_0}{\rho} u . \tag{2.57}$$

We are tempted to assume (as some texts do) that $u \sim \rho^{l+1}$ is fine for $l = 0$; viz., $u = C\rho$. If we succumb to this temptation, we find that the left hand side of Eq. (2.58) goes to 0 while the right hand side approaches a constant: $-C\rho_0$. So, we instead resort to Mathematica and find the general solution of Eq. (2.58) to be

$$u \approx \sqrt{\rho_0\rho}\,(C_1 BesselJ\,[1, 2\sqrt{\rho_0\rho}] + C_2 BesselY\,[1, 2\sqrt{\rho_0\rho}])\,. \qquad (2.58)$$

For small ρ, the first term is linear in ρ and the second term diverges. Hence, for both $\ell = 0$ and for $\ell \neq 0$ the behavior near the origin is given by $\rho^{\ell+1}$. The rest of the analysis is the same as for the $l > 0$ case.

The truncated v series is the set of associated Laguerre polynomials $L_{-n-l-1}^{2l+1}\left(\frac{2r}{na}\right)$, and the full radial function $R(r) = u(r)/r$ is given by

$$R_{nl}(r) = N_{nl}(a)e^{-r/na}\left(\frac{2r}{na}\right)^l L_{-n-l-1}^{2l+1}\left(\frac{2r}{na}\right)\,. \qquad (2.59)$$

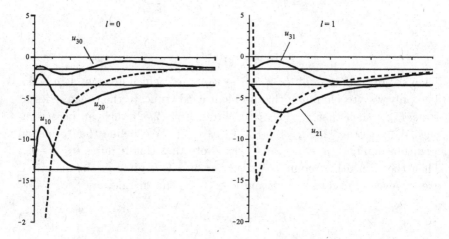

Fig. 2.6 The effective Coulomb potential for $l = 0$ and $l = 1$ and some energy levels and wavefunctions.

Figure 2.6 shows the effective potential including the centrifugal term for $l \geq 0$, the first three energy levels, and $u = rR$ wave functions. The full wave function is $\psi_{nlm}(r, \theta, \phi) = R_{nl}(r)Y_l^m(\theta, \phi)$.

We have used the definition $\rho \equiv \kappa r$. We have then used the standard definition of the Bohr radius a: $a \equiv 1/n\kappa$ to get the the function arguments r/na. (We have, as is customary, also absorbed powers of 2 into the arguments.) $N_{nl}(a)$ is the normalization constant.

Problem 2.14. *The radial functions $R_{nl}(r)$ are dependent on l. For example,*

$$R_{2s}(r) = \frac{1}{2a}\sqrt{\frac{2}{a}} \left(1 - \frac{r}{2a}\right) e^{-r/2a} \;\; ; \;\; R_{2p}(r) = \frac{1}{2a^2}\sqrt{\frac{1}{6a}} \; re^{-r/2a}$$

a) *In spherical coordinates, the probability density $P(r)$ of finding the electron within dr of r is proportional to $r^2 R^2(r)$. Calculate the value of radius r at which you are most likely to find a 2s and a 2p electron.*

b) *The labels s, p, etc. designate different values of angular momentum. Treating the values you have just found as the classical values of the radii of the electron orbits, explain why Bohr's quantization rule $L = n\hbar$ is inconsistent with classical mechanics. Hint: Relate energy to angular momentum for a circular orbit. That is what Bohr did.*

Problem 2.15. $R_{2s}(r) = \frac{1}{2a}\sqrt{\frac{2}{a}} \left(1 - \frac{r}{2a}\right) e^{-r/2a}$. *Recall that $u \equiv rR$. We found that for the s states ($l = 0$), $u \sim \left[\rho - n\rho^2\right]$ as $\rho \to 0$. Confirm this for u_{2s}. Hint: Be careful to keep all terms of the same order.*

This completes the analysis of our four examples using brute force solution of the Schrödinger equation. We hope that this chapter has been instructive, and tedious enough to whet your appetite for the more elegant SUSYQM approach.

Chapter 3

Operator Formalism in Quantum Mechanics

3.1 Operators and Representations

As we discussed in Chapter 1, the identification of the momentum with a differential operator led to the correspondence between the first term in the Schrödinger equation and the kinetic energy: $\mathbf{p} \equiv -i\hbar\nabla \rightarrow p^2/2m \equiv \frac{-\hbar^2}{2m}\nabla^2\Psi$, in our convention simply $-\nabla^2\Psi$. For simplicity we restrict ourselves to one dimension. p and x do not commute. The former is a differential operator, the latter simply a multiplier. Explicitly,

$$p\,x\,\psi(x) = -i\frac{d}{dx}x\psi(x) = -ix\frac{d}{dx}\psi(x) - i\psi(x) \,, \qquad (3.1)$$

which we can write as $(xp - p\,x)\psi(x) = i\psi(x)$, where the term in parentheses is called the commutator of x and p, and written $[x, p]$. More elegantly, we write this as an operator equation: $[x, p] = i$; i.e. applying the commutator of x and p to any wave function reproduces an extra term: i times that wave function.[1]

If we use the momentum space p rather than the coordinate space x, then the wave functions would become $\tilde{\phi}(p)$. In this case the momentum operator p would be a multiplier, and x would be the differential operator: $x \equiv i\frac{d}{dp}$. But the operation $[x, p]$ on $\tilde{\phi}$ would still yield $i\tilde{\phi}$. This indicates that the operator formalism is more general. It does not depend on which representation we use.

The hamiltonian in one-dimensional coordinate space is an operator comprising a differential operator $-\frac{d^2}{dx^2}$ and a multiplicative operator $V(x)$: $H = -\frac{d^2}{dx^2} + V(x)$.

---

[1]But one must take care to keep the function in while performing the operations, otherwise the extra term will tend to get lost.

3.2 Dirac Notation

Dirac invented a notation which not only simplified writing down the equations of quantum mechanics, but also was independent of representation. Its elements are brackets $\langle\ \rangle$, and he separated them into a left side $\langle\ |$ called a "bra" and a right side $|\ \rangle$ called a "ket". The kets represent the wave functions, the bras, their complex conjugates, and their bracket combination an integration of the two. The Dirac notation does not include the representation variable, in this case, x. It could just as well describe the integral in the momentum representation, or in any other. It contains only the physics. Thus $\langle m|n \rangle$ is the overlap between states $|m\rangle$ and $|n\rangle$. Note that $\langle m|n \rangle$ is a number, just like the dot product in the space of real vectors. Since the spaces here are made up of complex vectors, the dot product must be generalized. It is called the inner product, with the property that $\langle m|n \rangle = \langle n|m \rangle^*$, in the coordinate representation, suppressing the limits of integration, $\int \psi_m^*(x)\psi_n(x)dx = (\int \psi_n^*(x)\psi_m(x)dx)^*$ An "inner product space" is one for which the inner product exists. This is a bit more subtle than it seems. For a finite inner product space of dimensionality N, there are N linearly independent vectors that span the space. For infinite-dimensional vector spaces, one requires that in addition to the existence of an inner product, all sequences of vectors converge to a limit in the same vector space. That is, any vector can be expanded in terms of an infinite basis. These are called Hilbert spaces, and it is in them that quantum mechanics operates.

The application of some operator A on a state $|n\rangle$ is designated $A|n\rangle \equiv |An\rangle$. The time independent Schrödinger equation, for example, may be written as $|Hn\rangle = E_n|n\rangle$ irrespective of representation. However an inner product is written $\langle m|An \rangle$, to emphasize that the operator A acting on the ket $|n\rangle$ produces a new ket $|An\rangle$, whose overlap with the ket $|m\rangle$ is then computed.

The expectation value of an operator is written as $\langle n|An \rangle$, in the one-dimensional coordinate representation, $\int \psi_n^*(x)A\psi_n(x)dx$. More generally, $\langle m|An \rangle$, $\int \psi_m^*(x)A\psi_n(x)dx$ in the coordinate representation, represents a transition from an initial state $|n\rangle$ to a final state $|m\rangle$ due to the operator A. The only shortcoming of the Dirac notation is that it has no way to describe the probability of finding the particle in a given interval (a, b), $\int_a^b \psi^*(x)\,\psi(x)dx$. The bracket implies the integration over the entire domain.

Now how can we relate an operation on a ket to one on a bra? We define the "adjoint" or "hermitian conjugate" of operator A, $A^\dagger$:

$$\langle A^\dagger m | n \rangle \equiv \langle m | An \rangle , \tag{3.2}$$

for any $|m\rangle$ and $|n\rangle$. That is, the adjoint of an operator acting on the bra gives the same result as the original operator acting on the ket. In the coordinate representation, this is $\int (A^\dagger \psi_m(x))^* \psi_n(x) dx = \int \psi_m^*(x) A \psi_n(x) dx$.

In matrix representation, if a column vector represents a ket, its adjoint is the bra represented by a row vector, with components that are the complex conjugates of those in the column vector:

$$|n\rangle \equiv \begin{pmatrix} n_1 \\ n_2 \\ \vdots \end{pmatrix} \qquad \langle n| \equiv \begin{pmatrix} n_1^* & n_2^* \cdots \end{pmatrix} .$$

In two dimensions, $A \equiv \begin{pmatrix} a_{11} & a_{12} \\ a_{21} & a_{22} \end{pmatrix}$, $A^\dagger \equiv \begin{pmatrix} a_{11}^* & a_{21}^* \\ a_{12}^* & a_{22}^* \end{pmatrix}$, $|n\rangle \equiv \begin{pmatrix} n_1 \\ n_2 \end{pmatrix}$, and $\langle m| \equiv \begin{pmatrix} m_1^* & m_2^* \end{pmatrix}$, the adjoint of $|m\rangle \equiv \begin{pmatrix} m_1 \\ m_2 \end{pmatrix}$.

Problem 3.1. *In this two-dimensional matrix formulation, show that* $\langle A^\dagger m | n \rangle = \langle m | An \rangle$.

The rule for matrix multiplication $U = ST$ is $U_{ik} = \Sigma_j S_{ij} T_{jk}$. The hermitian conjugate of this product is the product taken in reverse order of the hermitian conjugates of the component matrices: $U^\dagger = T^\dagger S^\dagger$.

3.2.1 *Hermitian Operators*

We have seen several examples of physical quantities: position, momentum, energy, being represented by operators. In quantum mechanics all physical observables are represented by operators. Measurements of a physical observable yield the eigenvalues of its associated operator. Hence the eigenvalues of quantum operators must be real. This constrains the form of the operators to be self-adjoint: $A^\dagger = A$. Self-adjoint operators are also known as hermitian operators.

For states $|n\rangle$ which are eigenvectors[2] of a hermitian operator A: $A|n\rangle = \lambda_n |n\rangle$, we can deduce that the eigenvalues λ_n must be real numbers : $\langle n|An \rangle = \langle n|\lambda_n n \rangle = \lambda_n \langle n|n \rangle$. For a hermitian operator, $\langle n|An \rangle = \langle An|n \rangle = \langle \lambda_n n|n \rangle = \lambda_n^* \langle n|n \rangle$. Therefore $\lambda_n^* = \lambda_n$.

---

[2]We shall use the terms "eigenvector" and "eigenstate" interchangeably.

Problem 3.2. *In the one-dimensional coordinate representation, $p = -i\frac{d}{dx}$. Assuming that the wave function and its complex conjugate vanish at the boundaries, show using integration by parts that $p \equiv -i\frac{d}{dx}$ is hermitian. Is $\frac{d}{dx}$ hermitian?*

Problem 3.3.

     a) *Prove that the sum of two Hermitian operators is Hermitian.*

     b) *Prove that if operator A is Hermitian, then so is A^n.*

The particle wave function can be an eigenstate of some operator. But it can also be a mixture: a linear combination of eigenstates. In this case we are interested in two things. The first is the expectation value of a particular operator A; i.e, the average value of the associated observable after many (technically infinitely many) measurements, given by $\langle A \rangle \equiv \langle \psi | A \psi \rangle$. But since any measurement can only return an eigenvalue of the operator, we are also interested in the probability that a particular eigenvalue is returned. We would like to employ the eigenvectors of the operator as the basis set. We will require them to be orthonormal: $\langle m | n \rangle = \delta_{mn}$. If they are not, we can make them so by procedures such as Gram-Schmidt orthogonalization.[3]

Once the eigenvectors constitute an orthonormal set, we can show the following:

If a mixed state is given by $|\psi\rangle = \sum_n c_n |n\rangle$ where $A|n\rangle = \lambda_n |n\rangle$ then the probability of obtaining the result λ_n is $|\langle \psi | n \rangle|^2 = |c_n|^2$. We know that any state $|\psi\rangle$ must be normalized: the probability of finding the particle in that state somewhere in the domain must be 1. Exploiting the orthonormality of the eigenstates, we can without loss of generality write the mixed state as $|\psi\rangle = \sum_m c_m |m\rangle$.

Then $1 = \langle \psi | \psi \rangle = \sum_m c_m^* \langle m | \sum_n c_n | n \rangle = \sum_{mn} c_m^* c_n \langle m | n \rangle = \sum_{mn} c_m^* c_n \delta_{mn} = \sum_n |c_n|^2$.

Since the total probability of finding the particle is the sum of the probabilities for the particle to be in each of the eigenstates, $|c_n|^2$ represents the probability that a measurement of $|\psi\rangle$ will find it in state $|n\rangle$; i.e., the result of such a measurement will be the eigenvalue λ_n.

We can also think of this as follows: Since $|\psi\rangle = \sum_m c_m |m\rangle$, $\langle n | \psi \rangle =$

---

[3]The street grid of Chicago is orthonormal: there are eight blocks per mile north-south or east-west. New York's grid is orthogonal but needs to be normalized: there are 20 blocks per mile north-south, but 8 blocks per mile east-west. Paris on the other hand is crying out for the full Gram-Schmidt procedure.

$\langle n| \sum_m c_m |m\rangle = \sum_m c_m \delta_{mn} = c_n$. In words, the probability is the square of the absolute value of the $\langle n|\psi\rangle$ overlap.

Problem 3.4. *In Chapter 2 we showed that the wave functions and energies for an infinite square well of length π were given by $|n\rangle = \sqrt{\frac{2}{\pi}} \sin nx$, $E_n = n^2$. Show that the eigenfunctions constitute an orthonormal set.*

Problem 3.5. *A state of a particle in the well is in a mixture of the two lowest states: $\psi = \frac{1}{2}|1\rangle + \frac{\sqrt{3}}{2}|2\rangle$.*

    (a) Check that ψ is normalized.
    (b) What is the expectation value of the energy? Does this correspond to any single measurement?
    (c) What is the probability that a measurement of the energy returns the value 4?

Physical observables are of course not only the expectation values of operators for particles in a given state. They can also describe measurable quantities for particles changing state. For example, the "transition probability" is a measure of the likelihood of an electron jumping from one orbit to another, in Bohr model terminology. It is proportional to $|\langle m|An\rangle|^2$.

3.3 Constants of Motion in Quantum Mechanics

Constants of motion are conserved quantities. What does a conservation law mean in quantum mechanics? In classical mechanics the measurable quantities are dynamical variables such as position, momentum, angular momentum, and energy. In quantum mechanics, these dynamical quantities become operators. In particular, to say the physical observable associated with some operator A is "conserved" means that $d\langle A\rangle/dt = 0$, in which case we call the eigenvalues of A "good quantum numbers." We have

$$\frac{d\langle A\rangle}{dt} = \frac{d\langle \Psi|A\Psi\rangle}{dt} = \left\langle \frac{\partial \Psi}{\partial t} \,\middle|\, A\Psi \right\rangle + \left\langle \Psi \,\middle|\, \frac{\partial A}{\partial t}\Psi \right\rangle + \left\langle \Psi \,\middle|\, A\frac{\partial \Psi}{\partial t} \right\rangle.$$

The time dependent Schrödinger equation relates such conservation to the hamiltonian, as follows:

$$|H\Psi\rangle = \left| i\frac{\partial \Psi}{\partial t} \right\rangle,$$

where H is hermitian. So

$$\frac{d\langle A\rangle}{dt} = i\langle H\Psi|A\Psi\rangle + \left\langle \frac{\partial A}{\partial t} \right\rangle - i\langle \Psi|AH\Psi\rangle.$$

Rearranging,

$$\frac{d\langle A\rangle}{dt} = i\langle[H,A]\rangle + \left\langle\frac{\partial A}{\partial t}\right\rangle. \tag{3.3}$$

For A not explicitly time-dependent, $\langle\frac{\partial A}{\partial t}\rangle = 0$, and the time dependence of $\langle A\rangle$ is determined entirely from its commutator with H.

A simple example is the identity operator **1**. Since it commutes with everything, $\frac{d\langle\mathbf{1}\rangle}{dt} = 0$. But this is just $d\langle\Psi|\Psi\rangle/dt$. This is a statement of conservation of normalization; viz., the probability of finding a particle somewhere is 1 at all times. Another example is $A \equiv x$. Ignoring our convention, putting back in $2m$,

$$\frac{d\langle x\rangle}{dt} = i\langle[H,x]\rangle = i\left\langle\left[\frac{p^2}{2m} + V(x), x\right]\right\rangle = i\left\langle\left[\frac{p^2}{2m}, x\right]\right\rangle, \tag{3.4}$$

where we have used the fact that $[V(x), x] = 0$.

Problem 3.6. *Using the Taylor expansion of $V(x)$ prove that $[V(x), x] = 0$.*

But $[p^2, x] = ppx - xpp$, where $xp - px = i$. Then $ppx = p(xp - i)$ and $xpp = (xp)p = (px + i)p$. Canceling the terms pxp, $[H, x] = -2ip/2m$. Therefore

$$m\frac{d\langle x\rangle}{dt} = \langle p\rangle, \tag{3.5}$$

which is the quantum mechanical equivalent of the classical definition of momentum.

Problem 3.7. *(Ehrenfest's theorem) Show that*

$$\frac{d\langle p\rangle}{dt} = \left\langle-\frac{dV}{dx}\right\rangle.$$

Thus, we see the importance of the hamiltonian: if $[H, A] = 0$ then $\langle A\rangle$ is a constant of the motion. In retrospect, this comes as no surprise. The time dependent Schrödinger equation links H with the time evolution of the wave function, from which the time evolution of the expectation value of an operator is obtained.

Another way of considering the consequences of commutation with the hamiltonian is to recognize that the eigenstates of the hamiltonian are the same as those for the operator(s) with which it commutes. Thus, we write the wave functions for the hydrogen atom (ignoring spin) as $\psi_{n\ell m}$, indicating that the total angular momentum and its projection commute with the hamiltonian.

Let us prove a general result: *Commuting operators have the same eigenvectors.*

Proof: Suppose $[A, B] = 0$ and $|An\rangle = \lambda_n|n\rangle$. Then, since $BA = AB$, $AB|n\rangle = B\lambda_n|n\rangle = \lambda_n|Bn\rangle$. This says that $|Bn\rangle$ is an eigenstate of A with the same eigenvalue as $|n\rangle$ itself. This can only be the case if $|Bn\rangle$ is parallel to $|n\rangle$ itself; i.e. $|Bn\rangle = \mu_n|n\rangle$. Thus $|n\rangle$ is an eigenstate of B with eigenvalue μ_n.

While this is a general result, its application to the hamiltonian is the very definition of a good quantum number. It tells us which quantum numbers become "bad" as a hamiltonian is changed, for example as we go from the Schrödinger to the Dirac equation. Operator algebra has enabled us to discover which quantum numbers are good under which circumstances, without the need for choosing any particular representation. It is indispensable for SUSYQM.

3.4 Application: The Harmonic Oscillator

In Chapter 2 we have seen that rather tedious computations yield the solutions (eigenenergies and eigenfunctions) for some of the elementary systems we looked at. However, there is a different path to these results. The method we are about to show for the harmonic oscillator case doesn't involve solving the Schrödinger equation, while obtaining the same results. The way is provided by an algebraic method often attributed to Dirac.[4] At the core of this algebraic method is finding a set of specially crafted operators A^- and A^+ that will transform one eigenfunction of the hamiltonian into another eigenfunction with a different energy. In this way, starting from any eigenfunction ψ_n, one can obtain all the other eigenfunctions by successive application of A^- or A^+.

To begin, let us rewrite the Schrödinger equation for the harmonic oscillator, Eq. (2.20), as the eigenvalue problem for hamiltonian operator $H\psi(x) = E\psi(x)$, where

$$H = -\frac{d^2}{dx^2} + \frac{1}{4}\omega^2 x^2 \ . \tag{3.6}$$

Next, we define two operators

$$A^- \equiv \frac{d}{dx} + \frac{1}{2}\omega x \ , \quad A^+ \equiv -\frac{d}{dx} + \frac{1}{2}\omega x \ , \tag{3.7}$$

---

[4]P.A.M. Dirac, "Communications of the Dublin Institute of Advanced Study", **A1**, 5-7 (1943). Dirac references Fock, *Zeits. f. Phys.* **49** 339 (1928).

which are known respectively as *annihilation* and *creation* operators. (The reason for these names will become clear shortly.) We can immediately check, by operating on an arbitrary function ψ with them, that $A^- (A^+\psi) - A^+ (A^-\psi) = \psi$. This property of the operators is usually written as a commutation relation:

$$A^- A^+ - A^+ A^- \equiv [A^-, A^+] = \omega . \tag{3.8}$$

The operator $N \equiv \frac{1}{\omega} A^+ A^-$ is Hermitian.

Problem 3.8. *Using the results of Problem 3.2, prove N is Hermitian.*

Consequently, all its eigenvalues are real. Let us denote by $|n\rangle$ the eigenvectors of N. Then

$$N|n\rangle = n|n\rangle . \tag{3.9}$$

Furthermore, we can show that N is a linear function of the hamiltonian.

$$
\begin{aligned}
N\psi &= \frac{1}{\omega} A^+ A^- \psi \\
&= \frac{1}{\omega}\left(-\frac{d}{dx} + \frac{1}{2}\omega x\right)\left(\frac{d}{dx} + \frac{1}{2}\omega x\right)\psi \\
&= \frac{H}{\omega}\psi - \frac{1}{2}\psi .
\end{aligned}
\tag{3.10}
$$

Therefore the two operators N and H commute, and consequently, they have a common set of eigenvectors. Hence $|n\rangle$ is an energy eigenvector as well. From (3.10) we have

$$H = \omega\left(N + \frac{1}{2}\right) , \tag{3.11}$$

and therefore,

$$H|n\rangle = \left(n + \frac{1}{2}\right)\omega|n\rangle , \tag{3.12}$$

which means that the eigenenergies of the harmonic oscillator are given by

$$E_n = \left(n + \frac{1}{2}\right)\omega . \tag{3.13}$$

Problem 3.9. *Repeat the above analysis for the operator $M \equiv \frac{1}{\omega} A^- A^+$. How are its eigenvalues m related to n?*

At this stage, we should be careful when comparing Eq. (2.30) with Eq. (3.13). So far, all we know in Eq. (3.13) is that n is a non-negative real number. Indeed, from the positivity of the norm of $(A^-\,|\,n\rangle)$ one gets

$$n = \langle n|N|n \rangle = \frac{1}{\omega}\left(\langle n|\,A^+ \right)\left(A^-\,|\,n \rangle \right) \geq 0 \, . \qquad (3.14)$$

To find the nature of the positive number n, we will explore some properties of A^- and A^+ operators. Using the commutation relation (3.8) we note that

$$[N, A^+] = \frac{1}{\omega}[A^+A^-, A^+] = \frac{A^+}{\omega}[A^-, A^+] + [A^+, A^+]\frac{A^-}{\omega} = A^+ \, , \qquad (3.15)$$

and, similarly

$$[N, A^-] = -A^- \, . \qquad (3.16)$$

We can now prove that if $|\,n\rangle$ is a normalized eigenvector of N with eigenvalue n, then $A^+\,|\,n\rangle$ and $A^-\,|\,n\rangle$ are also eigenvectors of N. Indeed,

$$N\,A^+\,|\,n\rangle = \left([N, A^+] + A^+\,N \right)|\,n\rangle = (n+1)\,A^+\,|\,n\rangle \, , \qquad (3.17)$$
$$N\,A^-\,|\,n\rangle = \left([N, A^-] + A^-\,N \right)|\,n\rangle = (n-1)\,A^-\,|\,n\rangle \, . \qquad (3.18)$$

We observe that $A^+\,|\,n\rangle$ is an eigenvector of N with eigenvalue increased by 1, and we write accordingly

$$A^+\,|\,n\rangle = c_+\,|\,n+1\rangle \, , \qquad (3.19)$$

where c_+ is a constant to be determined using the normalization condition $\langle n+1\,|\,n+1\rangle = 1$. We have

$$\langle n|A^-A^+\,|\,n\rangle = |c_+|^2 \, . \qquad (3.20)$$

But $A^-A^+ = \omega + A^+A^- = \omega\,(1 + N)$, and therefore the left hand side of Eq. (3.20) is equal to $\omega\,(n + 1)$. Hence, $|c_+|^2 = \omega\,(n + 1)$, and Eq. (3.19) becomes

$$A^+\,|\,n\rangle = \sqrt{\omega\,(n+1)}\,|\,n+1\rangle = \sqrt{\omega\,(n+1)}\,|\,n+1\rangle \, . \qquad (3.21)$$

Similarly,

$$A^-\,|\,n\rangle = \sqrt{\omega\,n}\,|\,n-1\rangle \, . \qquad (3.22)$$

Now, we can see why A^+ (A^-) is called a *creation* (*annihilation*) operator. Acting on the energy eigenvector $|\,n\rangle$, this operator yields another energy eigenvector $|\,n+1\rangle$ $(|\,n-1\rangle)$ whose corresponding eigenenergy is increased (decreased) by one quantum of energy ω.

We are now ready to completely identify the positive real number n. Let us start from an arbitrary state $|n\rangle$ and successively apply the annihilation operator A^-. We obtain

$$A^-|n\rangle = \sqrt{n\omega}\,|n-1\rangle \,,$$
$$(A^-)^2|n\rangle = \sqrt{n(n-1)\,\omega^2}\,|n-2\rangle \,,$$
$$\vdots$$
$$(A^-)^k|n\rangle = \sqrt{n(n-1)\cdots(n-k+1)\,\omega^k}\,|n-k\rangle \,, \qquad (3.23)$$
$$\vdots$$

If n is an arbitrary *non-integer* positive real number, then after applying the annihilation operator A^- an integral number of times k, we must overshoot 0 and arrive at an eigenvector $|n-k\rangle$ whose eigenvalue $n-k$ is negative. This contradicts Eq. (3.14). Therefore, n must be a positive integer, and consequently, the sequence (3.23) terminates at the eigenvector of N with the lowest possible eigenvalue. We call this eigenvector $|0\rangle$. If we go any further, the sequence terminates, since from Eq. (3.23), the eigenvalue of $|0\rangle$ is 0.

$$A^-|0\rangle = 0 \,. \qquad (3.24)$$

The vector $|n-k\rangle = |0\rangle$ is the ground state vector, with the lowest energy $E_0 = \omega/2$. Now, we can finally conclude that the numbers n in Eq. (3.13) are non-negative integers. Hence, we recover the same result as in Eq. (2.30), but this time via the less cumbersome algebraic method.

Starting from the ground state $|0\rangle$ we can successively apply the creation operator A^+ to obtain any state $|n\rangle$. Using (3.21), we obtain

$$|1\rangle = \frac{A^+}{\sqrt{\omega}}|0\rangle \,,$$
$$|2\rangle = \frac{A^+}{\sqrt{2\omega}}|1\rangle = \frac{(A^+)^2}{\sqrt{2\omega^2}}|0\rangle \,,$$
$$|3\rangle = \frac{A^+}{\sqrt{3\omega}}|2\rangle = \frac{(A^+)^3}{\sqrt{3!\,\omega^3}}|0\rangle \,,$$
$$\vdots$$
$$|n\rangle = \frac{A^+}{\sqrt{n\omega}}|n-1\rangle = \frac{(A^+)^n}{\sqrt{n!\,\omega^n}}|0\rangle \,. \qquad (3.25)$$

So far we have used the abstract representation of the eigenvectors of the hamiltonian of the system. We make contact with the familiar coordinate representation, where the energy eigenvectors become simply the well-known eigenfunctions via $|n\rangle \rightarrow \psi_n(x)$. We can use the "differential realization" of the creation operator A^+ from Eq. (3.7) in conjunction with Eq. (3.25) to find all $\psi_n(x)$ if we know the ground state $\psi_0(x)$. To find $\psi_0(x)$, we rewrite Eq. (3.24) as a simple differential equation

$$\left(\frac{d}{dx} + \frac{1}{2}\,\omega\,x \right) \psi_0(x) = 0 \qquad (3.26)$$

using the differential realization of A^- from (3.7). Solving this first order homogeneous differential equation, one gets

$$\psi_0(x) = \left(\frac{\omega}{2\pi} \right)^{\frac{1}{4}} e^{-\frac{1}{4}\omega x^2}\,, \qquad (3.27)$$

where the integration constant was determined from the normalization condition $\int_{-\infty}^{+\infty} |\psi_0(x)|^2 = 1$. Note that this is the only place where we have to solve a differential equation, one which is much simpler that the Schrödinger equation.

Problem 3.10. *Obtain Eq. (3.27) from Eq. (3.26).*

Substituting Eq. (3.26) into Eq. (3.25) we obtain

$$\psi_1(x) = \frac{1}{\sqrt{\omega}} \left(-\frac{d}{dx} + \frac{1}{2}\,\omega\,x \right) \left(\frac{\omega}{2\pi} \right)^{\frac{1}{4}} e^{-\frac{1}{4}\omega x^2}$$

$$\psi_2(x) = \frac{1}{\sqrt{2!\,\omega^2}} \left(-\frac{d}{dx} + \frac{1}{2}\,\omega\,x \right)^2 \left(\frac{\omega}{2\pi} \right)^{\frac{1}{4}} e^{-\frac{1}{4}\omega x^2}$$

$$\vdots$$

$$\psi_n(x) = \frac{1}{\sqrt{n!\,\omega^n}} \left(-\frac{d}{dx} + \frac{1}{2}\,\omega\,x \right)^n \left(\frac{\omega}{2\pi} \right)^{\frac{1}{4}} e^{-\frac{1}{4}\omega x^2} \qquad (3.28)$$

which can be brought into the same form as Eq. (2.35) using Rodrigues formula for the Hermite polynomials.

We now see that recursion relations Eqs. (2.37) and (2.38) are nothing but the manifestation of the action of the annihilation and creation operators on the wave function $\psi_n(x)$. Because of their special action on the energy eigenfunctions, the operators A^- and A^+ are also known as "ladder operators." We will see in the next chapter that many potentials allow for a generalization of the algebraic method introduced here. The generalization of this method to other potentials is SUSYQM.

Chapter 4

Supersymmetric Quantum Mechanics

In Chapter 1, we gave a brief introduction to SUSYQM.[1,2] In this chapter we will elaborate on it further and use it to solve several problems. We will introduce it more formally in Chapter 6.

4.1   The Concept of Superpotential

As we saw in Chapter 2, interactions in quantum mechanics are described by a potential function $V(x, a)$, where x is the coordinate variable and the parameter a describes the "strength" of the interaction. For example, in the Coulomb problem, the effective one-dimensional potential $V(r, e, \ell) = -\frac{e^2}{r} + \frac{\ell(\ell+1)\hbar^2}{2mr^2}$ depends on the coordinate r and on two parameters: the electric charge e and the orbital angular momentum ℓ. The singularity at the origin, the depth of the potential, and the behavior of the potential at infinity are governed by these parameters, and appropriately, the energy of the system $E_n = \frac{me^4}{2\hbar^2}\left[\left(\frac{e^2}{\ell+1}\right)^2 - \left(\frac{e^2}{n+\ell+1}\right)^2\right]$ depends on them. It is in this sense that we say that these parameters, which we generically denote by a, define the "strength" of the potential.

Our objective is to solve the Schrödinger equation to determine the complete set of eigenvalues E_n and eigenfunctions $\psi_n(x, a)$ for a given hamiltonian H:

$$H\,\psi(x, a) \equiv \left[-\frac{\hbar^2}{2m}\frac{d^2}{dx^2} + V(x, a)\right]\psi(x, a) = E\,\psi(x, a) \ . \qquad (4.1)$$

---

[1] E. Witten, "Dynamical Breaking of Supersymmetry", *Nucl. Phys.* B, **185**, 513–554 (1981).

[2] F. Cooper and B. Freedman, "Aspects of Supersymmetric Quantum Mechanics", *Ann. Phys.* **146**, 262–288 (1983).

In Chapter 2, we considered several examples of eigenvalue equations and we solved them with varying degrees of effort. In Section 3.4 we solved one such example; namely, the harmonic oscillator, using a scheme based on algebra, rather than solving a differential equation. In this chapter we will show that the algebraic method used for the harmonic oscillator can be extended to other quantum mechanical systems. However, it is worth noting that there is a frequently overlooked benefit associated with the traditional method. In Chapter 2, after rather tedious calculations, we not only found all eigenvalues of the hamiltonian, we also described all eigenfunctions in closed form as in Eq. (2.35) for the harmonic oscillator. SUSYQM takes a different approach. It provides a simpler method of deriving the energy eigenvalues. However, it determines the eigenfunctions only recursively.

In the case of the harmonic oscillator, we were able to factorize the hamiltonian $H = \left(-\frac{\hbar^2}{2m} \frac{d^2}{dx^2} + \frac{1}{2} m\omega^2 x^2 \right)$ into a product of two first order operators and an additive constant: $A^+ A^- + \frac{1}{2}\hbar\omega$, where $A^- = \frac{\hbar}{\sqrt{2m}} \frac{d}{dx} + \sqrt{\frac{m}{2}} \omega x$ and $A^+ = -\frac{\hbar}{\sqrt{2m}} \frac{d}{dx} + \sqrt{\frac{m}{2}} \omega x$.

We will generalize this method to factorize a hamiltonian with an arbitrary potential $V(x,a)$. For the harmonic oscillator, the expression $\sqrt{\frac{m}{2}} \omega x$ in $A^\pm$ generated the term $\frac{1}{2} m\omega^2 x^2$ for the potential. Analogously, we define a real function $W(x,a)$, that would generate the potential $V(x,a)$. This progenitor $W(x,a)$ of the potential function is known as the superpotential. In fact a superpotential $W(x,a)$ generates two potentials $V_\pm(x,a)$ which are called partner potentials. Similar to the case of the harmonic oscillator, we introduce two operators $A^+ = -\frac{\hbar}{\sqrt{2m}} \frac{d}{dx} + W(x,a)$ and $A^- = \frac{\hbar}{\sqrt{2m}} \frac{d}{dx} + W(x,a)$. Their operator product $A^+ A^-$ is equivalent to $-\frac{\hbar^2}{2m} \frac{d^2}{dx^2} + \left[W^2(x,a) - \frac{\hbar}{\sqrt{2m}} \frac{dW(x,a)}{dx} \right]$. Hence the Schrödinger equation for a quantum mechanical system described by a potential $V(x,a) = W^2(x,a) - \frac{dW(x,a)}{dx}$ can be factorized as

$$\left[-\frac{\hbar^2}{2m} \frac{d^2}{dx^2} + W^2(x,a) - \frac{\hbar}{\sqrt{2m}} \frac{dW(x,a)}{dx} \right] \psi(x,a)$$

$$= \left(-\frac{\hbar}{\sqrt{2m}} \frac{d}{dx} + W(x,a) \right) \left(\frac{\hbar}{\sqrt{2m}} \frac{d}{dx} + W(x,a) \right) \psi(x,a)$$

$$\equiv A^+ A^- \psi(x,a) . \tag{4.2}$$

If we interchange the operators A^- and A^+, we obtain

$$A^- A^+ \psi(x,a) \equiv \left(\frac{\hbar}{\sqrt{2m}} \frac{d}{dx} + W(x,a) \right) \left(-\frac{\hbar}{\sqrt{2m}} \frac{d}{dx} + W(x,a) \right) \psi(x,a)$$

$$= \left[-\frac{\hbar^2}{2m} \frac{d^2}{dx^2} + W^2(x,a) + \frac{\hbar}{\sqrt{2m}} \frac{dW(x,a)}{dx} \right] \psi(x,a) \quad (4.3)$$

which is also a Schrödinger equation, but for a different potential: $W^2(x,a) + \frac{\hbar}{\sqrt{2m}} \frac{dW(x,a)}{dx}$. Thus, a superpotential $W(x,a)$ helps us factorize two hamiltonians, which we will denote by $H_\pm \equiv -\frac{\hbar^2}{2m} \frac{d^2}{dx^2} + V_\pm(x,a)$, associated with potentials

$$V_\pm(x,a) = W^2(x,a) \pm \frac{\hbar}{\sqrt{2m}} \frac{dW(x,a)}{dx}. \quad (4.4)$$

These are known as partner hamiltonians. We will denote eigenvalues and eigenfunctions of the $H_\pm$ by $E_n^\pm$ and $\psi_n^\pm(x,a)$, $n = 0,1,2,\cdots$. Before we proceed further, we will set $\hbar = 2m = 1$, which reduces $A^\pm$ to $\mp\frac{d}{dx}+W(x,a)$ respectively. We can make a few interesting observations here.

(1) Suppose $\psi_0^-(x,a)$, the solution of $A^- \psi_0^-(x,a) = 0$ is a normalizable function. Then $H_- \psi_0^-(x,a) \equiv A^+ A^- \psi_0^-(x,a) = 0$, i.e., $\psi_0^-(x,a)$ is an eigenstate of H_- with the eigenvalue zero, $E_0 = 0$.

(2) Since A^+ is the adjoint or hermitian conjugate[3] of A^-, from $H_- \psi_n^-(x,a) = A^+ A^- \psi_n^-(x,a) = E_n^- \psi_n^-(x,a)$, we obtain

$$E_n^- = \frac{\int \psi_n^{-*} A^+ A^- \psi_n^- \, dx}{\int \psi_n^{-*} \psi_n^- \, dx} = \frac{\int |A^- \psi_n^-|^2 \, dx}{\int |\psi_n^-|^2 \, dx} \geq 0. \quad (4.5)$$

Thus, the eigenvalues of the operator $H_- \equiv A^+ A^-$ are greater than or equal to zero. In such cases, we call the operator H_- a "semi-positive definite" operator.

Problem 4.1. *Show that the ground state eigenfunction of $\psi_0^-(x,a)$ can be written as $\psi_0^-(x,a) \sim e^{-\int_{x_0}^x W(x,a)\,dx}$.*

Analogously, if we had instead started with the solution of $A^+ \psi_0^+(x,a) = 0$, we would have gotten

$$H_+ \psi_0^+(x,a) \equiv A^- A^+ \psi_0^+(x,a) = 0 ; \quad (4.6)$$

---

[3]These two operators are hermitian conjugates of each other, i.e., for any two functions $\eta(x)$ and $\lambda(x)$, we have $\int [A^+\eta(x)]^* \lambda(x)\,dx = \int \eta(x)^* A^- [\lambda(x)]\,dx$ and $\int [A^-\eta(x)]^* \lambda(x)\,dx = \int \eta(x)^* A^+ [\lambda(x)]\,dx$.

thus, the zero energy ground state eigenfunction of H_+ is given by the solution of

$$\left(-\frac{d}{dx} + W(x,a)\right)\psi^+(x,a) = 0 ,$$

i.e.

$$\psi_0^+(x,a) \sim e^{\int_{x_0}^x W(x,a)\,dx} \sim \frac{1}{\psi_0^-(x,a)} .$$

From this reciprocal relationship between $\psi_0^-(x,a)$ and $\psi_0^+(x,a)$, we see that they cannot be normalized simultaneously. This leads to three different possibilities:

(1) $\psi_0^-(x,a)$ is normalizable and $\psi_0^+(x,a)$ is not;
(2) $\psi_0^+(x,a)$ is normalizable and $\psi_0^-(x,a)$ is not;
(3) neither $\psi_0^-(x,a)$ nor $\psi_0^+(x,a)$ is normalizable.

We will now analyze these three cases in the following section.

4.1.1 Broken and Unbroken Supersymmetry

Let us consider the above three possibilities. In the first two cases, the ground state energy of H_- or H_+ is zero. In the third case neither of the two hamiltonians has a zero-energy bound state. When either of the two partner hamiltonians has a zero energy bound state, the system is said to possess *unbroken* supersymmetry, and when neither holds a zero energy bound state the supersymmetry is said to be *spontaneously broken*. We will explain the reasons behind this nomenclature in Chapter 6.

Let us explore which properties of $W(x,a)$ are important in determining whether supersymmetry is broken or unbroken. Normalizability of $\psi_0^-(x,a)$ requires that the wave function vanish as $x \to \pm\infty$. This implies that $\psi_0^-(\pm\infty,a) \sim e^{-\int_{x_0}^{\pm\infty} W(x,a)\,dx} = 0$, and hence $\int_{x_0}^{\pm\infty} W(x,a)\,dx = \infty$. We consider the integrals separately. The integral $\int_{x_0}^{+\infty} W(x,a)\,dx$ is the area under the curve from x_0 to ∞ and is infinite. This implies that the superpotential $W(x,a)$ must have a positive value at $x = \infty$. Similarly, $\int_{x_0}^{-\infty} W(x,a)\,dx = -\int_{-\infty}^{x_0} W(x,a)\,dx = +\infty$ implies that the area under the curve from $-\infty$ to x_0 must be $-\infty$. Consequently, $W(x,a)$ must have a negative value at $x = -\infty$. Hence, $W(x,a)$ must cross the x-axis. This is illustrated in Fig. 4.1.

While any odd function $W(x)$ with the property $\int_{x_0}^{\pm\infty} W(x,a)\,dx = \infty$ will lead to an unbroken supersymmetry, odd parity is not necessary. For example, while the odd function $W(x, A) = A \tanh x$ leads to a system with unbroken SUSY, so does the function $W(x, A, B) = A \tanh x + B$ with $A > B > 0$ which has no particular parity. Note that for cases that have finite domain, $x = -\infty$ and $x = +\infty$ are to be replaced by x_L and x_R, the left and right boundary points, respectively. Normalization of ψ_0^- would then require that $W(x, a)$ diverge (positively at x_R and negatively at x_L) at both the finite boundary points.

It is important to point out that a superpotential that has a negative value at x_R and positive at x_L would also preserve SUSY. From here onward we will assume, without loss of generality, that for all cases of unbroken SUSY the superpotential is negative at the left and positive at the right boundary.

In short, a superpotential that has the same sign at both ends of the domain will generate a system with broken supersymmetry, and one that has opposite sign will lead to unbroken supersymmetry.

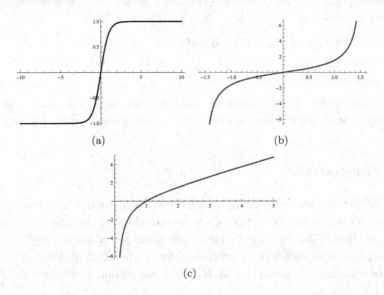

Fig. 4.1 Examples of superpotentials $W(x)$ that lead to unbroken supersymmetry for (a) an infinite domain, (b) a finite domain and (c) a semi-infinite domain.

Problem 4.2. *For the following three cases, compute ground state wave functions and show that they vanish at both end points of their domains:*

(a) Finite domain: $W(x, \alpha) = \alpha \tan x$ with $\alpha > 0$;

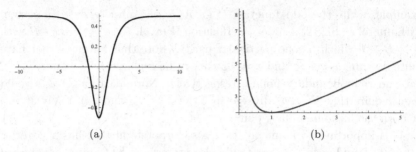

Fig. 4.2   Examples of superpotentials $W(x)$ that lead to broken supersymmetry for (a) an infinite domain and (b) a semi-infinite domain.

(b) *Semi-infinite domain:* $W(x, \alpha, \beta) = \alpha - \beta \coth x$ *with* $\alpha > \beta > 0$; *and*

(c) *Infinite domain:* $W(x, \gamma) = \gamma \tanh x$ *with* $\gamma > 0$.

Problem 4.3. *The superpotential $W(r, \omega, l)$ for the three-dimensional harmonic oscillator is given by $W(r, \omega, l) = \omega r - \frac{l}{r}$.*

(a) *Plot $W(r, 10, -1)$ as a function of r;*

(b) *Show that the function $\exp\left(-\int_0^\infty W(r, 10, -1)\, dr\right)$ does not converge at at least one boundary point.*

Since the two hamiltonians H_- and H_+ originate from one common superpotential, we will now see that they share many common properties.

4.2   Isospectrality

One of the interesting relationships between two partner hamiltonians is known as isospectrality. That is, except possibly for the ground state of H_-, both hamiltonians have exactly the same set of eigenvalues. Their corresponding eigenfunctions are related by an operation of A^- or A^+. Let $\psi_n^\pm$ denote the eigenfunctions of $H_\pm$ that correspond to eigenvalues $E_n^\pm$; $n = 1, 2, \ldots$

$$H_+\left(A^- \psi_n^-\right) = A^- A^+ \left(A^- \psi_n^-\right) = A^- \left(A^+ A^- \psi_n^-\right)$$
$$= A^- \left(H_- \psi_n^-\right) = E_n^- \left(A^- \psi_n^-\right) . \qquad (4.7)$$

Thus, for an eigenfunction ψ_n^- of H_-, the function $A^- \psi_n^-$ is an eigenfunction of H_+, and the corresponding eigenvalue is E_n^-. All excited states ψ_n^- have a one-to-one correspondence with ψ_{n-1}^+:

$$\psi_{n-1}^+ = c A^- \psi_n^- \quad \text{and} \quad E_{n-1}^+ = E_n^- . \qquad (4.8)$$

If the supersymmetry is unbroken, the ground state of H_- obeys $A^- \psi_0^- = 0$ and it does not connect to any state of H_+. This is why we did not include above the case $n = 0$. On the other hand, if supersymmetry is broken, $A^- \psi_0^- = 0$ does not have a normalizable solution, we would have a strict isospectrality between two hamiltonians; i.e., $E_n^+ = E_n^-$ for all values of n. Thus in each case, if we knew the eigenvalues and eigenfunctions of either of the two partner hamiltonians, we could determine the eigenvalues and the eigenfunctions of the other.

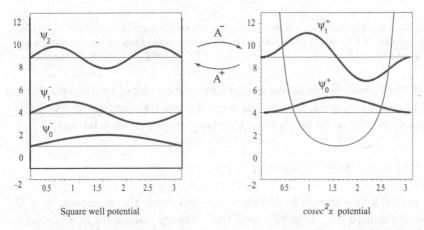

Square well potential $cosec^2 x$ potential

Fig. 4.3 The infinite square well and its $cosec^2 x$ partner potential, together with a few energy levels and the corresponding eigenfunctions. Since this is a case of unbroken supersymmetry, note that operators A^- and A^+ respectively destroy and create nodes of wave functions they act upon.

Problem 4.4. *Determine the normalization constants to show that normalized eigenfunctions $\psi_n^\pm$ are related by*

$$\psi_n^+ = \frac{1}{\sqrt{E_{n+1}^-}} A^- \psi_{n+1}^- ; \quad \text{and} \quad \psi_n^- = \frac{1}{\sqrt{E_{n-1}^+}} A^+ \psi_{n-1}^+ . \quad (4.9)$$

Here is an example specially selected to exhibit the efficacy of this formalism. A judiciously chosen superpotential in this example will yield two partner potentials: one the well known infinite well potential and the other, the less familiar $cosec^2 x$ (the trigonometric Pöschl-Teller or Pöschl-Teller I) potential. We determined spectra for both of these two potentials in Chapter 2, one very easily and the other with considerable difficulty, and found

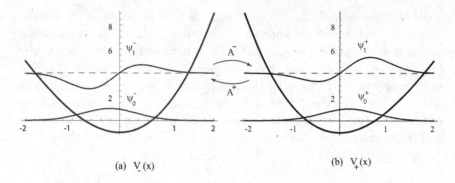

(a) $V_-(x)$ (b) $V_+(x)$

Fig. 4.4   A schematic diagram for two partner potentials for a broken SUSY case. Note that operators A^- and A^+ do not alter the nodal structure of eigenfunctions upon which they act. In addition, the ground state energies $E_0^{(-)}$ and $E_0^{(+)}$ are greater than zero.

that they were the same except for the ground state, for domain $(0, \pi)$. In this chapter, we will solve the second from the first, using SUSYQM formalism. We begin by showing that they are superpotential partners.

4.3   Pöschl-Teller Potential

Consider a superpotential $W(x) = -b \cot x$ with the parameter $b > 0$, and domain $(0, \pi)$. This falls into the category illustrated in Fig. 4.1.c, and thus the supersymmetry is unbroken. The supersymmetric partner potentials are:

$$V_-(x, b) = W^2(x) - \frac{dW}{dx} = b(b-1) \, \mathrm{cosec}^2 x - b^2$$

and   (4.10)

$$V_+(x, b) = W^2(x) + \frac{dW}{dx} = b(b+1) \, \mathrm{cosec}^2 x - b^2 \ .$$

Now for the special case of $b = 1$, the potential $V_-(x, 1)$ is a trivial constant function $-b^2 = -1$; i.e., just an infinite one-dimensional square well potential whose bottom is set to -1 and whose boundaries are at 0 and π, while its partner potential $V_+(x, 1)$ is given by $(2 \, \mathrm{cosec}^2 x - 1)$. Thus, $V_-(x, 1)$ and $V_+(x, 1)$, two very different potentials, have exactly the same eigenvalues except one. Since we can find out the eigenvalues and eigenfunctions of the infinitely deep square well potential from any book on modern physics, SUSYQM automatically gives us all desired properties of its partner, the very nontrivial $(2 \, \mathrm{cosec}^2 x - 1)$ potential. The eigenfunctions of the square well potential $V_-(x, 1)$, are given by $\psi_n^-(x, 1) \sim \sin(nx)$

and the corresponding eigenvalues by $E_n^- = n^2$, $n = 0, 1, 2, \ldots$. Hence, using $\psi_{n-1}^+ \sim A^- \psi_n^-$ and $A^- = \left(\frac{d}{dx} - \cot x\right)$, the eigenspectrum of the $\mathrm{cosec}^2 x$ potential is given by $\psi_{n-1}^+(x, 1) \sim \left(\frac{d}{dx} - \cot x\right) \sin(nx)$ and $E_n^+ = n^2$, $n = 1, 2, 3, \ldots$. Thus, this method gives a step-by-step simple method for recursively generating all eigenfunctions.

Problem 4.5. *Determine the following two functions and plot them:* $\psi_3^-(x, 1)$ *and* $\psi_2^+(x, 1)$.

Problem 4.6. *By explicitly operating with* A^- *and* A^+ *on eigenfunctions* $\psi_3^-(x, 1)$ *and* $\psi_2^+(x, 1)$, *verify that* A^- *removes a node while* A^+ *adds a node to the eigenfunctions of* H_- *and* H_+.

Problem 4.7. *A Dirac-delta function potential* $2\delta(x - x_0)$, *centered at the origin, can be described by a superpotential* $W(x) = 2\Theta(x - x_0) - 1$. *The function* $\Theta(x - x_0)$ *represents a step function at* $x = x_0$; *i.e.,*

$$\Theta(x - x_0) = \begin{cases} 0 & \text{for} \quad x < x_0; \\ \frac{1}{2} & \text{for} \quad x = x_0; \\ 1 & \text{for} \quad x > x_0. \end{cases}$$

(a) Show that this superpotential leads to unbroken supersymmetry.

(b) Determine the eigenfunction corresponding to the zero energy ground state.

(c) From the shape of the partner potential V_+, *determine the number of bound states supported by an attractive Delta function potential.*

Chapter 5

Shape Invariance

In Chapter 4, we learned that a quantum mechanical system generated by a superpotential $W(x, a)$ yields two partner hamiltonians, $H_-(x, a)$ and $H_+(x, a)$. SUSYQM formalism related these two hamiltonians in such a way that if we knew the spectrum of one, we could find the spectrum of the other.

5.1 Shape Invariant Potentials

There is a restricted class of superpotentials $W(x, a)$ for which the partner potentials $V_\pm(x, a)$ obey a special constraint known as "shape invariance." For this class we do not have to know the spectrum of one partner hamiltonian to know the other.

The shape invariance condition states that

$$V_+(x, a_0) = V_-(x, a_1) + R(a_0) , \qquad (5.1)$$

where a_1 is a function of a_0, and $R(a_0)$ is an additive constant. Two shape invariant partner potentials $V_+(x, a_0)$ and $V_-(x, a_1)$ have the same x-dependence but differ from each other in the value of parameters a_i.

In terms of hamiltonians, the shape invariance condition reads

$$H_+(x, a_0) + g(a_0) = H_-(x, a_1) + g(a_1) ; \qquad (5.2)$$

where, we have written $R(a_0) = g(a_1) - g(a_0)$. Since these two hamiltonians differ by a constant, their eigenvalues differ by the same constant and both have the same set of eigenfunctions. I.e., $E_n^+(a_0) = E_n^-(a_1) + g(a_1) - g(a_0)$ and $\psi_n^+(x, a_0) = \psi_n^-(x, a_1)$ for all values of n.

Now, let us see how shape invariance leads to exact solvability. Suppose we want to find the spectrum of $H_-(x, a_0)$. Let us assume that our superpotentials $W(x, a_i)$, $i = 0, 1, \ldots n$ are such that $\psi_0^-(x, a_i) \sim e^{-\int_{x_0}^{x} W(x, a_i)\, dx}$

are normalizable functions; i.e., we have unbroken supersymmetry for these values of parameters. If wave functions are not normalizable, supersymmetry will be broken. (Such systems will be considered in Chapter 6.)

Since supersymmetry is unbroken, we know that the ground state energy $E_0^-(a_i) = 0$, $i = 0, 1, \ldots, n$. From the shape invariance condition Eq. (5.2), we then know that $E_0^-(a_0) = E_0^-(a_1) + g(a_0) - g(a_1) = g(a_0) - g(a_1)$. From the isospectrality condition described in the previous chapter, we have $E_1^-(a_0) = E_0^+(a_0)$. Thus, we already have the two lowest eigenvalues of H_-: $E_0^-(a_0) = 0$ and $E_1^-(a_0) = g(a_0) - g(a_1)$.

Now let us compute the next eigenstate $E_2^-(a_0)$. From our discussion on isospectrality, we know that $E_2^-(a_0)$ is equal to $E_1^+(a_0)$. But from shape invariance we have $E_1^+(a_0) = E_1^-(a_1) + g(a_1) - g(a_0)$. Using isospectrality again, we have $E_1^-(a_1) = E_0^+(a_1)$, which in turn from shape invariance is equal to $E_0^-(a_2) + g(a_2) - g(a_1)$, where a_2 is another parameter that is related to a_1 in exactly the same way as a_1 is related to a_0; i.e., $a_1 = f(a_0)$ and $a_2 = f(a_1)$.

Thus we have

$$
\begin{aligned}
E_2^-(a_0) &\overset{\text{SUSYQM}}{=} E_1^+(a_0) \\
&\overset{\text{SI}}{=} E_1^-(a_1) + g(a_1) - g(a_0) \\
&\overset{\text{SUSYQM}}{=} E_0^-(a_2) + g(a_2) - g(a_1) + g(a_1) - g(a_0) \\
&= g(a_2) - g(a_0) \ ,
\end{aligned}
\tag{5.3}
$$

where we have assumed that the hamiltonian $H_-(x, a_2)$ also corresponds to unbroken SUSY and hence the ground state energy of the system $E_0^-(a_2) = 0$. The general expression for an eigenvalue of $H_-(x, a_0)$ is $E_n^-(a_0) = g(a_n) - g(a_0)$. Now let us find the eigenfunctions. The zero-energy ground state eigenfunction and the first excited state of H_- are given by

$$
\psi_0^-(x, a_0) \sim e^{-\int_{x_0}^x W(x, a_0)\, dx} \ , \quad \text{and}
$$
$$
\psi_1^-(x, a_0) \sim A^+(x, a_0)\, \psi_0^+(x, a_0) \ .
\tag{5.4}
$$

However, from shape invariance we have $\psi_0^+(x, a_0) = \psi_0^-(x, a_1) \sim e^{-\int_{x_0}^x W(x, a_1)\, dx}$. Thus, we have

$$
\psi_1^-(x, a_0) \sim A^+(x, a_0) e^{-\int_{x_0}^x W(x, a_1)\, dx} \ .
\tag{5.5}
$$

Next, let us find $\psi_2^-(x, a_0)$. The ground state eigenfunction of $H_-(x, a_2)$ is given by $\psi_0^-(x, a_2) \sim \exp\left(-\int_{x_0}^x W(x, a_2) dx\right)$. Using Eq. (5.4), the

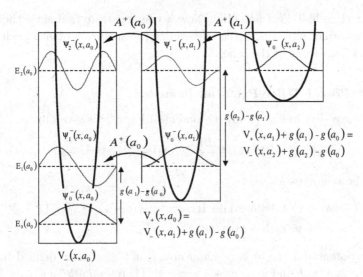

Fig. 5.1 Shape invariant potentials, their energy levels and the corresponding eigenfunctions.

corresponding eigenfunction is obtained as follows:

$$
\begin{aligned}
\psi_2^- (x, a_0) &= \frac{1}{\sqrt{E_1^+(a_0)}} A^+(x, a_0)\, \psi_1^+(x, a_0) \\
&= \frac{1}{\sqrt{E_1^+(a_0)}} A^+(x, a_0)\, \psi_1^-(x, a_1) \\
&= \frac{1}{\sqrt{E_1^+(a_0)\, E_0^+(a_1)}} A^+(x, a_0)\, A^+(x, a_1)\, \psi_0^+(x, a_1) \\
&= \frac{1}{\sqrt{E_1^+(a_0)\, E_0^+(a_1)}} A^+(x, a_0)\, A^+(x, a_1)\, \psi_0^-(x, a_2) \ . \quad (5.6)
\end{aligned}
$$

All eigenfunctions and eigenvalues of the hamiltonian $H_-(x, a_0)$ can be determined by this iterative algorithm. Thus, SUSYQM and shape invariance determine the entire spectrum of a potential without the need to know its partner *a priori*.

5.2 Examples of Shape Invariant Systems

In Chapter 4 we used the spectrum of the infinite square well, which we know from Modern Physics, to determine the spectrum of its partner, the cosec^2x (Pöschl-Teller) potential. Now we demonstrate that since they are shape invariant partners, we can determine the spectrum of either without *a priori* knowledge of the other.[1]

5.2.1 *Pöschl-Teller Potential Revisited*

As we have seen in Eq. (4.10) the potential $V_+(x, b)$ is given by

$$V_+(x, b) = W^2(x) + \frac{dW}{dx} = b(b + 1)\cosec^2 x - b^2 . \qquad (5.7)$$

It can be rewritten as

$$V_+(x, b) = (b + 1)[(b + 1) - 1]\cosec^2 x - (b + 1)^2 + (b + 1)^2 - b^2$$
$$= V_-(x, b + 1) + (b + 1)^2 - b^2 . \qquad (5.8)$$

So the potential $V_-(x, b)$ is a shape invariant potential as defined in Eq. (5.1), with $a_0 = b$ and $a_1 = a_0 + 1 = b + 1$. Therefore, $g(b) = b^2$.

As a special case of the Pöschl-Teller potential, we solve the infinite square well. As we saw earlier, setting $b = 1$ leads to $V_+(x, 1) = 2\cosec^2 x - 1$ and $V_-(x, 1) = -1$, with the latter representing an infinitely deep potential well in the region $0 < x < \pi$. The ground state eigenfunction of $H_-(x, 1)$ is $\psi_0^-(x, 1) \sim e^{-\int W(x,1)dx} \sim e^{\int \cot x dx} \sim \sin x$; its energy is, as usual, 0. Now, we use shape invariance to determine the excited states.

From Eq. (5.8), $V_+(x, 1) = V_-(x, 2) + 3$; thus, $g(2) - g(1) = 3$. Using the fact that the ground state energy $E_0^-(2)$ of $H_-(x, 2)$ is zero (like all H_- ground states where ψ_0^- is normalizable), we find that the ground state energy of $H_+(x, 1)$ is $E_0^+(1) = 3$. But this is just the energy of the first excited state of H_-.

The common ground state eigenfunction of $H_+(x, 1)$ and $H_-(x, 2)$ is given by $\psi_0^+(x, 1) = \psi_0^-(x, 2) \sim e^{-\int W(x,2)dx} \sim e^{\int 2\cot x dx} \sim \sin^2 x$. Then the first excited state of $H_-(x, 1)$ is given by $\psi_0^-(x, 1) \sim A^+(x, 1)\sin^2 x = \left(-\frac{d}{dx} - \cot x\right)\sin^2 x \sim \sin 2x$. Thus, we have derived the energy and the eigenfunction of the first excited state of $H_-(x, 1)$. By iterating this procedure, we can generate the entire spectrum. In particular, the eigenvalues are given by $E_n^-(b) = (b + n)^2 - b^2$.

In the above example, our parameters from Eqs. (5.1) and (5.2) were $a_0 = b$ and $a_1 = b + 1 = a_0 + 1$. This type of shape invariance, for which $a_n = a_{n-1} + \text{constant}$, is known as additive shape invariance. In contrast, shape invariance conditions in which parameters are related by $a_1 = q\, a_0$

---

[1] Note that the shape invariant partners need not look much alike.

are called multiplicative shape invariance. Potentials with multiplicative shape invariance are only known as series, not in closed form, and we will discuss them later.

5.2.2 Harmonic Oscillator

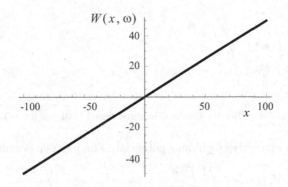

Fig. 5.2   Superpotential for the one dimensional harmonic oscillator for $\omega = 1$.

This system is described by a superpotential $W(x,\omega) = \frac{1}{2}\omega x$. First of all we see that the superpotential is an odd function of x, hence we expect the ground state wave function to be normalizable. The wave function $\psi_0(x)$ is now given by $N\exp\left(-\int_0^x W(x,\omega)\,dx\right) = N\exp\left(-\int_0^x \frac{1}{2}\omega x\,dx\right) = N\exp\left(-\frac{1}{4}\omega x^2\right)$, where N is the normalization constant. This function goes to zero at $\pm\infty$ and can be integrated over the entire domain to give a finite result. Now let us turn to shape invariance. The partner potentials are given by

$$V_\pm(x,\omega) = \frac{1}{4}\omega x^2 \pm \frac{1}{2}\omega\,.$$

Hence, we have

$$V_+(x,\omega) = V_-(x,\omega) + \omega\,.$$

In this case $g(a_1) - g(a_0) = \omega$, and $g(a_n) - g(a_0) = n\omega$. This is the origin of equal spacing of the eigenvalues of the harmonic oscillator.

5.2.3 Coulomb Potential

The superpotential for the Coulomb system is

$$W(r,\ell,e) = \frac{1}{2}\frac{e^2}{\ell+1} - \frac{\ell+1}{r}\,.$$

It represents a particle with angular momentum ℓ interacting with a force

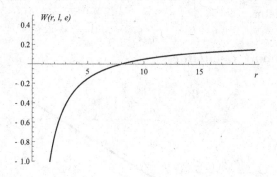

Fig. 5.3   Effective Coulomb superpotential with $\ell = 1, e = 1$.

center via an attractive Coulomb potential. The partner potentials are

$$V_+(r,\ell,e) = \frac{1}{4}\left(\frac{e^2}{\ell+1}\right)^2 - \frac{e^2}{r} + \frac{(\ell+1)(\ell+2)}{r^2}$$

and   (5.9)

$$V_-(r,\ell,e) = \frac{1}{4}\left(\frac{e^2}{\ell+1}\right)^2 - \frac{e^2}{r} + \frac{\ell(\ell+1)}{r^2} .$$

This leads to

$$V_+(r,\ell,e) = V_-(r,\ell+1,e) + \underbrace{\frac{1}{4}\left(\frac{e^2}{\ell+1}\right)^2}_{-g(\ell)} - \underbrace{\frac{1}{4}\left(\frac{e^2}{\ell+2}\right)^2}_{-g(\ell+1)} .$$ (5.10)

Thus, with $a_0 = \ell+1$ and $a_1 = \ell+2$, we have a shape invariant superpotential. $g(a_0)$ is $-\frac{1}{4}\left(\frac{e^2}{\ell+1}\right)^2$ and the energy eigenvalues are given by

$$E_n^{(-)} = g(a_n) - g(a_0) = \frac{1}{4}\left(\frac{e^2}{\ell+1}\right)^2 - \frac{1}{4}\left(\frac{e^2}{\ell+n+1}\right)^2 .$$

The ground state eigenfunction is

$$\psi_0^{(-)}(r,\ell,e) = r^{\ell+1}\exp\left(-\frac{1}{2}\frac{e^2}{\ell+1}r\right) .$$

Problem 5.1. *Compute the three lowest energy eigenfunctions of the Coulomb potential, and determine for what values of its parameters the supersymmetry is unbroken.*

5.2.4 *The Hulthen Potential*

Another example of a shape invariant potential is the Hulthen potential:

$$V(r,a) = -\frac{Z}{a}\frac{e^{-\frac{r}{a}}}{1-e^{-\frac{r}{a}}} . \tag{5.11}$$

It is a short-range potential. For small values of r, i.e., $r << a$, the potential behaves as $-\frac{Z}{r}$; i.e., like an attractive Coulomb potential. Thus, the constant Z can be identified with the atomic number. For large values of r, the potential decreases exponentially: $V(r,a) = -\frac{Z}{a}\exp\left(-\frac{r}{a}\right)$. The Hulthen potential is used in many branches of physics, such as nuclear physics, atomic physics, condensed matter physics, and chemical physics. The model of the three-dimensional delta-function could well be considered as a Hulthen potential with the radius of the force going down to zero.

To determine the spectrum of this potential, let us consider the following superpotential

$$W(r,l) = \frac{a}{l} - \frac{l}{a}\left[\frac{1+e^{-\frac{r}{a}}}{1-e^{-\frac{r}{a}}}\right] = \frac{a}{l} - \frac{l}{2a}\coth\left(r/2a\right) . \tag{5.12}$$

The corresponding partner potentials are

$$V_{\pm}(r,l) = \frac{l(l\pm1)}{4a^2}\operatorname{cosech}^2\left(\frac{r}{2a}\right) - 2\coth\left(\frac{r}{2a}\right) + \left(\frac{l}{2a}\right)^2 + \left(\frac{a}{l}\right)^2 .$$

The potential $V_-(r,l)$, apart from an additive constant, reduces to the Hulthen potential for $l = 1$. After some algebra, we find that

$$V_+(r,l) = V_-(r,l+1) + \underbrace{\left(\frac{l^2}{4a^2} + \frac{a^2}{l^2}\right)}_{-g(l)} - \underbrace{\left(\frac{(l+1)^2}{4a^2} + \frac{a^2}{(l+1)^2}\right)}_{-g(l+1)} .$$

Thus, we see that the superpotential given above is shape invariant, and hence the the spectrum of the Hulthen potential can be determined without recourse to the differential equation.

Problem 5.2. *Obtain expressions for the two lowest energy eigenfunctions and the general expression for eigenvalues of the Hulthen potential.*

Problem 5.3. *For $W(x,a) = a\tanh x$,*

(a) Show that it is a shape invariant potential;
(b) Determine its eigenenergies and first two eigenfunctions;
(c) Show that it holds a finite number of bound states.

Problem 5.4. *The following shape invariant superpotential*

$$W(x, a, b) = a \tan x - b \cot x ,$$

with $0 < x < \frac{\pi}{2}$ and $a > 0$, $b > 0$ is equivalent to

$$W(x, c, d) = c \csc z + d \cot z .$$

Determine parameters c, d and the new variable z in terms of a, b and x. What is the domain of the new superpotential?

5.3 List of Translational Shape Invariant Potentials

We list below several translational shape invariant potentials, and their corresponding potentials and parameters.

- **Harmonic oscillator**

Superpotential:	$W = \frac{1}{2}\omega x ,\quad -\infty < x < \infty$
Potential:	$V_- = \frac{1}{4}\omega x^2 - \frac{1}{2}\omega$
a_0:	a_0
a_1:	$a_0 + 1$
$g(a_0)$:	$g(a_0) = \omega a_0$
Eigenenergies:	$E_n^{(-)} = n\omega$

- **3-D Oscillator**

Superpotential:	$W = \frac{1}{2}\omega r - \frac{\ell+1}{r} ,\quad 0 < r < \infty$
Potential:	$V_- = \frac{\ell(\ell+1)}{r^2} - \frac{(2\ell+3)\omega}{2} + \frac{\omega^2 r^2}{4}$
a_0:	ℓ
a_1:	$\ell + 1$
$g(a_0)$:	$g(a_0) = 2\omega\ell$
Eigenenergies:	$E_n^{(-)} = 2n\omega$

- **Coulomb**

Superpotential:	$W = \frac{e^2}{2(\ell+1)} - \frac{\ell+1}{r} ,\quad 0 < r < \infty$
Potential:	$V_- = \frac{\ell(\ell+1)}{r^2} - \frac{e^2}{r} + \frac{e^4}{4(\ell+1)^2}$
a_0:	ℓ
a_1:	$\ell + 1$
$g(a_0)$:	$g(a_0) = -\frac{e^4}{4(a_0+1)^2}$
Eigenenergies:	$E_n^{(-)} = \frac{1}{4}\left(\frac{e^2}{\ell+1}\right)^2 - \frac{1}{4}\left(\frac{e^2}{\ell+n+1}\right)^2$

- **Morse**

Superpotential:	$W = A - e^{-x}$, $\quad -\infty < x < \infty$
Potential:	$V_- = A^2 + e^{-2x} - (2A+1)e^{-x}$
a_0:	$-A$
a_1:	$-A+1$
$g(a_0)$:	$g(a_0) = -a_0^2$
Eigenenergies:	$E_n^{(-)} = A^2 - (A-n)^2$

- **Scarf (hyperbolic)**

Superpotential:	$W = A \tanh x + B \operatorname{sech} x$, $\quad -\infty < x < \infty$
Potential:	$V_- = -\left[A(A+1) - B^2\right] \operatorname{sech}^2 x$
	$\qquad + (2A+1)B \tanh x \operatorname{sech} x + A^2$
a_0:	$-A$
a_1:	$-A+1$
$g(a_0)$:	$g(a_0) = -a_0^2$
Eigenenergies:	$E_n^{(-)} = A^2 - (A-n)^2$

- **Rosen-Morse (hyperbolic)**

Superpotential:	$W = A \tanh x + \frac{B}{A}$, $\quad -\infty < x < \infty$, $B < A^2$
Potential:	$V_- = -A(A+1) \operatorname{sech}^2 x + 2B \tanh x + A^2 + \frac{B^2}{A^2}$
a_0:	$-A$
a_1:	$-A+1$
$g(a_0)$:	$g(a_0) = -a_0^2 - \frac{B^2}{a_0^2}$
Eigenenergies:	$E_n^{(-)} = A^2 - (A-n)^2 - \frac{B^2}{(A-n)^2} + \frac{B^2}{A^2}$

- **Eckart**

Superpotential:	$W = -A \coth r + \frac{B}{A}$, $\quad 0 < r < \infty$, $B > A^2$
Potential:	$V_- = A(A-1) \operatorname{cosech}^2 r - 2B \coth r + A^2 + \frac{B^2}{A^2}$
a_0:	A
a_1:	$A+1$
$g(a_0)$:	$g(a_0) = -a_0^2 - \frac{B^2}{a_0^2}$
Eigenenergies:	$E_n^{(-)} = A^2 - (A+n)^2 + B^2 \left[\frac{1}{A^2} - \frac{1}{(A+n)^2}\right]$

- **Pösch-Teller (hyperbolic)**

Superpotential:	$W = A \coth r - B \operatorname{cosech} r$, $\quad 0 < r < \infty$, $A < B$
Potential:	$V_- = \left[A(A+1) + B^2\right] \operatorname{cosech}^2 r + A^2$
	$\qquad - (2A+1)B \coth r \operatorname{cosech} r$

a_0: $-A$
a_1: $-A + 1$
$g(a_0)$: $g(a_0) = -a_0^2$
Eigenenergies: $E_n^{(-)} = A^2 - (A - n)^2$

- **Scarf (trigonometric)**

   Superpotential: $W = A \tan x - B \sec x$, $-\frac{\pi}{2} < x < \frac{\pi}{2}$, $A > B$
   Potential: $V_- = \left[A(A-1) + B^2\right] \sec^2 x - A^2$
   $$-(2A-1)B \tan x \sec x$$

   a_0: A
   a_1: $A + 1$
   $g(a_0)$: $g(a_0) = a_0^2$
   Eigenenergies: $E_n^{(-)} = (A + n)^2 - A^2$

- **Rosen-Morse (trigonometric)**

   Superpotential: $W = -A \cot x - \frac{B}{A}$, $0 < x < \pi$
   Potential: $V_- = A(A-1) \operatorname{cosec}^2 x + 2B \cot x - A^2 + \frac{B^2}{A^2}$
   a_0: A
   a_1: $A + 1$
   $g(a_0)$: $g(a_0) = -\frac{B^2}{a_0^2} + a_0^2$
   Eigenenergies: $E_n^{(-)} = (A + n)^2 - A^2 + B^2 \left[\frac{1}{A^2} - \frac{1}{(A+n)^2}\right]$

Problem 5.5. *Determine relative values of parameters A and B such that the Eckart potential holds at least three bound states.*

5.4 Scattering in SUSYQM

So far, we had concentrated on shape invariance and its effect on bound systems; in particular, their eigenspectra. In this section, we will see that shape invariance has implications for scattering states as well.

Scattering plays a very important role in the probing and modeling of atomic and sub-atomic interactions. Our early understanding of the structure of atoms came from scattering experiments. In nuclear physics as well, much of the information available was obtained from scattering of judiciously chosen projectiles by nuclei.

In this section we will study scattering by a one-dimensional localized potential. In one dimension, as particles are hurled at the center of the interaction, some particles are reflected back and others are transmitted. The quantities of interest are the ratios of particles reflected and transmitted to the number of incident particles, observed very far from the localized potential where particles can be assumed to be in interaction free zones. These ratios are known respectively as the reflection and transmission amplitudes.

For particles to be free far away from a localized potential; i.e., described by a sinusoidal wavefunction, the potential must tend towards a finite asymptotic value as $x \to -\infty$, or $x \to \infty$, or both. Let us denote these finite values by $V_{\pm\infty}$. We assume that incident free particles approach the potential from the left, and hence are described by a plane wave[2] e^{ikx}. In such cases, the asymptotic limits of the wave functions with energy E are given by

$$\lim_{x \to -\infty} \psi(x, E) \sim e^{ikx} + r(k) e^{-ikx} ,$$

and

$$\lim_{x \to \infty} \psi(x, E) \sim t(k') e^{ik'x} , \tag{5.13}$$

where k and k' are the asymptotic momenta of the particle. They are related to the energy by $E = k^2 + V_{-\infty} = k'^2 + V_{+\infty}$. Reflection and transmission amplitudes $r(k)$ and $t(k)$ are related respectively to the coefficient of reflection by $\mathcal{R} = |r(k)|^2$ and the coefficient of transmission by $\mathcal{T} = |t(k)|^2$. These amplitudes are measures of the relative strengths of the reflected and transmitted waves, ($r(k) e^{-ikx}$ and $t(k') e^{ik'x}$), compared to the incident waves e^{ikx} from the left.

In supersymmetric quantum mechanics, since we have two partner potentials $V_{\pm} \equiv W^2(x) \pm W'(x)$, we will have to consider two different asymptotic limits of the type given in Eq. (5.13), one for each potential. Since the potentials reach finite asymptotic values at $\pm\infty$, we assume that so does the superpotential $W(x)$; i.e., $\frac{dW(x)}{dx}\Big|_{\pm\infty} = 0$. Thus, for a SUSYQM system, we obtain

$$\lim_{x \to -\infty} \psi^{\mp}(x, E) \sim e^{ikx} + r^{\mp}(k) e^{-ikx} ;$$

and

$$\lim_{x \to \infty} \psi^{\mp}(x, E) \sim t^{\mp}(k') e^{ik'x} . \tag{5.14}$$

Momenta k and k' are given by $\sqrt{E - (W_{-\infty})^2}$ and $\sqrt{E - (W_{+\infty})^2}$, where $W_{\pm\infty}$ are the asymptotic values of the superpotential as $x \to \pm\infty$.

---

[2]Note that this term is multiplied by a time-dependent term $e^{-i\omega t}$ and thus incident particles are appropriately represented by $e^{i(kx - \omega t)}$, a plane wave moving toward increasing x.

Now, using SUSYQM, we will show that r^+ and r^-, and similarly t^+ and t^- are related to each other.[3] The wave functions of H_+ and H_- are related by

$$\left(\frac{d}{dx} + W(x)\right) \psi_E^-(x) = N_1 \, \psi_E^+(x)$$

$$\left(-\frac{d}{dx} + W(x)\right) \psi_E^+(x) = N_2 \, \psi_E^-(x) \,,$$

(5.15)

where N_1 and N_2 are constants. Applying these conditions to the asymptotic forms given in Eq. (5.14), we obtain

$$\left(\frac{d}{dx} + W_{-\infty}\right) \left(e^{ikx} + r^-(k)\,e^{-ikx}\right) = N_1 \left(e^{ikx} + r^+(k)\,e^{-ikx}\right) \,,$$

which gives

$$(ik + W_{-\infty})\,e^{ikx} + r^- \left(-ik + W_{-\infty}\right) e^{-ikx} = N_1 \left(e^{ikx} + r^+ e^{-ikx}\right) \,,$$

(5.16)

where we have suppressed the k-dependence of $r^\pm$. Similarly, we obtain

$$(-ik + W_{-\infty})\,e^{ikx} + r^+ \left(ik + W_{-\infty}\right) e^{-ikx} = N_2 \left(e^{ikx} + r^- \, e^{-ikx}\right) \,.$$

(5.17)

Comparing the coefficients of $e^{\pm ikx}$ in Eq. (5.16), we obtain the following relationships:

$$(ik + W_{-\infty}) = N_1 \quad \text{and} \quad r^- \left(-ik + W_{-\infty}\right) = N_1 \, r^+ \,. \qquad (5.18)$$

Solving these two equations yields

$$r^+ = \left(\frac{W_{-\infty} - ik}{W_{-\infty} + ik}\right) r^- \,; \qquad\qquad r^- = \left(\frac{W_{-\infty} + ik}{W_{-\infty} - ik}\right) r^+ \,. \quad (5.19)$$

This identity, which follows from SUSYQM, has profound impact. Because $\left(\frac{W_{-\infty} + ik}{W_{-\infty} - ik}\right)$ is just a phase, the coefficients of reflection $\mathcal{R}^\pm \equiv |r^\pm(k)|^2$, are the same for partner potentials $V_\pm$. If $r^-(k) = 0$, then $r^+(k) = 0$; this describes reflectionless potentials. It also shows that if $W_{-\infty}$ is negative, then $r^-(k)$ has an additional pole (compared to $r^+(k)$) on the positive

---

[3]R. Adhikari, R. De and R. Dutt, "Supersymmetric WKB approach to scattering problems", *Phys. Lett. A* **152**, 381–387 (1991).

A. Khare and U.P. Sukhatme, "Scattering amplitudes for supersymmetric shape-invariant potentials by operator methods", *Jour. Phys. A-Math. & Gen.* **21**, L501 (1988).

imaginary axis of the k-plane. This pole corresponds to the zero-energy eigenstate of H_-. A similar analysis leads to

$$t^+(k) = \left(\frac{W_{-\infty} - ik}{W_{+\infty} - ik'}\right) t^-(k) . \qquad (5.20)$$

Problem 5.6. *Prove Eq. (5.20).*

Problem 5.7. *Show that Eq. (5.17) also generates the same relationship between $r^-(k)$ and $r^+(k)$ as Eq. (5.19).*

Problem 5.8. *Consider a system described by the shape invariant superpotential $W(x, a_0) = a_0 \tanh x$.*

(a) *Determine the partner potentials $V_\pm(x, a_0)$ generated by this superpotential.*

(b) *Using plotting software, plot superpotential $W(x, a_0)$ and partner potentials $V_\pm(x, a_0)$ for $a_0 = 1$ and $-5 < x < 5$.*

(c) *For $a_0 = 1$, determine the coefficients $r^-(k, a_0)$ and $t^-(k, a_0)$.*

So far we have focused on the implications of supersymmetry. Now let us turn to potentials with shape invariance. For such potentials, we have

$$V_+(x, a_i) = V_-(x, a_{i+1}) + R(a_0). \qquad (5.21)$$

Since these two potentials, $V_+(x, a_i)$ and $V_-(x, a_{i+1})$, differ only by a constant term $R(a_0)$, they share the same scattering states as well as bound states. This implies that they must also have the same reflection and transmission amplitudes. I.e., $r^+(k, a_i) = r^-(k, a_{i+1})$ and $t^+(k, a_i) = t^-(k, a_{i+1})$. Hence, we obtain the following recursion relation,

$$r^-(k, a_1) = \left(\frac{W_{-\infty}(a_0) - ik}{W_{-\infty}(a_0) + ik}\right) r^-(k, a_0) . \qquad (5.22)$$

At this point we will consider all translational shape invariant potentials that allow for scattering[4] and explore the consequences of the above recursion relation. But we must note at the outset that we cannot solve the problem entirely without eventual recourse to the Schrödinger equation.

(1) Scarf II Potential: The superpotential for this system is given by

$$W(x, A, B) = A \tanh x + B \operatorname{sech} x . \qquad (5.23)$$

---

[4]Shape invariant potentials such as the harmonic oscillator have no upper bounds. As a result, such potentials have no scattering states.

The asymptotic value of the superpotential W at $-\infty$ is $-A$. Hence, Eq. (5.22) leads to

$$r^-(k, A+1, B) = \frac{(A+ik)}{(A-ik)} \, r^-(k, A, B) \, . \tag{5.24}$$

Since any function can be written as a ratio of two functions, albeit there are an infinitely large number of ways to do so, let us write $r^-(k, A, B)$ as the ratio of two functions $u^-(k, A+ik, B)$ and $v^-(k, A-ik, B)$. Thus,

$$r^-(k, A, B) = \frac{u^-(k, A+ik, B)}{v^-(k, A-ik, B)} \equiv \frac{u^-(k, A', B)}{v^-(k, A'', B)} \, , \tag{5.25}$$

where $A' \equiv A + ik$ and $A'' \equiv A - ik$. Eq. (5.24) now implies

$$\frac{u^-(k, A'+1, B)}{v^-(k, A''+1, B)} = \frac{A'}{A''} \frac{u^-(k, A', B)}{v^-(k, A'', B)} \, . \tag{5.26}$$

This is a difference equation, akin to the recursion relations for special functions of mathematical physics. To solve it, let us compare it with the following property of the Gamma functions: $\Gamma(z+1) = z\,\Gamma(z)$. If a function $f(z)$ obeys $f(z+1) = z\,f(z)$, we have $\frac{f(z+1)}{\Gamma(z+1)} = \frac{f(z)}{\Gamma(z)}$ for all values of the variable z. This is possible only if these ratios are equal to a function $\alpha(z)$ with period one; i.e., $\alpha(z+1) = \alpha(z)$. Employing this property, we see that the dependence of $u^-(k, A', B)$ and $v^-(k, A'', B)$ on A' and A'' should be proportional to $\Gamma(A')$ and $\Gamma(A'')$ respectively. Thus, we can write

$$u^-(k, A', B) = \Gamma(A') \times \alpha_k^u(A, B)$$

and

$$v^-(k, A'', B) = \Gamma(A'') \times \alpha_k^v(A, B) \, ,$$

where $\alpha_k^u(A, B)$ and $\alpha_k^v(A, B)$ are functions of A, B, and k. As noted above, the dependence on A must be a periodic function of period one; i.e., $\alpha_k{}^{u,v}(A+1, B) = \alpha_k{}^{u,v}(A, B)$. Putting all of the above together, we obtain

$$r^-(k, A, B) = \frac{\Gamma(A+ik)}{\Gamma(A-ik)} \, \alpha_k(A, B) \, , \tag{5.27}$$

where α_k, as stated earlier, is a function of k, A and B.

This is as far as we can proceed with SUSYQM and shape invariance. The function α_k needs to be determined from the actual solution of the related Schrödinger equation.

(2) Rosen-Morse II Potential

The superpotential for this system is given by

$$A \tanh x + \frac{B}{A} \quad \text{with} \quad B < A^2 \ . \tag{5.28}$$

The asymptotic value of the superpotential W at $-\infty$ is $-A + B/A$. Hence, Eq. (5.22) leads to

$$r^-_{\text{RM\,II}}(k, A+1) = \left(\frac{-A + \frac{B}{A} - ik}{-A + \frac{B}{A} + ik} \right) r^-_{\text{RM\,II}}(k, A) \tag{5.29}$$

$$= \left(\frac{A^2 + ikA - B}{A^2 - ikA - B} \right) r^-_{\text{RM\,II}}(k, A)$$

$$= \frac{\left[A + \frac{ik}{2} + \frac{i}{2}\sqrt{4B - k^2} \right] \left[A + \frac{ik}{2} - \frac{i}{2}\sqrt{4B - k^2} \right]}{\left[A - \frac{ik}{2} + \frac{i}{2}\sqrt{4B - k^2} \right] \left[A - \frac{ik}{2} - \frac{i}{2}\sqrt{4B - k^2} \right]}$$

$$\times \ r^-_{\text{RM\,II}}(k, A) \ .$$

Using arguments similar to these of the last example, we obtain

$$r^-_{\text{RM\,II}}(k, A) =$$

$$\frac{\Gamma\left[A + \frac{ik}{2} + \frac{i}{2}\sqrt{4B - k^2} \right] \Gamma\left[A + \frac{ik}{2} - \frac{i}{2}\sqrt{4B - k^2} \right]}{\Gamma\left[A - \frac{ik}{2} + \frac{i}{2}\sqrt{4B - k^2} \right] \Gamma\left[A - \frac{ik}{2} - \frac{i}{2}\sqrt{4B - k^2} \right]} \ \alpha_k(A, B) \ . \tag{5.30}$$

Again, this is as far as we can go with SUSYQM and shape invariance. We have extracted information about the structure of these coefficients that follow from SUSYQM. We still need to employ the Schrödinger equation to get complete information on reflection and transmission coefficients. This will be the case for all shape invariant potentials.

Chapter 6

Supersymmetry and its Breaking

In this chapter we will describe the concept of supersymmetry in detail and discuss its implications for SUSYQM.

6.1 Supersymmetry

We have seen that in supersymmetric quantum mechanics the partner hamiltonians are given by $H_- = A^+A^-$ and $H_+ = A^-A^+$, where A^+ and A^- are hermitian conjugate operators of each other. The structure of these hamiltonians guarantees that their eigenenergies are never negative.

$$\langle\phi|H_\mp|\phi\rangle \equiv \langle\phi|A^\pm A^\mp|\phi\rangle = \left(\langle\phi|A^\pm\right)\left(A^\mp|\phi\rangle\right)$$
$$= \left|A^\mp|\phi\rangle\right|^2 \geq 0 \ . \tag{6.1}$$

Since the above statement is true for an arbitrary state of the system, for the ground state we know that the energy is either zero or positive. We will soon see the significance of these two possibilities.

Let us explore the conditions under which we have unbroken or broken supersymmetry. For that let us introduce two new operators $Q^\pm$ that will allow us to express many of the equations we saw earlier in a matrix form:

$$Q^- \equiv \begin{pmatrix} 0 & 0 \\ A^- & 0 \end{pmatrix} \quad \text{and} \quad Q^+ \equiv \begin{pmatrix} 0 & A^+ \\ 0 & 0 \end{pmatrix} \ . \tag{6.2}$$

Multiplying these operators we obtain

$$Q^- Q^+ = \begin{pmatrix} 0 & 0 \\ 0 & A^-A^+ \end{pmatrix} \quad \text{and} \quad Q^+ Q^- = \begin{pmatrix} A^+A^- & 0 \\ 0 & 0 \end{pmatrix} \ . \tag{6.3}$$

We can now define a hamiltonian H in matrix form with $H_\mp$ as the diagonal components:

$$H \equiv \begin{pmatrix} H_- & 0 \\ 0 & H_+ \end{pmatrix} = \begin{pmatrix} A^+A^- & 0 \\ 0 & A^-A^+ \end{pmatrix} \ . \tag{6.4}$$

However, the above hamiltonian is just $Q^-Q^+ + Q^+Q^-$. This expression is known as the anti-commutator of operators $Q^\pm$ and is denoted by $\{Q^-, Q^+\}$. Thus, we have

$$\{Q^-, Q^+\} = H \ . \tag{6.5}$$

Problem 6.1.

(a) Show that $(Q^\pm)^2 = 0$.
(b) Show that operators $Q^\pm$ commute with hamiltonian H; i.e.,

$$[Q^\pm, H] \equiv Q^\pm H - H Q^\pm = 0 \ .$$

The first property

$$[Q^\pm, H] = 0 \ , \tag{6.6}$$

signals the presence of an underlying symmetry in the system. Explicitly it implies that $Q^- H |\psi\rangle - H Q^- |\psi\rangle = 0$, where $|\psi\rangle$ is an eigenstate of the hamiltonian. Thus $H (Q^- |\psi\rangle) = E (Q^- |\psi\rangle)$; i.e., the operation of $Q^\pm$ on state $|\psi\rangle$ does not change the energy of the state, and hence represents a symmetry of the system. This is the supersymmetry of SUSYQM, and $Q^\pm$ are called its generators.

The second property of operators

$$(Q^-)^2 = (Q^+)^2 = 0 \ , \tag{6.7}$$

implies that the action of operators $Q^\pm$ twice on any state does not generate a new state, as its magnitude is zero. This reminds one of the Pauli exclusion principle where any attempt to create two fermions in a state is bound to be futile. Because of this behavior of $Q^\pm$ we term them fermionic.

In Chapter 4 we learned that supersymmetry operators $A^\pm$ interconnect eigenstates of $H_\pm$. Let us now see how these interconnections are represented in terms of operators $Q^\pm$ and H. Eigenfunctions $\psi_n^-(x)$ and $\psi_{n-1}^+(x)$ of hamiltonians $H_\mp$ can be recast into column-vector form, $\begin{pmatrix} \psi_n^-(x) \\ 0 \end{pmatrix}$ and $\begin{pmatrix} 0 \\ \psi_{n-1}^+(x) \end{pmatrix}$, and they are then eigenstates of the hamiltonian H.

Problem 6.2. *Verify that vectors $\begin{pmatrix} \psi_n^-(x) \\ 0 \end{pmatrix}$ and $\begin{pmatrix} 0 \\ \psi_{n-1}^+(x) \end{pmatrix}$ are degenerate eigenstates of the hamiltonian H with eigenvalue $E_n^- = E_{n-1}^+$.*

These column vectors are related through the charges $Q^{\mp}$ as follows:

$$Q^- \begin{pmatrix} \psi_n^-(x) \\ 0 \end{pmatrix} \equiv \begin{pmatrix} 0 & 0 \\ A^- & 0 \end{pmatrix} \begin{pmatrix} \psi_n^-(x) \\ 0 \end{pmatrix}$$

$$= \begin{pmatrix} 0 \\ A^- \psi_n^-(x) \end{pmatrix} = \begin{pmatrix} 0 \\ \psi_{n-1}^+(x) \end{pmatrix} , \qquad (6.8)$$

and

$$Q^+ \begin{pmatrix} 0 \\ \psi_n^+(x) \end{pmatrix} \equiv \begin{pmatrix} 0 & A^+ \\ 0 & 0 \end{pmatrix} \begin{pmatrix} 0 \\ \psi_n^+(x) \end{pmatrix}$$

$$= \begin{pmatrix} A^+ \psi_n^+(x) \\ 0 \end{pmatrix} = \begin{pmatrix} \psi_{n+1}^-(x) \\ 0 \end{pmatrix} . \qquad (6.9)$$

Since $Q^\pm$ do not change the energy of a state, these transformations relate two states with the same eigenvalue of H. Thus, while $A^\pm$ connected eigenstates of two different hamiltonians, operators $Q^\pm$ transform two states of H into each other and hence demonstrate the degeneracy of the system.

We know from spin-orbit coupling that integer orbital angular momentum ℓ when added to spin angular momentum $s = \frac{1}{2}$, gives us total angular momentum $j = \ell \pm \frac{1}{2}$. Recalling that integer momentum states are associated with bosons and half-integer states with fermions, we recognize that a coupling between a boson and a fermion always leads to a fermion. Analogously, we expect that operators $Q^\pm$, which we established to have fermionic behavior, applied to a state would change its character from bosonic to fermionic and vice-versa. However, here the concept of bosonic or fermionic is rather abstract. We consider, arbitrarily, the states of the form $\begin{pmatrix} \psi_n^-(x) \\ 0 \end{pmatrix}$ to be bosonic. Then $Q^- \begin{pmatrix} \psi_n^-(x) \\ 0 \end{pmatrix} = \begin{pmatrix} 0 \\ \psi_{n-1}^+(x) \end{pmatrix}$ are fermionic states. Similarly, when Q^+ is applied to a fermionic state $\begin{pmatrix} 0 \\ \psi_n^+(x) \end{pmatrix}$, it gives a state of the form $\begin{pmatrix} \psi_{n+1}^-(x) \\ 0 \end{pmatrix}$, a bosonic state. In short, vectors of the form $\begin{pmatrix} \phi(x) \\ 0 \end{pmatrix}$ and $\begin{pmatrix} 0 \\ \phi(x) \end{pmatrix}$ represent a bosonic and a fermionic state respectively.

In summary, we have defined supersymmetry generators $Q^\pm$. Their anticommutator generates the hamiltonian of the system. Since these operators commute with H, there is supersymmetry in the system.

6.2   Breaking of Supersymmetry

Despite this symmetry structure at the level of the operators, it is possible to have a system where the symmetry is broken. Let us consider the case that supersymmetry is the symmetry of the hamiltonian; i.e., $HQ^\pm - Q^\pm H = 0$. If this were not so we would say that supersymmetry is explicitly broken. But even if H commutes with $Q^\pm$, there is another kind of symmetry breaking known as the spontaneous breaking of supersymmetry. Let us see how that works.

A unitary operator[1] that generates the supersymmetry transformation of a state is of the form $e^{i(\epsilon Q^- + \bar\epsilon Q^+)}$, where ϵ is a parameter that measures how far the state deviates from the original state[2] and $\bar\epsilon$ is its complex conjugate. For $\epsilon = 0$, we obtain the identity, as expected.

The ground state of the system (also sometimes known as the vacuum state), represented by $|0\rangle$, must be unique. It must not change when operated upon by $e^{i(\epsilon Q^- + \bar\epsilon Q^+)}$. In other words, the ground state must be invariant under this symmetry operation. I.e., if supersymmetry is to remain a good symmetry of the system, for an arbitrary value of the parameter ϵ, we must have

$$e^{i(\epsilon Q^- + \bar\epsilon Q^+)}|0\rangle = |0\rangle \ . \tag{6.10}$$

This implies that for unbroken SUSY we must have $Q^\pm |0\rangle = 0$. If $Q^\pm |0\rangle \neq 0$, i.e., if the ground state does not respect the symmetry, we have a case of spontaneous breakdown of supersymmetry.

Problem 6.3. *Show that* $\langle 0|H|0\rangle \neq 0$ *implies that either* $|Q^-|0\rangle| \neq 0$ *or* $|Q^+|0\rangle| \neq 0$, *and hence supersymmetry is broken.*

Since the hamiltonian H is given by the sum $(Q^-Q^+ + Q^+Q^-)$, the supersymmetry is unbroken if and only if the energy of the ground state of hamiltonian H is zero. Non-vanishing of the ground state energy signals the breaking of SUSY.

If the supersymmetry is unbroken, we have

$$\begin{pmatrix} A^+A^- & 0 \\ 0 & A^-A^+ \end{pmatrix} \begin{pmatrix} \phi^-(x) \\ \phi^+(x) \end{pmatrix} = 0 \ . \tag{6.11}$$

---

[1]Unitary operators play a very important role in quantum mechanics. Their operation on a vector does not change the length of that vector. A unitary operator U obeys $U^\dagger U = U U^\dagger = 1$. Thus, the action by a unitary operator can be viewed as a generalization of multiplication by a complex number of modulus unity.

[2]Here the exponentiation of an operator is to be understood in terms of a power series. I.e. $e^A = 1 + A + \frac{1}{2!}A^2 + \frac{1}{3!}A^3 + \cdots$.

I.e., we must have $A^+A^-\phi^-(x) = 0$, and $A^-A^+\phi^+(x) = 0$. These in turn imply

$$A^-\phi^-(x) = 0, \quad \text{and} \quad A^+\phi^+(x) = 0 . \tag{6.12}$$

Problem 6.4. *Show that Eq. (6.11) leads to Eqs. (6.12).*

The solutions of Eqs. (6.12) are

$$\phi^\pm(x) = N_\pm e^{\pm \int_{x_0}^x W(x',a_0)\, dx'} ;$$

where $N_\pm$ are constants of normalization. This implies $\phi^+(x) \sim 1/\phi^-(x)$. Hence, if $\phi^-(x)$ is normalizable; i.e., it goes to zero at infinity, the function $\phi^+(x)$ would not be a normalizable function as it would blow up at infinity. The only way to have a normalizable zero energy eigenstate is for either $\phi^-(x)$ or $\phi^+(x)$ to vanish identically. Let us assume that it is the function $\phi^+(x)$ that is zero. Then the ground state of the system has the form:

$$\begin{pmatrix} \phi^-(x) \\ 0 \end{pmatrix} .$$

Since the function $\phi^-(x)$ is an eigenstate of the operator A^+A^- with eigenvalue zero, we will denote it by $\psi_0^{(-)}(x)$. This function is then given by

$$\psi_0^{(-)}(x) = N e^{-\int_{x_0}^x W(x',a_0)\, dx'} .$$

As we saw in Chapter 4 normalizability of the state requires that wave function $\psi_0^{(-)}(x)$ vanishes at $x = \pm\infty$. This implies that $\int_{x_0}^{\pm\infty} W(x',a_0)\, dx'$ must be ∞. For the integral $\int_{x_0}^\infty W(x',a_0)\, dx'$ to be unbounded requires that function $W(x',a_0)$ must have a positive value as $x \to \infty$. Similarly, for $\int_{x_0}^{-\infty} W(x',a_0)\, dx'$ to be unbounded requires that function $W(x',a_0)$ must have a negative value as $x \to -\infty$. Thus, the function $W(x',a_0)$ must have an odd number of zeroes on the real axis for the normalizability of the ground state function $\psi_0^{(-)}(x)$, guaranteeing the existence of a zero energy state for H and the presence of supersymmetry. An odd function automatically guarantees all of above. In Fig. 6.1, we show two examples of W's, one for infinite domain and other for a finite domain, that correspond to a system with zero energy ground state and hence unbroken SUSY.

On the other hand, if $W(x',a_0)$ has an even number of zeroes, or no zeroes at all, then SUSY is automatically broken spontaneously. In Fig. 6.2 we depict some examples of $W(x',a_0)$ that lead to a system with spontaneously broken supersymmetry.

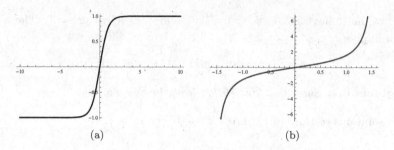

Fig. 6.1   Examples of superpotential $W(x', a_0)$ related to unbroken SUSY.

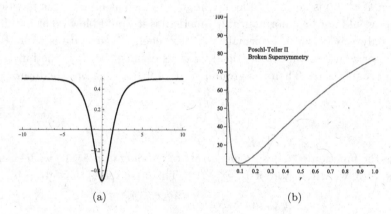

Fig. 6.2   Examples of superpotential $W(x', a_0)$ related to broken SUSY.

6.3   The Witten Index

In broken supersymmetric cases, there is no zero energy ground state. Hence the number of fermionic states N_F (eigenstates of H with upper component zero) is equal in number to those of the bosonic type N_B (eigenstates of H with lower component zero). In the unbroken case the number of bosonic states is one greater than that of fermionic states. Let us define an operator F that gives zero acting on a bosonic state and returns a fermionic state unchanged. Thus, bosonic and fermionic states would be eigenstates of operator $(-)^F$ with eigenvalues 1 and -1 respectively. Edward Witten[3] devised an index[4] $\Delta(\beta) = Tr\,(-)^F\,e^{-\beta H}$ that is zero if and only if the difference $N_B - N_F$ is zero. This index determines whether SUSY is

---

[3] E. Witten, "Constraints on supersymmetry breaking", *Nucl. Phys. B* **202**, 253 (1982).
[4] Trace Tr is the sum of all eigenvalues of an operator.

broken or not. For the broken case it returns 0; for the unbroken case it gives 1. Nontrivial applications of this index are found in field theory.

Problem 6.5. *Show that* $N_B - N_F = 0$ *indeed implies that the index* $\Delta(\beta) = Tr\,(-)^F\,e^{-\beta H} = 0$.

Problem 6.6. *Show that* $N_B - N_F = 1$ *indeed implies that the index* $\Delta(\beta) = Tr\,(-)^F\,e^{-\beta H} = 1$.

6.4 Two-Step Shape Invariance for Systems with Broken Supersymmetry

In Chapter 5, the shape invariance of potentials was utilized to determine eigenvalues and eigenfunctions of various hamiltonians. These depend crucially on the system having a zero energy ground state; we could not have done this without supersymmetry. In this chapter we analyze shape invariant systems with broken supersymmetry. I.e., they are described by shape invariant potentials where values of the parameters leave the supersymmetry spontaneously broken. For these systems, the usual step-by-step procedure of Chapter 5 does not work.

At the outset, let us be clear that we are only interested in studying those potentials that admit bound states. In the list of potentials given in Chapter 5, only three of them hold bound states in the broken supersymmetric phase. For example, the Morse potential, described by the superpotential $W(x, A, B) = A - B e^{-x}$, does not hold bound states in the broken SUSY phase given by $A < 0$. This can be verified by drawing partner potentials for this superpotential for various values of A. It is important to note that since $A - B e^{-x} \equiv A - e^{-(x-x_0)}$, where $x_0 = \ln B$, the parameter B only moves the origin for x. It cannot affect the nature of supersymmetry. This is why we leave it set to 1.

We will now work with three potentials that hold bound states in their broken SUSY phase. These are: the three dimensional harmonic oscillator, the Pöschl-Teller I, and the Pöschl-Teller II.

There exists a variant of shape invariance that is referred to as a two-step shape invariance[5] that will help us determine the spectrum and eigenstates for these systems with broken SUSY. The first step converts the initial superpotential with broken SUSY into one where SUSY is not broken. The

---

[5] A. Gangopadhyaya, J. V. Mallow, U. P. Sukhatme, "Broken supersymmetric shape invariant systems and their potential algebras", *Phys. Lett. A* **283**, 279–284 (2001).

second step uses the method of Chapter 5 to determine the spectrum. We will demonstrate this by examples.

6.4.1   Examples of Two Step Shape Invariance

We consider first the example of the Pöschl-Teller I superpotential

$$W(x, A, B) = A \tan x - B \cot x \; ; \quad 0 < x < \pi/2 \; . \qquad (6.13)$$

The supersymmetric partner potentials are given by

$$
\begin{aligned}
V_-(x, A, B) &= A(A - 1) \; \sec^2 x \\
&\quad + B(B - 1) \; \mathrm{cosec}^2 x - (A + B)^2 \; ; \qquad (6.14) \\
V_+(x, A, B) &= A(A + 1) \; \sec^2 x + B(B + 1) \; \mathrm{cosec}^2 x - (A + B)^2 \; .
\end{aligned}
$$

The zero energy ground state of the system, if normalizable, would be given by $\exp\left(-\int^x W(y, A, B) \, dy\right) \equiv \cos^A x \sin^B x$. For $A > 0, B > 0$, the function $\psi_0^{(-)}(x, A, B)$ is normalizable and proportional to the ground state wave function, implying unbroken SUSY. Similarly, for $A < 0, B < 0$, the function $1/\psi_0^{(-)}(x, A, B)$ is normalizable, and thus has unbroken SUSY.

For $A > 0$, and $B < 0$, $\psi_0^{(-)}(x, A, B)$ is not normalizable due to its divergent behavior at $x = 0$ while its reciprocal $\frac{1}{\psi_0^{(-)}}$ is not normalizable due to its divergent behavior at $x = \frac{\pi}{2}$, and hence SUSY is necessarily broken. Likewise, for $A < 0$, and $B > 0$, once again the SUSY is broken. As mentioned before, the broken SUSY case has no zero energy state, and the standard shape invariance procedure of Chapter 5 cannot be used to obtain the eigenstates and eigenvalues.

Let us focus on the case $A > 0$, $B < 0$. The eigenstates of $V_\pm(x, A, B)$, even without the presence of SI, are related by

$$
\begin{aligned}
\psi_n^{(+)}(x, a_0) &= A(x, a_0) \, \psi_n^{(-)}(x, a_0) \; ; \\
\psi_n^{(-)}(x, a_0) &= A^+(x, a_0) \, \psi_n^{(+)}(x, a_0) \qquad (6.15)
\end{aligned}
$$

$$\text{and}$$

$$E_n^{(-)}(a_0) = E_n^{(+)}(a_0) \; .$$

The potentials of Eq. (6.14) are shape invariant. In fact there are two possible relations between parameters such that these two potentials exhibit shape invariance. One of them is the conventional $(A \to A+1, \; B \to B+1)$. The shape invariance condition is given by

$$V_+(x, A, B) = V_-(x, A + 1, B + 1) + (A + B + 2)^2 - (A + B)^2 \; . \quad (6.16)$$

For B sufficiently large and negative, $B+1$ is also negative; thus the super-potential resulting from this change of parameters still falls in the broken SUSY category, and hence $E_0^{(-)}(a_0) \neq 0$. In the absence of this crucial result, even with shape invariance, we are not able to proceed further.

The second possibility is $(A \to A+1, \quad B \to -B)$. The corresponding relationship is given by

$$V_+(x, A, B) = V_-(x, A+1, -B) + (A - B + 1)^2 - (A + B)^2 . \quad (6.17)$$

This change of parameters $(A \to A+1, \quad B \to -B)$ leads to a system with unbroken SUSY, since the parameter B changes sign. Hence the ground state of the system with potential $V_-(x, A+1, -B)$ is guaranteed to be at zero energy. From Eq. (6.17) we see that potentials $V_+(x, A, B)$ and $V_-(x, A+1, -B)$ differ only by a constant, hence we have

$$\psi_+(x, A, B) = \psi_-(x, A+1, -B) , \quad \text{and}$$
$$E_n^{(+)}(A, B) = E_n^{(-)}(A+1, -B) + (A+1-B)^2 - (A+B)^2 . \quad (6.18)$$

Thus, if we knew the spectrum of the unbroken SUSY $H_-(x, A+1, -B)$, we would be able to determine the spectrum of $H_+(x, A, B)$ with broken SUSY. Since the parameters of the potential $V_-(x, A+1, -B)$ lie in the region necessary for unbroken SUSY, we have machinery at hand to determine the eigenstates of this potential. The results are

$$E_n^{(-)}(A+1, -B) = (A+1-B+2n)^2 - (A+1-B)^2 . \quad (6.19)$$

When combined with Eqs. (6.15) and (6.17), we obtain

$$E_n^{(-)}(A, B) = (A+1-B+2n)^2 - (A+B)^2 ; \quad \text{and}$$
$$\psi_n^{(-)}(y, A, B) = (1+y)^{(1-A)/2}(1-y)^{B/2} P_n^{(B-1/2, 1/2-A)}(y) , \quad (6.20)$$

where $y = \cos 2x$ and $P_n^{(\alpha, \beta)}(y)$ are Jacobi polynomials.

Similar analyses can be used for determining the eigenvalues and eigenstates of the three dimensional harmonic oscillator and the Pöschl-Teller II potentials.

6.4.1.1 *Three-Dimensional Harmonic Oscillator with Broken SUSY*

The three-dimensional harmonic oscillator with broken SUSY is described by the superpotential

$$W(r, \ell, \omega) = \frac{1}{2}\,\omega r - \frac{\ell+1}{r} ; \quad \ell < -1 . \quad (6.21)$$

The function $\psi_0^{(-)}(r, \ell, \omega) = \exp\left(-\int^r W(r', \ell, \omega)\, dr'\right) = r^{\ell+1}e^{-\omega r^2}$ diverges near $r \to 0$ for $\ell < -1$ and hence corresponds to the case of broken SUSY. The supersymmetric partner potentials are

$$V_+(r, \ell, \omega) = \frac{\omega^2 r^2}{4} + \frac{(\ell+1)(\ell+2)}{r^2} - \left(\ell + \frac{1}{2}\right)\omega , \quad \text{and}$$

$$V_-(r, \ell, \omega) = \frac{\omega^2 r^2}{4} + \frac{\ell(\ell+1)}{r^2} - \left(\ell + \frac{3}{2}\right)\omega . \tag{6.22}$$

These two partner potentials are shape invariant since

$$V_+(r, \ell, \omega) = V_-(r, \ell+1, \omega) + 2\omega . \tag{6.23}$$

For sufficiently large negative values of ℓ, the potential $V_-(r, \ell+1, \omega)$ also lies in the realm of broken SUSY. However, there is another change of parameters that maintains shape invariance

$$V_+(r, \ell, \omega) = V_-(r, -\ell-2, \omega) - (2\ell+1)\omega . \tag{6.24}$$

Since $\ell < -1$, the potential $V_-(r, -\ell-2, \omega)$ has unbroken SUSY and thus has a zero energy ground state. Combining the results of Eqs. (6.22), (6.23) and (6.24), we obtain

$$E_n^{(+)}(\ell, \omega) = (2n - 2\ell - 1)\omega . \tag{6.25}$$

6.4.1.2 *Pöschl-Teller II Potential with Broken SUSY*

Our last example is the Pöschl-Teller II potential described by

$$W(r, A, B) = A \tanh r - B \coth r ; \quad 0 < r < \infty . \tag{6.26}$$

The function $\psi_0^{(-)}(r, A, B) \equiv \exp\left(-\int^r W(r', A, B)\, dr'\right)$ is given by $\psi_0^{(-)}(r, A, B) = \cosh^{-A} r \, \sinh^B r$. Here, for $A > 0$ and $B < 0$, neither $\psi_0^{(-)}(r, A, B)$ nor its inverse are normalizable, and hence we have a system with broken SUSY. The supersymmetric partner potentials are given by

$$V_+(r, A, B) = -A(A-1)\operatorname{sech}^2 r + B(B-1)\operatorname{cosech}^2 r + (A+B)^2 ,$$

$$\text{and} \tag{6.27}$$

$$V_-(r, A, B) = -A(A+1)\operatorname{sech}^2 r + B(B+1)\operatorname{cosech}^2 r + (A+B)^2 .$$

Here too we have two possible relations between parameters for these potentials to be shape invariant. They are

$$V_+(r, A, B) = V_-(r, A-1, B-1) + (A+B)^2 - (A+B-2)^2 , \tag{6.28}$$

and

$$V_+(r, A, B) = V_-(r, A - 1, -B) + (A + B)^2 - (A - B - 1)^2 . \quad (6.29)$$

As in the previous two examples, the first transformation does not lead to the determination of the spectrum since the new parameters $(A - 1, B - 1)$ still leave the system with broken SUSY for sufficiently large positive values of A. However, in the second transformation, the new values of the parameters $(A - 1, \ -B)$ lie in the domain of unbroken SUSY and hence the spectrum of $V_-(r, A - 1, -B)$ can be determined. The resulting spectrum is given by

$$E_n^{(-)}(A, B) = (A + B)^2 - (A - 1 - B - 2n)^2 .$$

Problem 6.7. *Show that a similar procedure yields the eigenstates of the Pöschl-Teller II and harmonic oscillator potentials in the broken SUSY phase.*

Chapter 7

Potential Algebra

In this chapter we will show that starting from the commutation relations for the angular momentum operators we arrive at a Schrödinger equation for the $\text{sech}^2 x$ potential, and thus we will see that there is a connection between the algebra of angular momentum operators and the $\text{sech}^2 x$ shape invariant potential. We will further show that such a connection can be generalized to all shape invariant potentials. This mathematical structure is called potential algebra.

7.1 The Algebra of Angular Momentum Operators

Let us recall from Chapter 2, in our discussion of the hydrogen atom, that for the time-independent Schrödinger equation in spherical coordinates, the angular solutions are the same for all spherically symmetric $V(r)$; namely, the spherical harmonics $Y(\theta, \phi)$, Eq. (2.44). This is the quantum mechanical version of angular momentum conservation for spherically symmetric potentials. $Y(\theta, \phi)$ is the solution to

$$\frac{1}{\sin\theta}\frac{\partial}{\partial\theta}\left(\sin\theta\frac{\partial Y(\theta,\phi)}{\partial\theta}\right) + \frac{1}{\sin^2\theta}\left(\frac{\partial^2 Y(\theta,\phi)}{\partial\phi^2}\right) + l(l+1)Y(\theta,\phi) = 0 .$$

If we multiply by $\sin^2\theta$ and divide by $Y(\theta, \phi) \equiv \Theta(\theta)\Phi(\phi)$, we separate the angular parts:

$$\frac{1}{\Theta}\left[\sin\theta\frac{\partial}{\partial\theta}\left(\sin\theta\frac{\partial}{\partial\theta}\Theta\right) + l(l+1)\sin^2\theta\right] + \frac{1}{\Phi}\frac{\partial^2\Phi}{\partial\phi^2} = 0 . \qquad (7.1)$$

Then the θ-dependent and ϕ-dependent terms must be separately constant. We thus define

$$\frac{1}{\Theta}\left[\sin\theta\frac{\partial}{\partial\theta}\left(\sin\theta\frac{\partial}{\partial\theta}\Theta\right) + l(l+1)\sin^2\theta\right] \equiv m^2 ,$$

whence

$$\frac{1}{\Phi}\frac{\partial^2 \Phi}{\partial \phi^2} \equiv -m^2 \, ,$$

whose solution is $e^{im\phi}$.

If we demand single-valuedness[1] for $\Phi(\phi)$ over the range 0 to 2π then $m = 0, \pm1, \pm2...$ Substituting this into the Θ equation yields a differential equation whose solutions are the associated Legendre functions $P_l^m(\cos\theta)$, themselves obtainable from the Legendre polynomials $P_l(\cos\theta)$ via

$$P_l^m(x) \equiv (1 - x^2)^{|m|/2}\left(\frac{d}{dx}\right)^m P_l(x) \, .$$

$P_l(x)$ is itself generated by

$$P_l(x) = \frac{1}{2^l l!}\frac{d}{dx}^l (x^2 - 1)^l \, .$$

Note that these definitions require that m and l be integers. The spherical harmonics $Y_l^m(\theta, \phi) = \Theta(\theta)\Phi(\phi)$ have only integer values of m and l.

All of these arguments for integer quantum numbers are self-consistent. The question is, do they cover all possibilities? Is it possible to have an angular momentum described by non-integer m and l? We will examine this question by starting first with the standard definition for angular momentum: $\mathbf{L} \equiv \mathbf{r} \times \mathbf{p}$. We will then write out its components $\mathbf{r} \equiv (x_1, x_2, x_3)$ and $\mathbf{p} \equiv (p_1, p_2, p_3)$, where $p_j \equiv -i\frac{\partial}{\partial x_j}$. (As usual, we take $\hbar = 1$.) From the commutation relations $[x_j, p_k] = i\,\delta_{jk}$, and the commutator theorem $[AB, C] = A[B, C] + [A, C]B$ we can obtain commutation relations between the components (L_1, L_2, L_3) of $\mathbf{L}$.

Problem 7.1. *From the components of $\boldsymbol{L} \equiv \boldsymbol{r} \times \boldsymbol{p}$, show that*

(a) *for i, j, k cyclic indices,* $[L_i, x_i] = 0$, $[L_i, x_j] = ix_k$, $[L_i, x_k] = -ix_j$, $[L_i, p_i] = 0$, $[L_i, p_j] = ip_k$, *and* $[L_i, p_k] = -ip_j$.
(b) $[L_i, L_j] = iL_k$ *for i, j, k cyclic indices.*
(c) $[L^2, L_i] = 0$.

Since L^2 commutes with each of its components, it shares the same set of eigenfunctions with each component. But since the components do not

---

[1]See the following two books for better arguments:

A modern approach to quantum mechanics by John S. Townsend, 2nd Edition, 2012 University Science Books.

Quantum Mechanics: Fundamentals by Kurt Gottfried, 1966, W. A. Benjamin.

commute with each other, *they* cannot have the same set of eigenfunctions. We can thus choose only one of the components at a time and examine its relationship to L^2. As is customary, we will select L_3. If we call the common eigenfunctions f, then $L^2 f = \lambda f$ and $L_3 f = \mu f$.[2]

Now recall that in Chapter 3 we defined raising and lowering operators $A^\pm$ from which we were able to obtain the energy eigenvalues of the hamiltonian H. We shall use the same technique here, to obtain the respective eigenvalues λ and μ of L^2 and L_3, and relate them to each other. We define raising and lowering operators $L_\pm \equiv L_1 \pm iL_2$. We of course need to show that these are indeed raising and lowering operators, as we did for $A^\pm$. Recall that in that case, A^+ and A^- did not commute with each other: $[A^-, A^+] = 1$, nor with H: $[H, A^\pm] = \pm\omega A^\pm$. Indeed that is precisely what allowed us to build the ladder.

In the present case, L^2 will not do to build a ladder, since it commutes with each of its components, and therefore with any combination of them. We are then left with examining the commutators of $L_\pm$ with L_3.

$$[L_3, L_+] = [L_3, L_1] + i[L_3, L_2]$$

$$= iL_2 + i(-i)L_1$$

$$= iL_2 + L_1 = L_+ .$$

Similarly, $[L_3, L_-] = -L_-$ and $[L_+, L_-] = 2L_3$. We refer to this as "closing the algebra," since their commutators generate each other and no other operators.

Then for any eigenfunction of L_3,

$$L_+ f = (L_3 L_+ - L_+ L_3)f = (L_3 L_+ f - \mu L_+ f) .$$

$$L_3(L_+ f) = (\mu + 1)(L_+ f) .$$

So $L_+ f$ is an eigenfunction of L_3 with eigenvalue $\mu + 1$. Similarly, $L_- f$ is an eigenfunction of L_3 with eigenvalue $\mu - 1$. We have thus built a ladder of eigenfunctions and eigenvalues for L_3.

Now $\langle L^2 \rangle = \langle L_1^2 \rangle + \langle L_2^2 \rangle + \langle L_3^2 \rangle$ so $\langle L_3^2 \rangle \leq \langle L^2 \rangle$. This means that there is a μ_{max}. Similarly there exists a μ_{min}. That is, we have a ladder bounded at top and bottom. If we call f_{top} the eigenfunction corresponding to μ_{max}, and f_{bottom} the eigenfunction corresponding to μ_{min}, then

---

[2]The choice of these particular Greek letters is not arbitrary: λ will be related to l and μ to m.

"bounded" means $L_+ f_{\text{top}} = 0$ and $L_- f_{\text{bottom}} = 0$. We will use these relationships to obtain λ.

$$L_- L_+ = (L_1 - iL_2)(L_1 + iL_2) = L_1^2 + L_2^2 + i(L_1 L_2 - L_2 L_1) = L_1^2 + L_2^2 - L_3.$$

But $L_1^2 + L_2^2 = L^2 - L_3^2$. So

$$L_- L_+ = L^2 - L_3^2 - L_3 .$$

Similarly

$$L_+ L_- = L^2 - L_3^2 + L_3 .$$

We apply $L_- L_+$ to f_{top}, obtaining

$$0 = \left(L^2 f_{\text{top}} - L_3^2 - L_3 \right) f_{\text{top}} = L^2 f_{\text{top}} - \mu_{\text{max}}^2 f_{\text{top}} - \mu_{\text{max}}^2 f_{\text{top}} .$$

Collecting terms,

$$L^2 f_{\text{top}} = \mu_{\text{max}} (\mu_{\text{max}} + 1) f_{\text{top}} .$$

By tradition we define $l \equiv \mu_{\text{max}}$, thus

$$L^2 f_{\text{top}} = l(l+1) f_{\text{top}} .$$

Similarly, operating on f_{bottom} with $L_+ L_-$ and defining $\bar{l} \equiv \mu_{\text{min}}$,

$$L^2 f_{\text{bottom}} = \bar{l}(\bar{l} - 1) f_{\text{bottom}} .$$

All functions f are degenerate with respect to L^2, since it played no role in the generation of the ladder. Thus $l(l+1) = \bar{l}(\bar{l} - 1)$, whose solutions are $l = \bar{l} - 1$ and $l + \bar{l} = 0$. The first cannot be, since by definition, $l > \bar{l}$, so $\bar{l} = -l$. This sets the limits on μ, which by tradition we now call m: $-l \le m \le l$. We have also seen that $L_\pm$ changes m by integer steps. Since we defined l and $\bar{l}(= -l)$ as the maximum and minimum values of m, the range of the ladder $-l \to l$ must be an integer $N = 2l$. So we now discover that $l = N/2$; i.e., l (and therefore m) can take on half-integer as well as integer values.

Let us review how this happened. We started with the definition of angular momentum: $\mathbf{L} \equiv \mathbf{r} \times \mathbf{p}$. We used it to obtain the commutation relations between the components of $\mathbf{L}$. We then started with those commutation relations. Did they lead to $\mathbf{L} \equiv \mathbf{r} \times \mathbf{p}$ and only to that? We have now discovered that while they can of course do so: integer l, m, they also lead to another possibility as well: half-integer l, m. This possibility can be said to produce a new angular momentum, which we call spin.

Let us now return to the case of the orbital angular momentum, $\mathbf{L} = \mathbf{r} \times \mathbf{p}$, and investigate its hitherto abstract operators in their coordinate form, called their "realization". We do this for several reasons. One reason is to show that the eigenfunctions f are indeed what we expect from the analysis of Chapter 2 summarized at the beginning of this chapter; viz., the spherical harmonics $Y_l^m(\theta, \phi)$. The second reason is to connect, somewhat surprisingly, the angular momentum equations with the Schrödinger equation for a particular potential.

We first write the angular momentum in its quantum mechanical form: $\mathbf{L} \equiv \mathbf{r} \times \frac{1}{i}\nabla$. Since we are working with spherically symmetric potentials, we use spherical polar coordinates.

$$\nabla = \hat{r}\frac{\partial}{\partial r} + \hat{\theta}\frac{1}{r}\frac{\partial}{\partial \theta} + \hat{\phi}\frac{1}{r\sin\theta}\frac{\partial}{\partial \phi} \ .$$

Using the cross product relationships between the unit vectors $\hat{r}$, $\hat{\theta}$, $\hat{\phi}$:

$$\hat{r} \times \hat{r} = 0, \ \hat{r} \times \hat{\theta} = \hat{\phi}, \ \hat{r} \times \hat{\phi} = -\hat{\theta} \ ,$$

we obtain

$$\mathbf{L} = \frac{1}{i}\left(\hat{\phi}\frac{\partial}{\partial \theta} - \hat{\theta}\frac{1}{\sin\theta}\frac{\partial}{\partial \phi}\right) \ .$$

Now, however, we need the Cartesian components L_1, L_2, L_3 and the combinations $L_\pm \equiv L_1 \pm iL_2$ in terms of θ and ϕ. We use $\hat{\theta} = (\cos\theta\cos\phi)\hat{x_1} + \cos\theta\sin\phi)\hat{x_2} - (\sin\theta)\hat{x_3}; \phi = -(\sin\phi)\hat{x_1} + (\cos\phi)\hat{x_2}$, and after collecting terms in $\hat{x_1}$, $\hat{x_2}$, $\hat{x_3}$, we obtain

$$L_1 = \frac{1}{i}\left(-\sin\phi\frac{\partial}{\partial \theta} - \cos\phi\cot\theta\frac{\partial}{\partial \phi}\right)$$

$$L_2 = \frac{1}{i}\left(\cos\phi\frac{\partial}{\partial \theta} - \sin\phi\cot\theta\frac{\partial}{\partial \phi}\right)$$

$$L_3 = \frac{1}{i}\frac{\partial}{\partial \phi}$$

$$L_\pm \equiv L_1 \pm iL_2 = \pm e^{\pm i\phi}\left(\frac{\partial}{\partial \theta} \pm i\cot\theta\frac{\partial}{\partial \phi}\right) \ .$$

Writing the eigenfunction as $f = \Theta(\theta)\Phi(\phi)$, $L_3 f = mf$ yields $\Phi(\phi) = e^{im\phi}$. As we recall, $L^2 = L_+L_- + L_3^2 - L_3$. Applying the differential forms of these operators to f yields, after some algebra, the associated Legendre equation for Θ; i.e., Θ is the associated Legendre function P_l^m. Therefore $f = Y_l^m(\theta, \phi)$ are the spherical harmonics.

Problem 7.2.

(a) Apply L_- to Y_2^2 to obtain Y_2^1 up to a normalization constant.
(b) Apply the differential forms of $L+$, L_-, and L_3 to Y_2^2.
(c) Show that Y_2^2 is an eigenfunction of L^2 with the correct eigenvalue.

Now, as promised, we will show that there is a way, using only the angular momentum commutation relations, to get a Schrödinger equation for a particular potential. In so doing, we will obtain the energy eigenvalues simply by use of these relations, rather than by the tedious brute-force analytic method. Note that this is analogous to the harmonic oscillator method, but run backwards. There, one knows the potential, uses it to define raising and lowering operators, and then obtains their commutation relations, from which the ladder of energies can be built. Here we will obtain a potential, knowing the commutation relations: $[L_i, L_j] = iL_k$ for i, j, k cyclic indices, and $[L^2, L_i] = 0$ or alternatively, by $L_\pm = L_1 \pm iL_2$: $[L_3, L_\pm] = \pm L_\pm$. and $[L^2, L_i] = 0$. We have seen that the raising and lowering operators are what we used to construct the ladder of states, as, for example in the above problem, so we suspect (correctly) that the latter form will be more useful.

We want a Schrödinger equation $H\psi = E\psi$; i.e., an eigenvalue equation. Then the operator we construct from the angular momentum operators must be the product of raising and lowering operators to insure that we obtain the same eigenfunction with which we started. The realization of $L_\pm$ is

$$L_\pm \equiv L_1 \pm iL_2 = \pm e^{\pm\phi}\left[\frac{\partial}{\partial\theta} \pm \cot\theta\left(i\frac{\partial}{\partial\phi}\right)\right]. \qquad (7.2)$$

We also want an equation that can be cast as a Schrödinger equation; i.e., with no first derivative.

We choose the product L_+L_-.

$$L_+L_- = -e^{i\phi}\left[\frac{\partial}{\partial\theta} + \cot\theta\left(i\frac{\partial}{\partial\phi}\right)\right] \cdot e^{-i\phi}\left[\frac{\partial}{\partial\theta} - \cot\theta\left(i\frac{\partial}{\partial\phi}\right)\right]$$

$$= -\left[\frac{\partial^2}{\partial\theta^2} + \cot\theta\frac{\partial}{\partial\theta} - \cot^2\theta\left(i\frac{\partial}{\partial\phi}\right)^2 + \left(i\frac{\partial}{\partial\phi}\right)\right]. \qquad (7.3)$$

Acting on an eigenfunction $\Theta(\theta)e^{im\phi}$, the above operator gives:

$$L_+L_-\Theta(\theta)e^{im\phi} = -\left[\frac{\partial^2}{\partial\theta^2} + \cot\theta\frac{\partial}{\partial\theta} - m^2\cot^2\theta + m\right]\Theta(\theta)e^{im\phi}. \qquad (7.4)$$

I.e., the action of the operator on $\Theta(\theta)$ would be very similar to that of the Schrödinger operator if only we did not have the first order derivative $\cot\theta\frac{\partial}{\partial\theta}$.

To accomplish this we change variables to $z = z(\theta)$ where this functional dependence will be determined by demanding that we obtain a Schrödinger type operator in the new variable. This substitution gives $\frac{\partial}{\partial\theta} = \frac{\partial z}{\partial\theta}\frac{\partial}{\partial z}$ and $\frac{\partial^2}{\partial\theta^2} = \left(\frac{\partial z}{\partial\theta}\right)^2\frac{\partial^2}{\partial z^2} + \left(\frac{\partial^2 z}{\partial\theta^2}\right)\frac{\partial}{\partial z}$. We obtain

$$\frac{\partial^2}{\partial\theta^2} + \cot\theta\frac{\partial}{\partial\theta} = \left(\frac{\partial z}{\partial\theta}\right)^2\frac{\partial^2}{\partial z^2} + \left[\left(\frac{\partial^2 z}{\partial\theta^2}\right) + \cot\theta\frac{\partial z}{\partial\theta}\right]\frac{\partial}{\partial z}.$$

If we choose z such that the expression within square bracket vanishes, we will only have a second derivative term in the operator of Eq. (7.4). Solving the constraint $\frac{\partial^2 z}{\partial\theta^2} + \cot\theta\frac{\partial z}{\partial\theta} = 0$, yields $\frac{\partial z}{\partial\theta} = \operatorname{cosec}\theta$; i.e.,

$$z = \ln\left[\tan\left(\frac{\theta}{2}\right)\right]. \tag{7.5}$$

This change of variable maps the domain $(0,\pi)$ to $(-\infty,\infty)$: the domain of the one-dimensional Schrödinger equation. In these new coordinates L_+ and L_- become

$$L_\pm = e^{\pm i\phi}\left(-i\sinh z\frac{\partial}{\partial\phi} \pm \cosh z\frac{\partial}{\partial z}\right), \tag{7.6}$$

while L_3 remains unchanged. Like the A^+A^- operator we have seen before, we would like to see if we can associate the operator L_+L_- with a quantum mechanical hamiltonian. The eigenvalue equation $L_+L_-\psi = \eta\psi$ yields,

$$\left[-\sinh^2 z\frac{\partial^2}{\partial\phi^2} - i\frac{\partial}{\partial\phi} - \cosh^2 z\frac{\partial^2}{\partial z^2}\right]\psi(z,\phi) = \eta\psi(z,\phi). \tag{7.7}$$

Dividing this equation by $\cosh^2 z$, using $\operatorname{sech}^2 z = 1 - \tanh^2 z$, and rearranging the terms, we get

$$\left[-\frac{\partial^2}{\partial z^2} + \left(-i\frac{\partial}{\partial\phi} + \frac{\partial^2}{\partial\phi^2} - \eta\right)\operatorname{sech}^2 z\right]\psi(z,\phi) = \frac{\partial^2}{\partial\phi^2}\psi(z,\phi). \tag{7.8}$$

Analogous to what we did for the spherical harmonics earlier, we separate ψ into its dependence on z and on ϕ, $\psi(z,\phi) = g(z)e^{im\phi}$. Then Eq. (7.8) yields

$$\left[-\frac{\partial^2}{\partial z^2} + \left(m - m^2 - \eta\right)\operatorname{sech}^2 z\right]g(z) = -m^2 g(z). \tag{7.9}$$

Note that this is now a one-dimensional Schrödinger equation.

Since $L_+L_- = L^2 - L_3^2 + L_3$, the eigenvalue η from Eq. (7.7) is given by $\eta = l(l+1) - m^2 + m$. Substituting this expression in Eq. (7.9) leads to the following Schrödinger equation

$$\left[-\frac{\partial^2}{\partial z^2} - l(l+1)\operatorname{sech}^2 z \right] g(z) = -m^2 g(z) \qquad (7.10)$$

where $V(z, m, l) = -l(l+1)\operatorname{sech}^2 z$ is the potential and $-m^2$ is the corresponding energy.

7.2 Algebra of Shape Invariance

In the previous section we learned that there is a connection between the algebra of angular momentum operators and the $\operatorname{sech}^2 x$ shape invariant potential. In this section we will show that all shape invariant potentials can be connected to a potential algebra.

Let us start by observing that the shape invariance condition (5.2) written in terms of A^+ and A^-

$$A^+(z, a_0) A^-(z, a_0) - A^-(z, a_1) A^+(z, a_1) = R(a_0) , \qquad (7.11)$$

is almost a commutation relation, if A^+ and A^- would have the same arguments. This suggests that we could use A^+ and A^- to build some new operators, which we denote by J_+, J_- and J_3, which in analogy with their angular momentum counterparts L_+, L_- and L_3, satisfy similar commutations relations

$$[J_3, J_+] = J_+ , \quad [J_3, J_-] = -J_- , \quad [J_+, J_-] = f(J_3) . \qquad (7.12)$$

When $f(J_3) = 2J_3$, we recover the usual angular momentum algebra case. In general, we design the operators $J_\pm$ and J_3 such that $f(J_3)$ in Eq. (7.12) is connected to $R(a_0)$ and therefore with the shape invariance condition (7.11). Our new algebra will be designed in such a way that it will encode in its third commutation relation an analog of the shape invariance formula (7.11).

Using Eq. (7.11) for guidance, we see that it would resemble the third equation in (7.12) if somehow we make the connection $a_0 \propto J_3$. Let us define

$$J_+ = e^{is\phi} \mathcal{A}^+ , \quad J_- \equiv (J_+)^\dagger = \mathcal{A}^- e^{-is\phi} , \qquad (7.13)$$

where s is a constant parameter. The operator $\mathcal{A}^-$ is obtained from $A^- \equiv A^-(z, a_0)$ by introducing an auxiliary variable ϕ independent of z and replacing the parameter a_0 by a function $\chi(i\partial_\phi)$

$$A^- \equiv A^-(z, a_0) \xrightarrow{a_0 \mapsto \chi(i\partial_\phi)} \mathcal{A}^- \equiv \mathcal{A}^-(z, \chi(i\partial_\phi)) . \qquad (7.14)$$

Similarly $\mathcal{A}^+ = (\mathcal{A}^-)^\dagger$. The reason for the substitution (7.14) will become evident after this short calculation.

$$[J_+, J_-] = J_+ J_- - J_- J_+ \tag{7.15}$$
$$= e^{is\phi} \mathcal{A}^+ (z, \chi(i\partial_\phi)) \, \mathcal{A}^- (z, \chi(i\partial_\phi)) \, e^{-is\phi} - \mathcal{A}^- (z, \chi(i\partial_\phi)) \, \mathcal{A}^+ (z, \chi(i\partial_\phi))$$
$$= \mathcal{A}^+ (z, \chi(i\partial_\phi + s)) \, \mathcal{A}^- (z, \chi(i\partial_\phi + s)) - \mathcal{A}^- (z, \chi(i\partial_\phi)) \, \mathcal{A}^+ (z, \chi(i\partial_\phi)) \, .$$

In the last line, we have used the fact that the operator $\partial_\phi \phi$ acting on an arbitrary function $f(\phi)$ yields $\partial_\phi \phi \, f(\phi) = \phi(\partial_\phi + 1) \, f(\phi)$. This implies $(i\partial_\phi) \, e^{-is\phi} = e^{-is\phi} \, (i\partial_\phi + s)$.

Observe now that the right hand side of Eq. (7.15) is similar to the left hand side of Eq. (7.11) provided that we make the following mappings:

$$a_0 \mapsto \chi(i\partial_\phi) \, , \qquad a_1 \mapsto \chi(i\partial_\phi + s) \, . \tag{7.16}$$

The parameters of shape invariance a_0 and a_1, can always be connected via a function η such that $a_1 = \eta(a_0)$. For example in the case of the Pöschl-Teller potential Eq. (5.8) the change of parameters is a simple translation: $a_1 \equiv \eta(a_0) = a_0 + 1$ where $a_0 = b$. We should remember that the identification, Eq. (7.16) is valid as long as we judiciously choose the function $\chi(i\partial_\phi)$ such that

$$\chi(i\partial_\phi + s) = \eta\left(\chi(i\partial_\phi)\right) \, . \tag{7.17}$$

Equation (7.17) is a rather generous constraint. In the above example, where the change of parameters was the simple translation, the function modeling this translation was the identity function $\chi(z) = z$. Indeed we have in general for translational shape invariance $\chi(z+s) = z+s = \chi(z)+s$, which models perfectly the desired change of parameters for the infinite well and $\mathrm{cosec}^2 x$ partner potentials if $s = 1$.

Next, we define the operator J_3 such that $[J_3, J_\pm] = \pm J_\pm$. Since $\left[-\frac{i}{s} \partial_\phi, e^{\pm is\phi}\right] = \pm e^{\pm is\phi}$, by direct calculation we can check that

$$J_3 = k - \frac{i}{s} \partial_\phi \tag{7.18}$$

satisfies these requirements. Here k is an arbitrary constant.

The last step in constructing the algebra for shape invariant systems is to identify the function $F(J_3)$ from Eq. (7.12). Comparing Eq. (7.11) with its extended form Eq. (7.15), and writing Eq. (7.18) as $sk - sJ_3 = i\partial_\phi$, we have

$$f(a_0) \mapsto f\left(\chi(i\partial_\phi)\right) = f\left(\chi(sk - sJ_3)\right) \equiv F(J_3) \, . \tag{7.19}$$

Putting together all of the above observations, we observe that in fact, we have proved the following theorem:

To any shape invariant system characterized by

$$A^+(z,a_1)A^-(z,a_1) - A^-(z,a_0)A^+(z,a_0) = f(a_0) , \quad a_1 = \eta(a_0) \quad (7.20)$$

we can associate an algebra generated by

$$J_+ = e^{is\phi} \, \mathcal{A}^+ \, (z, \chi(i\partial_\phi)) \quad (7.21)$$

$$J_- = \mathcal{A}^- \, (z, \chi(i\partial_\phi)) \, e^{-is\phi} \quad (7.22)$$

$$J_3 = k - \frac{i}{s} \, \partial_\phi \quad (7.23)$$

satisfying the commutation relations

$$[J_3, J_+] = J_+ , \quad [J_3, J_-] = -J_- , \quad [J_+, J_-] = F(J_3) , \quad (7.24)$$

where

$$F(J_3) = f\left(\chi(s\,k - s J_3)\right) . \quad (7.25)$$

The function f in Eq. (7.25) is given by the shape invariance condition (7.20), while the function χ must satisfy the compatibility equation: $\chi(i\partial_\phi + s) = \eta\left(\chi(i\partial_\phi)\right)$, where η models the change of parameter $a_1 = \eta(a_0)$ of Eq. (7.20).

As another example, let us find the algebra for the cyclic shape-invariant potentials.

7.2.1   Cyclic Potentials

Let us consider a particular change of parameters given by the following cycle (or chain):

$$a_0, \ a_1 = f(a_0), \ a_2 = f(a_1) ,\ldots, \ a_k = f(a_{k-1}) = a_0 , \quad (7.26)$$

and choose $R(a_i) = a_i \equiv \omega_i$. This choice generates a particular class of potentials known as cyclic potentials.[3]

Cyclic potentials form a series of shape invariant potentials; the series repeats after a cycle of k iterations. In Fig. 7.1 we show the first potential $V(x, a_0)$ from a 3-chain ($k = 3$) of cyclic potentials, corresponding to $\omega_0 = 0.15, \omega_1 = 0.25, \omega_2 = 0.60$.

Such potentials have an infinite number of periodically spaced eigenvalues. More precisely, the level spacings are given by $\omega_0, \omega_1, \ldots, \omega_{k-1}, \omega_0, \omega_1, \ldots, \omega_{k-1}, \omega_0, \omega_1, \ldots$, as illustrated in Fig. 7.2.

---

[3]U. Sukhatme, C. Rasinariu, A. Khare, "Cyclic Shape Invariant Potentials," *Phys. Lett. A* **234**, 401–409 (1997).

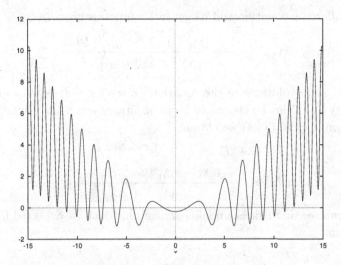

Fig. 7.1   First potential $V(x, a_0)$ from a 3-chain ($k = 3$).

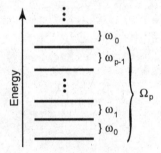

Fig. 7.2   Energy levels of a cyclic shape invariant potential of order p. The lowest level is at zero energy.

In order to generate the change of parameters (7.26) the function f should satisfy $f(f(\dots f(x)\dots)) \equiv f^k(x) = x$. The projective map

$$f(y) = \frac{\alpha y + \beta}{\gamma y + \delta} \quad , \tag{7.27}$$

with specific constraints on the parameters $\alpha, \beta, \gamma, \delta$, satisfies such a condition.

The next step is to identify the algebra behind this model. For this, we need to find the function χ satisfying the equation

$$\chi(z + p) = f(\chi(z)) \equiv \frac{\alpha\chi(z) + \beta}{\gamma\chi(z) + \delta} \quad . \tag{7.28}$$

It is a difference equation and its general solution is given by

$$\chi(z) = \frac{(\lambda_1 - \delta)\lambda_1^{z/p} + (\lambda_2 - \delta)\lambda_2^{z/p}B(z)}{\gamma\left[\lambda_1^{z/p} + \lambda_2^{z/p}B(z)\right]} , \tag{7.29}$$

where $\lambda_{1,2}$ are solutions of the equation $(x - \alpha)(x - \delta) - \beta\gamma = 0$. For simplicity $B(z)$ can be chosen to be an arbitrary constant. Plugging this expression in Eqs. (7.24) we obtain:

$$[J_3, J_\pm] = \pm J_\pm \ ;$$

$$[J_+, J_-] = -\frac{1}{c}\frac{A(\lambda_1 - \delta)\lambda_1^{-J_3} + B(\lambda_2 - \delta)\lambda_2^{-J_3}}{A\lambda_1^{-J_3} + B\lambda_2^{-J_3}} . \tag{7.30}$$

We can see that this is a nonlinear algebra, which we expected from the formalism.

Chapter 8

Special Functions and SUSYQM

An interesting application of shape invariance is the investigation of polynomial functions and their recursion relations.[1]

The process is as follows: We start with a second order differential equation whose solution is a known polynomial. This will in general contain a first order term. We need to recast the differential equation as a Schrödinger type equation in order to apply SUSYQM. That means we must eliminate the first-order term by a suitable transformation. Once this is done, SUSYQM does the rest.

Before we begin, however, let us remind ourselves of the shape invariance relations between adjacent wave functions

$$\psi_{n+1}^{(-)}(x, a_0) = \frac{1}{\sqrt{E_n^{(+)}(a_0)}} A^+(x, a_0)\psi_n^{(-)}(x, a_1) \tag{8.1}$$

$$\psi_n^{(-)}(x, a_1) = \frac{1}{\sqrt{E_{n+1}^{(-)}(a_0)}} A^-(x, a_0)\psi_{n+1}^{(-)}(x, a_0) . \tag{8.2}$$

---

[1]R. Dutt, A. Gangopadhyaya and U. P. Sukhatme, "Noncentral potentials and spherical harmonics using supersymmetry and shape invariance", *Am. Jour. Phys.* **65**, 400 (1997),

H. Rosu and J. R. Guzman, "Gegenbauer polynomials and supersymmetric quantum mechanics", *Nuovo Cimento della Societa Italiana di Fisica* **B 112**, 941 (1997).

C. Husko, B. Knufman, A. Gangopadhyaya, and J. V. Mallow, "Shape Invariance in Supersymmetric Quantum Mechanics and its Application to Selected Special Functions of Modern Physics", *Journal of Young Investigators*, **12**, (2005).

8.1   Hermite Polynomials

As an example, let us start with the Hermite differential equation:

$$\frac{d^2 H_n(\xi)}{d\xi^2} - 2\xi \frac{dH_n(\xi)}{d\xi} + 2n H_n(\xi) = 0 . \qquad (8.3)$$

In order to eliminate the first-order term we write $H_n(\xi) = f_n(\xi)\psi_n(\xi)$. Substituting this into the Hermite equation, we get:

$$\psi_n'' f_n + \psi_n' \left[2f_n' - 2f_n\xi\right] + \psi_n \left[f_n'' - 2f_n'\xi + 2nf_n\right] = 0 . \qquad (8.4)$$

Setting the coefficient of the $\psi_n'(\xi)$ term equal to 0 gives $2f_n' - 2f_n\xi = 0$, which gives $f_n(\xi) - c_n e^{\frac{\xi^2}{2}}$. Replacing $f_n(\xi)$ in Eq. (8.4) and making the substitution $K_n \equiv 2n + 1$ yields:

$$-\psi_n'' + \xi^2 \psi_n = K_n \psi_n . \qquad (8.5)$$

This equation is just the Schrödinger equation for the harmonic oscillator, and the corresponding superpotential is $W(\xi) = a\xi$, where a is a constant. The SUSYQM partner potentials are given by $V_-(\xi, a) = a^2\xi^2 - a$ and $V_+(\xi, a) = a^2\xi^2 + a$. First, note that this set of potentials is shape invariant since

$$V_+(\xi, a) = V_-(\xi, a) + 2a .$$

Note that the parameter a has no subscript j: the shift is the same for all partner potentials.[2] This is why the harmonic oscillator "collapses" into one ladder, as a special case of shape invariance.

Comparison with the potential term in Eq. (8.5) requires that we set $a = 1$. Now, the potential $V_-(\xi, a)$ differs from the traditional harmonic oscillator potential by an additive constant of -1. This is needed to make the SUSYQM ground state equal to 0.

The ground state eigenfunction can be determined using the differential realization of $A^- |0\rangle = 0$.

$$\psi_0^{(-)} = e^{-\int \xi d\xi} = e^{-\frac{\xi^2}{2}} .$$

Using shape invariance, we obtain the higher states by successive application of $A^+(\xi, a_n)$, where $A^+ = \left[-\frac{d}{d\xi} + \xi\right]$.

We can now obtain recursion relations for the Hermite polynomials. Since $H_n(\xi) = f_n(\xi)\psi_n(\xi) = c_n e^{\frac{\xi^2}{2}} \psi_n(\xi)$, inverting this relation gives

$$\psi_n(\xi) = N_n H_n(\xi) e^{-\frac{\xi^2}{2}} , \qquad (8.6)$$

---

[2] In standard texts, $\xi \equiv \sqrt{\left(\frac{2m\omega}{\hbar}\right)}x$, which in our convention is $\sqrt{\omega}x$. Thus, $a = \frac{\omega}{2}$.

where $N_n = (\frac{1}{\pi})^{\frac{1}{4}} \frac{1}{\sqrt{2^n n!}}$ in keeping with the traditional normalization of the Hermite polynomials.

Using Eqs. (8.1) and (8.2) we obtain the following recursion relations:

$$H_n(\xi) = \left[2\xi - \frac{d}{d\xi}\right] H_{n-1}(\xi)$$

and

$$2nH_{n-1}(\xi) = \left[\frac{d}{d\xi}\right] H_n(\xi) .$$

Problem 8.1. *Work out the details for the two recursion relations.*

These are two of the familiar Hermite polynomial recursion relations.

Comparing what we have done here with Chapter 2, in a sense we have run the harmonic oscillator scheme backwards. There, we began with the Schrödinger equation and obtained the wave functions as products of the decaying Gaussian functions and the Hermite polynomials. The recursion relations were not derived, although they might have been found by careful inspection. In the example here, we began with the Hermite Equation, turned it into a Schrödinger equation, then used SUSYQM and shape invariance to derive the recursion relations.

8.2 Associated Legendre Polynomials

This time we will start with the Schrödinger equation for a free particle and connect it to the associated Legendre polynomial. In three dimensions, the Schrödinger equations is given by

$$-\frac{1}{r}\frac{\partial^2}{\partial r^2}(r\psi) - \frac{1}{r^2 \sin\theta}\frac{\partial}{\partial \theta}\left(\sin\theta\frac{\partial\psi}{\partial\theta}\right) - \frac{1}{r^2 \sin^2\theta}\frac{\partial^2\psi}{\partial\phi^2} = E\psi . \tag{8.7}$$

Equation (8.7) permits a solution via separation of variables, if one writes the wave function in the form

$$\psi(r,\theta,\phi) = R(r)P(\theta)e^{im\phi} . \tag{8.8}$$

The equations satisfied by the quantities $R(r)$ and $P(\theta)$ are:

$$\frac{d^2R}{dr^2} + \frac{2}{r}\frac{dR}{dr} + \left(E - \frac{\ell(\ell+1)}{r^2}\right)R = 0 , \tag{8.9}$$

$$\frac{d^2P_{\ell,m}}{d\theta^2} + \cot\theta\frac{dP_{\ell,m}}{d\theta} + \left[\ell(\ell+1) - \frac{m^2}{\sin^2\theta}\right]P_{\ell,m} = 0 . \tag{8.10}$$

Since Eq. (8.10) will be transformed into the associated Legendre differential equation, our object of interest, we will ignore Eq. (8.9) for R. In addition, we note that we explicitly exhibited the ℓ, m-dependence in Eq. (8.10) as these parameters will become the indices for the associated Legendre polynomials P. To solve this equation by the SUSYQM method, we need to first recast it into a one-dimensional Schrödinger-like equation. In order to do this we substitute $\theta = f(z)$, where $f(z)$ is an arbitrary function to be soon determined. This substitution yields

$$\frac{d^2 P_{\ell,m}}{dz^2} + \left[-\frac{f''}{f'} + f' \cot f \right] \frac{dP_{\ell,m}}{dz} + f'^2 \left[\ell(\ell+1) - \frac{m^2}{\sin^2 f} \right] P_{\ell,m} = 0 \ . \tag{8.11}$$

Problem 8.2. *Show that Eq. (8.10) leads to Eq. (8.11).*

In order for Eq. (8.11) to be a Schrödinger-like equation, the first derivative term in will need to vanish. This requires

$$\frac{f''}{f'} = f' \cot f \ . \tag{8.12}$$

Solving the above equation, we obtain

$$\theta \equiv f = 2 \tan^{-1}(e^z) \quad \text{or} \quad e^z = \tan \frac{\theta}{2} \ . \tag{8.13}$$

The range of the new variable z is $-\infty < z < \infty$. From Eq. (8.13), we find $\sin\theta = \text{sech}\, z$ and $\cos\theta = -\tanh z$. Eq. (8.10) now reads

$$-\frac{d^2 P_{\ell,m}}{dz^2} + \left(\ell^2 - \ell(\ell+1) \ \text{sech}^2 z \right) P_{\ell,m} = \left(\ell^2 - m^2 \right) P_{\ell,m} \ . \tag{8.14}$$

As we have seen in Chapter 5, this is the Schrödinger equation for the Pöschl-Teller potential $V_-(z,\ell)$ generated by the superpotential $W(z,\ell) = \ell \tanh z$. Corresponding partner potentials $V_\pm(z,\ell)$ are $\ell^2 - \ell(\ell \mp 1) \ \text{sech}^2 z$, and the shape invariance condition relating them is

$$V_+(z,\ell) = V_-(z,\ell-1) + \ell^2 - (\ell-1)^2 \ .$$

Since the shape invariance condition reduces the value of ℓ in each step, and since ℓ must be positive for $V_-(z,\ell)$ to have the zero-energy ground state, this potential can only support a finite number of bound states. The eigenvalues are

$$E_n = \ell^2 - (\ell-n)^2 \ ; \quad n = 0, 1, 2, \ldots N \ , \tag{8.15}$$

where N is the number of bound states this potential holds, and is equal to the largest integer contained in ℓ. The eigenfunctions $\psi_n(z, \ell)$ are obtained by using supersymmetry operators:

$$\psi_n(z; \ell) \sim A^+(z; \ell) A^+(z; \ell - 1) \cdots A^+(z; \ell - n + 1) \psi_0(z; \ell - n) , \quad (8.16)$$

where $A^+(z; \ell) \equiv \left(-\frac{d}{dz} + \ell \tanh z\right)$. The ground state wave function is $\psi_0(z; \ell) \sim (\mathrm{sech}\, z)^\ell$. From Eq. (8.14), we have $E_n = \ell^2 - m^2$. Comparison with Eq. (8.15) yields

$$n = \ell - m .$$

The corresponding eigenfunctions $\psi_{\ell-m}(z; \ell)$ are, from Eq. (8.10), the associated Legendre polynomials of degree ℓ, and are denoted by $P_{\ell,m}(\tanh z)$. Now that these $P_{\ell,m}(\tanh z)$ functions can be viewed as solutions of a Schrödinger equation, we can apply all the machinery one uses for a quantum mechanical problem. For example, we know that the parity of the n-th eigenfunction of a symmetric potential is given by $(-1)^n$, we readily deduce the parity of $P_{\ell,m}$ to be $(-1)^{\ell-m} = (-1)^{\ell+m}$. Also, the application of supersymmetry algebra results in identities that are either not very well known or not easily available. With repeated application of the A^+ operators, we can determine $P_{\ell,m}$ for a fixed value $(\ell - m)$. As an illustration, we explicitly work out all the polynomials for $\ell - m = 2$. The lowest polynomial corresponds to $\ell = 2, m = 0$. For a general ℓ, using Eq. (8.16), one gets

$$\begin{aligned}
P_{\ell,\ell-2}(\tanh z) &\sim \psi_2(z; \ell) \\
&\sim A^+(z; \ell) A^+(z; \ell - 1) \psi_0(z; \ell - 2) \\
&\sim \left(-\frac{d}{dz} + \ell \tanh z\right) \left(-\frac{d}{dz} + (\ell - 1) \tanh z\right) \mathrm{sech}^{\ell-2} z \\
&\sim \left(-1 + (2\ell - 1) \tanh^2 z\right) \mathrm{sech}^{\ell-2} z .
\end{aligned} \quad (8.17)$$

I.e.,

$$P_{\ell,\ell-2}(x) \sim \left(-1 + (2\ell - 1)x^2\right) \left(1 - x^2\right)^{\frac{\ell-2}{2}} .$$

A similar procedure is readily applicable for other values of n. When one converts back to the original variable θ ($\mathrm{sech}\, z = \sin\theta$; $\tanh z = -\cos\theta$), the results are:

$$P_{\ell,\ell}(\cos\theta) \sim \sin^\ell\theta \quad (8.18)$$

$$P_{\ell,\ell-1}(\cos\theta) \sim \sin^{\ell-1}\theta \cos\theta$$

$$P_{\ell,\ell-2}(\cos\theta) \sim \left[-1 + (2\ell - 1)\cos^2\theta\right] \sin^{\ell-2}\theta$$

$$P_{\ell,\ell-3}(\cos\theta) \sim \left[-3 + (2\ell - 1)\cos^2\theta\right] \sin^{\ell-3}\theta \cos\theta$$

$$P_{\ell,\ell-4}(\cos\theta) \sim \left[3 - 6(2\ell - 3)\cos^2\theta + (2\ell - 3)(2\ell - 1)\cos^4\theta\right] \sin^{\ell-4}\theta .$$

Likewise, there are several recursion relations that are easily obtainable via SUSYQM methods. In particular, by applying A^- or A^+ once, we generate recursion relations of varying degrees (differing in the value of ℓ). From $\psi_{\ell-m}(z, \ell) = A^+(z, \ell)\psi_{\ell-m-1}(z, \ell-1)$, we get

$$P_{\ell,m}(x) = \left((1 - x^2)\frac{d}{dx} + \ell\, x \right) P_{\ell-1,m}, \qquad (8.19)$$

and from $\psi_{\ell-m}(z, \ell) = A^-(z, \ell-1)\psi_{\ell-m+1}(z, \ell+1)$, we obtain

$$P_{\ell,m}(x) = \left((1 - x^2)\frac{d}{dx} + (\ell-1)\, x \right) P_{\ell+1,m}. \qquad (8.20)$$

8.3   Associated Laguerre Polynomials

We now consider the associated Laguerre polynomials, which as we mentioned earlier are the radial parts of the hydrogen atom wave functions. These polynomials are solutions of the equation:

$$x\frac{d^2 L_n^\alpha(x)}{dx^2} + (\alpha + 1 - x)\frac{d L_n^\alpha(x)}{dx} + n L_n^\alpha(x) = 0. \qquad (8.21)$$

Here α is a constant parameter. To cast the above differential equation as a Schrödinger-like equation, we write $L_n^\alpha(x) = f_n^\alpha(x)U_n^\alpha(x)$. Elimination of the first derivative term leads to

$$f_n^\alpha(x) = c_n^\alpha x^{-\frac{(\alpha+1)}{2}} e^{\frac{x}{2}}.$$

Problem 8.3. *Work out the details.*

This substitution gives

$$-U'' + \left[\frac{(\alpha + 1)(\alpha - 1)}{4x^2} - \frac{(2n + \alpha + 1)}{2x} + \frac{1}{4} \right] U = 0. \qquad (8.22)$$

As we shall see in Chapter 11, the coefficient of $\frac{1}{x^2}$; viz., $\frac{(\alpha+1)(\alpha-1)}{4}$ has to be larger than $-\frac{1}{4}$, otherwise the wave function would diverge too strongly at the origin. The spectrum would not be bounded from below; i.e., there could be no $E = 0$ ground state. This constrains the coefficient α to be real.

Equation (8.22) can be generated from the superpotential

$$W(x, n, \alpha) = \frac{2n + \alpha + 1}{2(\alpha + 1)} - \frac{(\alpha + 1)}{2x}.$$

The SUSYQM operators are given by $A = \left(\frac{d}{dx} + \frac{(2n+\alpha+1)}{2(\alpha+1)} - \frac{\alpha+1}{2x} \right)$ and $A^+ = \left(-\frac{d}{dx} + \frac{(2n+\alpha+1)}{2(\alpha+1)} - \frac{\alpha+1}{2x} \right)$ and the partner potentials are given by:

$$V_-(x,n,\alpha) = W^2 - W'$$

$$= \frac{(\alpha+1)(\alpha-1)}{4x^2} - \frac{(2n+\alpha+1)}{4x} + \left[\frac{(2n+\alpha+1)}{2(\alpha+1)}\right]^2 \quad (8.23)$$

$$V_+(x,n,\alpha) = W^2 + W'$$

$$= \frac{(\alpha+1)(\alpha+3)}{4x^2} - \frac{(2n+\alpha+1)}{4x} + \left[\frac{(2n+\alpha+1)}{2(\alpha+1)}\right]^2 . \quad (8.24)$$

They are related by the following shape invariance condition:

$$V_+(x,n,\alpha) = V_-(x,n-1,\alpha+2)$$

$$+ \left[\left(\frac{2n+\alpha+1}{2(\alpha+1)}\right)^2 - \left(\frac{2n+\alpha+1}{2(\alpha+3)}\right)^2\right] . \quad (8.25)$$

The solutions $U_n^\alpha(x)$ of the Schrödinger equation are then related to each other via Eq. (8.1) and Eq. (8.2).

From our definition of U, we can write

$$U_n^\alpha(x) = K_n^\alpha x^{\frac{(\alpha+1)}{2}} e^{-\frac{x}{2}} L_n^\alpha(x) , \quad (8.26)$$

where K_n^α is a constant of normalization. To ensure that the $L_n^\alpha(x)$ have their traditional normalization, we choose

$$K_n^\alpha = \sqrt{\frac{n!}{(2n+\alpha+1)[(n+\alpha)!]^3}} .$$

Substitution of U into Eq. (8.1) gives

$$U_n^\alpha(x) = \frac{1}{\sqrt{E_{n,\alpha}^{(-)}}} A^+(x,n,\alpha) U_{n-1}^{\alpha+2} .$$

A tedious but straightforward simplification yields the following recursion relation among Laguerre polynomials:

$$L_n^a(x) = -\frac{(\alpha+1)}{n(n+\alpha+1)^2} \left[\left(\frac{n+\alpha+1}{\alpha+1}\right) x - (\alpha+2) - x\frac{d}{dx}\right] L_{n-1}^{\alpha+2}(x) . \quad (8.27)$$

An additional recursion relation is obtained via the lowering operation A, Eq. (8.2): $U_{n-1}^{\alpha+2}(x) = \frac{1}{\sqrt{E_{n,a}^{(-)}}} A(x,n,\alpha) U_n^\alpha(x)$. This gives

$$L_{n-1}^{\alpha+2}(x) = -(n+\alpha+1)\left[\frac{n}{x} + \frac{(\alpha+1)}{x}\frac{d}{dx}\right] L_n^\alpha(x) . \quad (8.28)$$

These examples: the Hermite, the associated Legendre, and the associated Laguerre polynomials show the power of SUSYQM with shape invariance. They provide an algorithm for deriving recursion relations for polynomial functions. The technique can be applied to other sets of functions, provided that their differential equations can be recast as Schrödinger-like equations, whose superpotentials can then be determined.

Chapter 9

Isospectral Deformations

Given any one-dimensional potential with n bound states, one can ask the following question: Are there any other potentials that have the same spectrum and the same reflection and transmission coefficients as the original potential? This question is motivated by practical applications. For example, in the case of $\alpha - \alpha$ scattering it was observed that there are ambiguities involved in determining the potential even when the phase shifts and bound states are precisely known.

In this chapter we will show how we can use SUSYQM to build a family of new potentials $\hat{V}_-(\lambda; x)$ that have the same spectrum and the same reflection and transmission coefficients as the "original" potential $V_-(x)$. We will also show that the family of potentials $\hat{V}_-(\lambda; x)$ is unique. That means that starting from $\hat{V}_-(\lambda; x)$ and using the same technique, we do not arrive at a new family of isospectral potentials.

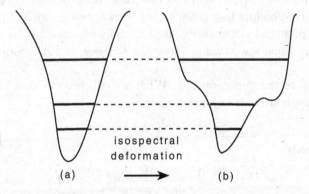

Fig. 9.1 Isospectral deformation. The potentials in (a) and (b) have the same spectrum and the same reflection and transmission coefficients.

The potential family described by $\hat{V}_-(\lambda; x)$ is known as the *isospectral deformation* of $V_-(x)$.

9.1   One-Parameter Isospectral Deformations

In the previous chapters we have seen that for a given potential $V_- = W^2 - W'$ its supersymmetric partner potential $V_+ = W^2 + W'$ has the same energy spectrum with the exception of the ground state $E_0^{(-)}$. We have

$$E_0^{(-)} = 0 \qquad\qquad E_{n+1}^{(-)} = E_n^{(+)} \quad \text{(isospectrality)}$$

$$\psi_n^{(+)} = \frac{d/dx + W(x)}{\sqrt{E_n^{(+)}}}\,\psi_{n+1}^{(-)} \qquad \psi_{n+1}^{(-)} = \frac{-d/dx + W(x)}{\sqrt{E_n^{(+)}}}\,\psi_n^{(+)} \qquad (9.1)$$

where the superpotential $W(x)$ is given by

$$W(x) = -\psi_0^{(-)}(x)'/\psi_0^{(-)}(x) \equiv -\frac{d}{dx}\ln\psi_0^{(-)}(x)\ . \qquad (9.2)$$

Also, as we have seen in Section 5.4, the reflection and transmission amplitudes for the partner potentials $V_\pm(X)$ are related by

$$r_-(k) = \frac{W_- + ik}{W_- - ik}\,r_+(k)\ , \qquad t_-(k) = \frac{W_+ - ik'}{W_- - ik}\,t_+(k)\ , \qquad (9.3)$$

where $k = \sqrt{E - W_-^2}$, $k' = \sqrt{E - W_+^2}$, with $W_\pm = \lim_{x\to\pm\infty} W(x)$.

The almost isospectrality of $V_-(x)$ and $V_+(x)$ suggests that starting from $V_+(x)$ we can investigate whether the corresponding partner potential $V_-(x)$ is unique or not. We will find that there are other partner potentials $\hat{V}_-(\lambda, x)$ corresponding to a given $V_+(x)$. Consequently, we will prove that for a given potential $V_-(x)$ there exists a family of potentials $\hat{V}_-(\lambda, x)$ that all have the same spectrum. These are the isospectral deformations of $V_-(x)$.

In terms of the superpotential $W(x)$ we will assume that there exist other superpotentials $\hat{W}(x)$ which yield the same $V_+(x)$:

$$V_+(x) = \hat{W}^2(x) + \hat{W}'(x)\ , \qquad (9.4)$$

or equivalently

$$W^2(x) + W'(x) = \hat{W}^2(x) + \hat{W}'(x)\ . \qquad (9.5)$$

A trivial solution of Eq. (9.5) is $\hat{W} = W$. In general, if we set

$$\hat{W}(x) = W(x) + f(x) \qquad (9.6)$$

and substitute it back into Eq. (9.5) we obtain

$$2 W(x) f(x) + f(x)^2 + f'(x) = 0 .$$ (9.7)

Making the substitution $f(x) = 1/y(x)$ we obtain the differential equation

$$y'(x) = 1 + 2 W(x) y(x)$$ (9.8)

that can be directly integrated. Its solution is

$$y(x) = \left(\int_{-\infty}^{x} e^{-2 \int_{-\infty}^{t} W(u) du} dt + \lambda \right) e^{\int_{-\infty}^{x} 2 W(t) dt} ,$$ (9.9)

where λ is an integration constant. Now, we use Eq. (9.2) to rewrite (9.9) as

$$y(x) = \frac{\int_{-\infty}^{x} \left[\psi_0^{(-)}(t) \right]^2 dt + \lambda}{\left[\psi_0^{(-)}(x) \right]^2} .$$ (9.10)

Therefore,

$$f(x) \equiv 1/y(x) = \frac{\left[\psi_0^{(-)}(x) \right]^2}{\int_{-\infty}^{x} \left[\psi_0^{(-)}(t) \right]^2 dt + \lambda}$$

$$= \frac{d}{dx} \ln[I(x) + \lambda]$$ (9.11)

where $I(x) = \int_{-\infty}^{x} [\psi_0^{(-)}(t)]^2 dt$. We conclude that the most general $\hat{W}(x)$ satisfying (9.5) is

$$\hat{W}(x) = W(x) + \frac{d}{dx} \ln[I(x) + \lambda] .$$ (9.12)

Problem 9.1. *Prove that $\hat{W}(x)$ and $W(x)$ have the same asymptotic limits.*

Finally, we are in a position to find the potential $\hat{V}_-(\lambda, x)$ corresponding to $\hat{W}(x)$. We find that all potentials given by

$$\hat{V}_-(\lambda, x) = \hat{W}(x)^2 - \hat{W}(x)' = V_-(x) - 2 \frac{d^2}{dx^2} \ln[I(x) + \lambda]$$ (9.13)

have the same $V_+(x)$ and the same spectrum.

Problem 9.2. *Prove Eq. (9.13).*

Observe that when $\lambda \to \pm\infty$, $\hat{V}_-(\lambda, x) \to V_-(x)$. Therefore, $V_-(x)$ itself is a member of this one-parameter family.

Problem 9.3. *Using Eqs. (9.3) and (9.13), prove that the reflection and the transmission coefficients of $\hat{V}_-(\lambda, x)$ don't change: $\hat{r}_-(k) = r_-(k)$, $\hat{r}_-(k) = r_-(k)$.*

We have found that Eq. (9.13) generates an infinite family of isospectral potentials for $V_-(x)$. Next, we need to find the acceptable range of values for the parameter λ. In order to see if there are any constraints on λ, we consider the ground state wave function $\hat{\psi}_0^{(-)}(x)$ corresponding to the deformed potential $\hat{V}_-(\lambda, x)$ and require that it be square integrable. Using Eqs. (9.12) and (9.2) we have successively:

$$\hat{\psi}_0^{(-)}(x) = N\, e^{-\int_{-\infty}^x \hat{W}(t)dt}$$

$$= N\, e^{-\int_{-\infty}^x \left(W(t) + \frac{d}{dt}\ln[I(t)+\lambda]\right)\,dt}$$

$$= N\, \frac{\psi_0^{(-)}(x)}{I(x) + \lambda}\,, \tag{9.14}$$

where N is a normalization constant. After some calculations we find $N = \sqrt{\lambda(\lambda+1)}$. Therefore, the normalized ground state wave function for $\hat{V}_-(\lambda, x)$ is

$$\hat{\psi}_0^{(-)}(x) = \frac{\sqrt{\lambda(\lambda+1)}}{I(x) + \lambda}\, \psi_0^{(-)}(x)\,. \tag{9.15}$$

Problem 9.4. *Calculate the normalization constant of Eq. (9.15).*

Note that $I(x)$ is a bounded quantity constrained to the interval $(0, 1)$. Consequently, we have three different possibilities for the range of values of λ:

(1) $\lambda < -1$ or $\lambda > 0$. In this case $\lambda(\lambda+1) > 0$ and the wave function (9.15) is a well behaved function which is square integrable. Thus the spectrum of $\hat{V}_-(\lambda, x)$ is identical with the spectrum of $V_-(x)$. Furthermore, we can see that in this case all the members of the family behave the same for $x \to \pm\infty$. If we know all the energy levels and the wave functions for $V_-(x)$ (i.e., if the potential is exactly solvable), then $\hat{V}_-(\lambda, x)$ is exactly solvable as well.
(2) $\lambda = -1$ or $\lambda = 0$. Because $I(x)$ vanishes at $x = \pm\infty$, in this case, the wave function $\hat{\psi}_0^{(-)}(x)$ is not square integrable. Thus SUSY is broken,

and $\hat{V}_-(\lambda, x)$ loses a bound state. We will not explore this case any further. We simply mention that $\lambda = 0$ yields the Pursey potential and $\lambda = -1$ yields the Abraham-Moses potential. For more details the reader should consult the reference cited below.[1]

(3) $-1 < \lambda < 0$. Because $I(x)$ is between 0 and 1, in this case, the potential $\hat{V}_-(\lambda, x)$ will be singular for certain values of x. These singularities raise significant questions about the physical meaning of such potentials, and we will not study them any further here.

9.2 Examples

As an example let us consider the Morse potential generated by the super-potential

$$W(x) = A - e^{-x} , \quad A > 0 . \tag{9.16}$$

Then, the potential is

$$V_-(x) = A^2 + e^{-2x} - 2(A + 1/2)e^{-x} , \tag{9.17}$$

with energy levels

$$E_n = A^2 - (A - n)^2 , \quad n = 0, 1, 2, \dots \tag{9.18}$$

and the corresponding wave functions given by

$$\psi_n = y^{s-n} e^{-y/2} L_n^{2s-2n}(y) ; \quad y = 2 e^{-x} , \quad s = A , \tag{9.19}$$

where $L_n^k(x)$ are the usual Laguerre polynomials.

Now we have all the tools we need to construct the family of isospectral deformations of the Morse potential. Note that the calculations become a little tedious, but they present no conceptual difficulty. The reader is encouraged to use a symbolic algebra system such as Mathematica or Maple to speed up the computational process. We begin by calculating the ground state wave function. From Eq. (9.19) we obtain

$$\psi_0^{(-)}(x) = 2^A e^{-e^{-x}} (e^{-x})^A . \tag{9.20}$$

Therefore we obtain

$$I(x) \equiv \int_{-\infty}^x \left[\psi_0^{(-)}(t) \right]^2 dt = \int_{-\infty}^x 2^{2A} e^{-2e^{-t}} \left(e^{-t} \right)^{2A} dt , \tag{9.21}$$

---

[1]A. Khare and U. Sukhatme, "Phase equivalent potentials obtained from supersymmetry." *J. Phys. A* **22**, 2847 (1989).

and consequently

$$\hat{V}_-(\lambda, x) = A^2 - (2A+1)e^{-x} + e^{-2x}$$
$$- 2\left(\frac{2^{2A+1}e^{-x-2e^{-x}}\left(e^{-x}\right)^{2A}}{\int_{-\infty}^{x} 2^{2A}e^{-2e^{-t}}\left(e^{-t}\right)^{2A} dt + \lambda} \right.$$
$$- \frac{2^{4A}e^{-4e^{-x}}\left(e^{-x}\right)^{4A}}{\left(\int_{-\infty}^{x} 2^{2A}e^{-2e^{-t}}\left(e^{-t}\right)^{2A} dt + \lambda\right)^2}$$
$$\left. - \frac{2^{2A+1}Ae^{-x-2e^{-x}}\left(e^{-x}\right)^{2A-1}}{\int_{-\infty}^{x} 2^{2A}e^{-2e^{-t}}\left(e^{-t}\right)^{2A} dt + \lambda} \right). \tag{9.22}$$

Despite the apparent complexity of Eq. (9.22), we can recognize on its first line the usual Morse potential, and in the subsequent lines the terms characterizing the isospectral deformation. These terms vanish as $\lambda \to \pm\infty$ leaving us with the original Morse potential, as expected. To simplify calculations (and the visualization of the results) we choose $A = 1$. We obtain successively:

$$\psi_0^{(-)}(x) = 2e^{-x-e^{-x}} \tag{9.23}$$
$$I(x) = e^{-x-2e^{-x}}\left(e^x + 2\right) \tag{9.24}$$
$$\hat{V}_-(\lambda, x) = e^{-2x} - 3e^{-x} + 1 - 2\left(-\frac{16e^{-4x-4e^{-x}}}{\left(\lambda + e^{-x-2e^{-x}}(e^x+2)\right)^2} \right.$$
$$\left. - \frac{8e^{-2x-2e^{-x}}}{\lambda + e^{-x-2e^{-x}}(e^x+2)} + \frac{8e^{-3x-2e^{-x}}}{\lambda + e^{-x-2e^{-x}}(e^x+2)} \right). \tag{9.25}$$

In Fig. 9.2 we illustrate the isospectral deformations of the Morse potential for three particular values of the λ parameter: $\lambda = 0.2$, $\lambda = 1.0$, and $\lambda = \infty$. The last case corresponds to the initial, undeformed Morse potential.

Problem 9.5. *Writing the Morse superpotential as $W(x) = A - Be^{-x}$ construct the normalized ground state wave function corresponding to the deformed Morse potential and represent it graphically for $\lambda = 0.2$, $\lambda = 1.0$, and $\lambda = \infty$.*

Problem 9.6. *Repeat this procedure to construct the isospectral deformations and the corresponding ground state wave functions for the Coulomb potential.*

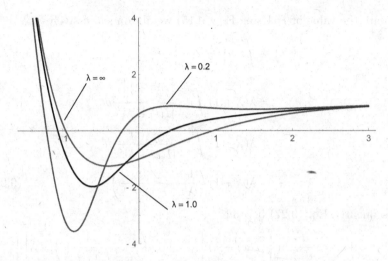

Fig. 9.2 Isospectral deformations of the Morse potential. The case $\lambda = \infty$ represents the original potential.

9.3 The Uniqueness of the Isospectral Deformation

Finally, we will address the question of the uniqueness of the isospectral deformation. If we start with any particular member $\hat{V}_-(\lambda, x)$ and again construct a one-parameter family of isospectral potentials, will this generate another distinct family? The answer is no, as we will see in the following proof.

For a potential $V_-(x)$, the one-parameter potential family is given by

$$\hat{V}_-(\lambda, x) = V_-(x) - 2 \frac{d^2}{dx^2} \ln[I(x) + \lambda] \, , \qquad (9.26)$$

where we take λ to be outside the $[-1, 0]$ interval. If we start with this $\hat{V}_-(\lambda, x)$ and deform it again, we obtain

$$\hat{\hat{V}}_- = \hat{V}_- - 2 \frac{d^2}{dx^2} \ln[\hat{I} + \hat{\lambda}] = V_- - 2 \frac{d^2}{dx^2} \ln[(I + \lambda)(\hat{I} + \hat{\lambda})] \, , \qquad (9.27)$$

where for simplicity we dropped the explicit λ and x dependence. Let us estimate the value of $\hat{I}$. Using Eq. (9.15) we obtain successively

$$\hat{I}(x) = \int_{-\infty}^{x} \left[\hat{\psi}_0^{(-)}(t) \right]^2 dt$$

$$= \lambda(\lambda + 1) \int_{-\infty}^{x} \frac{\left[\psi_0^{(-)}(t) \right]^2}{[I(t) + \lambda]^2} dt$$

$$= \lambda(\lambda + 1) \int_{-\infty}^{x} \frac{\frac{d}{dt} I(t)}{[I(t) + \lambda]^2} dt$$

$$= \lambda(\lambda + 1) \left(\frac{1}{\lambda} - \frac{1}{I(x) + \lambda} \right) . \tag{9.28}$$

Consequently, Eq. (9.27) becomes

$$\hat{V}_- = V_- - 2 \frac{d^2}{dx^2} \ln \left\{ (I(x) + \lambda) \left[\lambda(\lambda + 1) \left(\frac{1}{\lambda} - \frac{1}{I(x) + \lambda} \right) + \hat{\lambda} \right] \right\}$$

$$= V_- - 2 \frac{d^2}{dx^2} \ln[I(x) + \mu] , \tag{9.29}$$

where

$$\mu = \frac{\hat{\lambda} \lambda}{\hat{\lambda} + \lambda + 1} . \tag{9.30}$$

We observe that $\hat{V}_-$ doesn't yield anything new. The only difference is that the deformation parameter λ was replaced by μ. One can prove that if both λ and $\hat{\lambda}$ are in the same range outside $[-1, 0]$ then μ is in the same range outside of this interval as well. Therefore, the one-parameter deformation yields a unique isospectral family.

Chapter 10

Generating Additive Shape Invariant Potentials

In this chapter we will present a novel method for generating additive shape invariant potentials, where the parameters differ by an additive constant: $a_{i+1} = a_i +$ constant.

Shape invariance requires

$$V_+(x, a_i) + g(a_i) = V_-(x, a_{i+1}) + g(a_{i+1}) , \qquad (10.1)$$

where $g(a_i)$ are constants and $a_{i+1} = a_i + \frac{\hbar}{\sqrt{2m}}$. We will set $2m = 1$, but we retain $\hbar$. In terms of superpotentials, the shape invariance reads

$$W^2(x, a_i) + \hbar \frac{dW(x, a_i)}{dx} + g(a_i)$$
$$= W^2(x, a_i + \hbar) - \hbar \frac{dW(x, a_i + \hbar)}{dx} + g(a_i + \hbar) . \quad (10.2)$$

Thus, our defining equation for shape invariance, Eq. (10.2), is a difference-differential equation, and it will prove instrumental in deriving all additive shape invariant potentials. As shown in Fig. 10.1, it relates the superpotential W and its spatial derivative computed at two different parameter values a_i and $a_{i+1} = a_i + \hbar$; i.e., separated by a finite distance $\hbar$ from each other.

Our objective is to obtain all solutions of this difference-differential equation; viz., the entire set of solvable superpotentials $W(x, a)$. It is very important to observe that Eq. (10.2) is valid for all non-zero values of $\hbar$. Thus, if we expand it in powers of $\hbar$, the coefficient of each power must independently vanish. This transforms the difference-differential equation into a set of partial differential equations. Solutions of these equations will generate all possible additive shape invariant superpotentials.

103

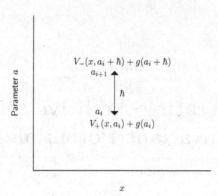

Fig. 10.1   Shape Invariance relates partner potentials and their derivatives defined at two different points on this "x-a" plane: $V_+(x, a_i) + g(a_i) = V_-(x, a_i + \hbar) + g(a_i + \hbar)$.

10.1   Conventional Shape Invariant Superpotentials

Conventional shape invariant superpotentials are those superpotentials that do not depend explicitly on $\hbar$. In order to derive these superpotentials, we expand $W(x, a_i + \hbar)$ in powers of $\hbar$ and substitute back into Eq. (10.2). This equation must hold for an arbitrary value of $\hbar$. Thus the coefficient of each power of $\hbar$ must independently vanish. Hence, for various powers of $\hbar$ we get

$$W \frac{\partial W}{\partial a} - \frac{\partial W}{\partial x} + \frac{1}{2} \frac{dg(a)}{da} = 0 \qquad \mathcal{O}(\hbar) \qquad (10.3)$$

$$\frac{\partial}{\partial a} \left(W \frac{\partial W}{\partial a} - \frac{\partial W}{\partial x} + \frac{1}{2} \frac{dg(a)}{da} \right) = 0 \qquad \mathcal{O}(\hbar^2) \qquad (10.4)$$

$$\frac{\partial^n}{\partial a^{n-1} \partial x} W(x, a) = 0 , \qquad n \geq 3 \qquad \mathcal{O}(\hbar^n) . \qquad (10.5)$$

Although this represents an infinite set, if equations of $\mathcal{O}(\hbar)$ and $\mathcal{O}(\hbar^3)$ are satisfied, all others automatically follow. Therefore, to find the complete set of solutions, it is sufficient to solve:

$$W \frac{\partial W}{\partial a} - \frac{\partial W}{\partial x} + \frac{1}{2} \frac{dg(a)}{da} = 0 \qquad (10.6)$$

and

$$\frac{\partial^3}{\partial a^2 \partial x} W(x, a) = 0 . \qquad (10.7)$$

Table 10.1 lists all known additive shape invariant superpotentials.[1] We will prove[2] that these are the only solutions of Eqs. (10.6) and (10.7).

Table 10.1 The complete family of $\hbar$-independent additive shape-invariant superpotentials.

Name	Superpotential	a	$g(a)$
Harmonic Oscillator	$\frac{1}{2}\omega x$	$\frac{\partial W}{\partial a} = 0$	$g(a) = \omega\,a$
3-D oscillator	$\frac{1}{2}\omega r - \frac{\ell+1}{r}$	$a = \ell$	$g(a) = 2\omega\,a$
Coulomb	$\frac{e^2}{2(\ell+1)} - \frac{\ell+1}{r}$	$a = \ell$	$g(a) = -\frac{e^4}{4(a+1)^2}$
Morse	$A - e^{-x}$	$a = -A$	$g(a) = -a^2$
Eckart	$-A\coth r + \frac{B}{A}$	$a = -A$	$g(a) = -a^2 - \frac{B^2}{a^2}$
Pöschl-Teller (hyperbolic)	$A\coth r - B\,\mathrm{cosech}\,r$	$a = -A$	$g(a) = -a^2$
Rosen-Morse (hyperbolic)	$A\tanh x + \frac{B}{A}$	$a = -A$	$g(a) = -a^2 - \frac{B^2}{a^2}$
Scarf (hyperbolic)	$A\tanh x + B\,\mathrm{sech}\,x$	$a = -A$	$g(a) = -a^2$
Rosen-Morse (trigonometric)	$-A\cot x - \frac{B}{A}$	$a = A$	$g(a) = -\frac{B^2}{a^2} + a^2$
Scarf (trigonometric)	$A\tan x - B\sec x$	$a = A$	$g(a) = a^2$

The general solution of Eq. (10.7) is of the form

$$W(x,a) = a \cdot X_1(x) + X_2(x) + u(a) , \qquad (10.8)$$

where $X_1(x), X_2(x)$ are arbitrary functions of x and $u(a)$ is an unknown function of a. We need to determine all possible combinations of $u(a)$, $X_1(x)$, and $X_2(x)$ that satisfy Eq. (10.6). We will ignore the trivial case when both $X_1(x)$ and $X_2(x)$ are constants. We will also ignore the case in which $X_1(x)$ and $X_2(x)$ are linearly dependent; i.e., $X_2(x) = \alpha X_1(x) + \beta$. In this case, $W(x) = aX_1(x) + \alpha X_1(x) + \beta + u(a) = (a+\alpha)X_1(x) + \beta + u(a)$. With a redefinition of a to $\tilde{a} \equiv a + \alpha$, this type of superpotentials reduces to the case where X_2 is a constant. This will be one of the cases we will consider shortly. Note that from here onward, we will use lower case Greek letters to denote constants that are independent of both a and x.

To determine $W(x,a)$, we first focus on determining $u(a)$. To do so, we take two derivatives of (10.6) with respect to a. This leads to the following

[1]R. Dutt, A. Khare, U. Sukhatme, "Supersymmetry, Shape Invariance and Exactly Solvable Potentials", *Am. J. Phys.*, **56**, 163 (1988),

[2]J. Bougie, A. Gangopadhyaya, J.V. Mallow, "Generation of a Complete Set of Additive Shape-Invariant Potentials from an Euler Equation", *Phys. Rev. Lett.*, 210402:1–210402:4 (2010),

J. Bougie, A. Gangopadhyaya, J.V. Mallow, C. Rasinariu, "Supersymmetric Quantum Mechanics and Solvable Models", *Symmetry*, **4**(3), 452–473 (2012).

differential equation:

$$3\dot{W}\ddot{W} + W\dddot{W} + \ddot{W}' + \frac{1}{2}\dddot{g} = 0 ,$$

where dots and primes represent derivatives taken with respect to a and x respectively. Since $\ddot{W}' \equiv \frac{\partial^3}{\partial a^2 \partial x} W(x,a) = 0$, this simplifies to

$$3\dot{W}\ddot{W} + W\dddot{W} + \frac{1}{2}\dddot{g} = 0 . \tag{10.9}$$

Inserting the form of the general solution (10.8) into (10.9) yields

$$X_1 \left(3\ddot{u} + a\dddot{u}\right) + X_2 \dddot{u} = H(a) , \tag{10.10}$$

where $H(a) = -\left(3\dot{u}\ddot{u} + u\dddot{u} + \frac{1}{2}\dddot{g}\right)$ is a function of a, and is independent of x. Since X_1 and X_2 are linearly independent, we find that there are only three possible ways for the left hand side of Eq. (10.10) to be independent of x:

- Case 1. X_1 is a constant and $\dddot{u} = 0$;
- Case 2. X_2 is a constant and $3\ddot{u} + a\dddot{u} = 0$;
- Case 3. Neither X_1 nor X_2 is a constant, but $3\ddot{u} + a\dddot{u} = 0$ and $\dddot{u} = 0$.

For each of these cases we determine the form of $u(a)$. Next, we determine $X_1(x)$ and $X_2(x)$. This we do by taking two derivatives of Eq. (10.6), one with respect to a and another with respect to x. We obtain

$$2\dot{W}\dot{W}' + W'\ddot{W} - \dot{W}'' = 0 . \tag{10.11}$$

Inserting the form of the general solution (10.8) into (10.11) yields

$$2\left(X_1 + \dot{u}\right)X_1' + \left(aX_1' + X_2'\right)\ddot{u} - X_1'' = 0 . \tag{10.12}$$

Now we analyze each of the three cases in detail.

10.1.0.1　*Case 1. X_1 is a constant and $\dddot{u} = 0$*

Let $X_1 = \alpha$. Since X_2 cannot be a constant as well, Eq. (10.10) requires $\dddot{u} = 0$. This leads to $u(a) = \gamma a^2 + \mu a + \nu$ for some arbitrary constants α, γ, μ, and ν. Inserting X_1 and u into Eq. (10.8) yields $W(x,a) = X_2(x) + \gamma a^2 + \eta a + \nu$, where $\eta = \alpha + \mu$.

We now find X_2 by inserting the above W into Eq. (10.6). This yields

$$\left(X_2 + \gamma a^2 + \eta a + \nu\right)\left(2a\gamma + \eta\right) - \frac{dX_2}{dx} = -\frac{1}{2}\frac{dg}{da}$$

or equivalently,

$$2a^3\gamma^2 + 3a^2\eta\gamma$$
$$+ a\left(2X_2\gamma + 2\nu\gamma + \eta^2\right) + X_2\eta + \eta\nu - X_2' = h(a) , \tag{10.13}$$

where $h(a) = -\frac{1}{2}\frac{dg}{da}$.

Since $X_2(x)$ is independent of a, and the left hand side of Eq. (10.13) is a sum of four linearly independent functions of a $\left(a^0, a^1, a^2, \text{and } a^3\right)$, and the term $h(a)$ on the right hand side is independent of x, the coefficient of each power of a must separately be independent of x. The linear term in a therefore requires that $2X_2\gamma + 2\nu\gamma + \eta^2$ be independent of x. Since a constant X_2 leads to a trivial solution, we must have $\gamma = 0$. The remaining x-dependent terms on the left hand side of Eq. (10.13), $X_2\eta - X_2'$ must be a constant:

$$\eta X_2 - X_2' = \beta . \tag{10.14}$$

The solution depends on the value of the constants η and β.

- Case 1A. $\eta = 0$, $\beta = 0$. This is not allowed as this leads to $X_2' = 0$, i.e., X_2 is constant;
- Case 1B. $\eta = 0$, $\beta \neq 0$. In this case, $X_2 = -\beta x + \xi$, so $W = -\beta x + \xi + \nu$. Defining $\beta = -\frac{1}{2}\omega$ yields the harmonic oscillator superpotential;
- Case 1C. $\eta \neq 0$. The solution is then $X_2(x) = -\frac{\beta}{\eta} + \frac{1}{\eta}e^{\eta(x-x_0)}$, $u = \mu a + \nu$, $X_1 = \alpha$. Therefore, $W = \left(\nu - \frac{\beta}{\eta} + \eta a\right) + \frac{1}{\eta}e^{\eta(x-x_0)}$. For $\eta = -1$, this yields $W = A - Be^{-x}$, where $A \equiv \nu + \beta - a$ and $B \equiv e^{x_0}$. This is the Morse superpotential. Note that A decreases as a increases, and hence signals a finite number of eigenstates.[3]

10.1.0.2 *Case 2. X_2 is constant*

In this case, let $X_2 = \alpha$; then Eq. (10.10) requires $3\ddot{u} + a\dddot{u} = 0$. This yields $u(a) = \mu a + \nu + \frac{\gamma}{a}$. We now insert this form of u and $X_2 = \alpha$ into Eq. (10.12) to get an ordinary differential equation in x for X_1:

$$2\left(X_1 + \mu - \frac{\gamma}{a^2}\right)X_1' + 2X_1'\frac{\gamma}{a^2} - X_1'' = 0 ,$$

or equivalently,

$$2X_1X_1' + 2\mu X_1' - X_1'' = 0 .$$

Integrating it once, we get

$$X_1^2 + 2\mu X_1 - X_1' = \beta . \tag{10.15}$$

This equation can be simplified by setting $\tilde{X}_1 = X_1 + \mu$. This leads to

$$\tilde{X}_1^2 - \tilde{X}_1' = \left(\beta + \mu^2\right) \equiv \theta . \tag{10.16}$$

The solutions for $\tilde{X}_1$ depend on the constant θ.

---

[3]The ground state wave function is given by $\psi_0(x) \sim e^{-\int W(x)dx}$. If A turns negative, $\psi_0(x)$ becomes unbounded as $x \to \infty$.

- Case 2A. $\theta = 0$, In this case, $\tilde{X}_1 = -\frac{1}{x - x_0}$. The whole superpotential is then given by

$$W(x, a) = a\left(-\frac{1}{x - x_0} - \mu\right) + \alpha + \left(\mu a + \nu + \frac{\gamma}{a}\right)$$

$$= -\frac{a}{x - x_0} + (\alpha + \nu) + \frac{\gamma}{a} . \tag{10.17}$$

Setting $\alpha + \nu = 0$ and $x_0 = 0$, and identifying $x \equiv r$, $a \equiv \ell + 1$ and $\gamma = e^2/2$, we get $W(r, \ell) = \frac{e^2}{2(\ell+1)} - \frac{\ell+1}{r}$, the Coulomb superpotential.

- Case 2B. $\theta > 0$. In this case, we have either $\tilde{X}_1 = -\sqrt{\theta}\coth(\sqrt{\theta}x - \eta)$ or $\tilde{X}_1 = -\sqrt{\theta}\tanh(\sqrt{\theta}x - \eta)$. In the first case, the superpotential is given by $W(x, a) = a\left(-\sqrt{\theta}\coth(\sqrt{\theta}x) - \mu\right) + \alpha + \left(\mu a + \nu + \frac{\gamma}{a}\right) \equiv$ $-A\coth(\sqrt{\theta}x) + \frac{B}{a}$, where we have set $\gamma = B$, $(\alpha + \nu) = 0$, $\theta = 1$ and $\eta = 0$. This is the Eckart potential. Similarly, the other solution with $\tanh x$ generates Rosen-Morse II.

- Case 2C. $\theta < 0$. In this case, we obtain $\tilde{X}_1 = -\sqrt{|\theta|}\cot(\sqrt{|\theta|}\,x - \eta)$. An analysis similar to the previous case generates the superpotential for Rosen-Morse I.

10.1.0.3 *Case 3. X_1 and X_2 are not constant, but $3\ddot{u} + a\dddot{u} = 0$ and $\dddot{u} = 0$*

In this case, since $\dddot{u} = 0$, and $3\ddot{u} + a\dddot{u} = 0$, we have $\ddot{u} = 0$. Therefore $u(a) = \mu a + \nu$. In this case, Eq. (10.12) yields

$$2\left(X_1 + \mu\right)X_1' - X_1'' = 0 .$$

Integrating,

$$X_1^2 + 2\mu X_1 - X_1' = \beta . \tag{10.18}$$

Thus, again we have

$$\tilde{X}_1^{\,2} - \tilde{X}_1' = (\beta + \mu^2) \equiv \theta . \tag{10.19}$$

Note that this is the same differential equation as (10.16) and will therefore give the same solutions for X_1 as Case 2. However, in this case, $u = \mu a + \nu$ (this is equivalent to choosing $\gamma = 0$ in Case 2) and X_2 is not constant. Instead, in each case we must plug our solutions for $u(a)$ and $X_1(x)$ into Eq. (10.6), which yields

$$(aX_1 + X_2 + \mu a + \nu)(X_1 + \mu) - aX_1' - X_2' = h(a) .$$

This equation is again simplified by setting $\tilde{X}_1 = X_1 + \mu$, which yields

$$\left(a\tilde{X}_1 + X_2 + \nu\right)\tilde{X}_1 - a{\tilde{X}_1}' - X_2' = h(a) .$$

Since $\tilde{X}_1$ and X_2 are independent of a, the terms linear in a and the terms independent of a on the left side of this equation must each separately be independent of x. Therefore,

$$\left(\tilde{X}_1\right)^2 - {\tilde{X}_1}' = \theta , \tag{10.20}$$

and

$$X_2\tilde{X}_1 + \nu\tilde{X}_1 - X_2' = \xi . \tag{10.21}$$

For different values of θ, we get different superpotentials:

- Case 3A. $\theta = 0$. We again get $\tilde{X}_1 = -\frac{1}{x}$, where with an appropriate choice for the origin we have set $x_0 = 0$. Equation (10.21) for X_2 becomes

$$(X_2 + \nu) + x\, X_2' = -\xi x . \tag{10.22}$$

Its solution is $X_2 = -\frac{1}{2}\xi\, x + \frac{\sigma}{x} - \nu$. With the identification $x \to r$, $\xi \to -\omega$, $a-\sigma \to \ell+1$, we get $W(r,\ell) = \frac{1}{2}\omega r - \frac{\ell+1}{r}$, the superpotential for the 3D-harmonic oscillator.

- Case 3B. $\theta > 0$. As seen before, $\theta > 0$ implies that $\tilde{X}_1 = -\sqrt{\theta}\coth(\sqrt{\theta}x - \eta)$ or $\tilde{X}_1 = -\sqrt{\theta}\tanh(\sqrt{\theta}x - \eta)$. By translation and scaling of x, we can simplify the first solution to $\tilde{X}_1 = -\coth x$. Substituting $\tilde{X}_1$ in Eq. (10.21), we get

$$-\coth x\, \tilde{X}_2 - {\tilde{X}_2}' = \xi \tag{10.23}$$

where we have set $\tilde{X}_2 = X_2 + \nu$. The solution to the homogeneous equation is $\tilde{X}_2 = \frac{\sigma}{\sinh x}$, and the particular solution is $\xi \coth x$. Hence $X_2 = \frac{\sigma}{\sinh x} + \xi\coth x - \nu$. Thus, the superpotential is given by $W(a,x) = (\xi - a)\coth x + \frac{\sigma}{\sinh x}$, the General Pöschl-Teller potential. The second solution generates the Scarf II potential.

- Case 3C. $\theta < 0$. A similar analysis for this case leads to Scarf I as the corresponding shape invariant superpotential.

Thus, we have generated all the superpotentials of Table 10.1 and shown that these are the only possible $\hbar$-independent solutions to the additive shape invariant condition.

10.2   Extended Potentials

In the last section we computed all solutions of the shape invariance condition Eq. (10.2), using Eqs. (10.6) and (10.7), assuming that the superpotentials did not have an explicit dependence on $\hbar$, and we were able to show that the superpotentials listed in Table 10.1 indeed formed a complete set.

What about the potentials that are explicitly dependent on $\hbar$? The authors in Ref. [4][4] found a new set of shape invariant potentials that depend explicitly on $\hbar$, and thus are not included in Table 10.1.

To generate these new potentials through our formalism, we will allow superpotentials to have an inherent dependence on $\hbar$. Thus, these superpotentials can now be expanded in terms of powers of $\hbar$:

$$W(x, a, \hbar) = \sum_{j=0}^{\infty} \hbar^j W_j(x, a) . \tag{10.24}$$

Using $a_0 = a$ and $a_1 = a + \hbar$, this power series yields

$$W(x, a_1, \hbar) = \sum_{j=0}^{\infty} \sum_{k=0}^{j} \frac{\hbar^j}{k!} \frac{\partial^k W_{j-k}}{\partial a^k} ,$$

$$W^2(x, a_1, \hbar) = \sum_{j=0}^{\infty} \sum_{s=0}^{j} \sum_{k=0}^{s} \frac{\hbar^j}{(j-s)!} \frac{\partial^{j-s} (W_k W_{s-k})}{\partial a^{j-s}} ,$$

and

$$\left. \frac{\partial W}{\partial x} \right|_{a=a_1} = \sum_{j=0}^{\infty} \sum_{k=0}^{j} \frac{\hbar^j}{k!} \frac{\partial^{k+1} W_{j-k}}{\partial a^k \partial x} .$$

Substituting these into Eq. (10.2) and setting the coefficients of each power of $\hbar$ equal to zero we obtain the following:
for $j = 1$

$$2 \frac{\partial W_0}{\partial x} - \frac{\partial}{\partial a} \left(W_0^2 + g \right) = 0 , \tag{10.25}$$

---

[4]C. Quesne, "Exceptional orthogonal polynomials, exactly solvable potentials and supersymmetry", *J. Phys. A*, **41**, 392001:1–392001:6 (2008); Solvable rational potentials and exceptional orthogonal polynomials in supersymmetric quantum mechanics. *Sigma*, **5**, 084:1–084:24 (2009);

S. Odake, R. Sasaki, "Infinitely many shape invariant discrete quantum mechanical systems and new exceptional orthogonal polynomials related to the Wilson and Askey-Wilson polynomials", *Phys. Lett. B*, **682**, 130–136 (2009); "Another set of infinitely many exceptional (X_ℓ) Laguerre polynomials", *Phys. Lett. B*, **684**, 173–176 (2010).

for $j = 2$

$$\frac{\partial W_1}{\partial x} - \frac{\partial}{\partial a}(W_0\, W_1) = 0 \,, \tag{10.26}$$

and for $j \geq 3$

$$2\frac{\partial W_{j-1}}{\partial x} - \sum_{s=1}^{j-1}\sum_{k=0}^{s}\frac{1}{(j-s)!}\frac{\partial^{j-s}}{\partial a^{j-s}}W_k\,W_{s-k}$$

$$+ \sum_{k=2}^{j-1}\frac{1}{(k-1)!}\frac{\partial^k W_{j-k}}{\partial a^{k-1}\partial x} + \left(\frac{j-2}{j!}\right)\frac{\partial^j W_0}{\partial a^{j-1}\partial x} = 0 \,. \tag{10.27}$$

We first observe that Eq. (10.25) is identical to Eq. (10.6); i.e., all conventional shape invariant potentials automatically satisfy Eq. (10.25). Hence, conventional superpotentials can be used as the base to erect the tower of extended superpotentials. If W_0 is a conventional superpotential, then Eq. (10.5) is automatically satisfied for W_0, meaning that the last term of Eq. (10.27) disappears.

As an example, we will now derive a potential that appears in the first paper of Ref. [4] as the first known extended potential. This potential will be built upon the conventional superpotential for the 3-D oscillator, $W_0 = \frac{1}{2}\omega x - \frac{a}{x}$, which obeys Eq. (10.25). The higher order terms of the superpotential W can be generated from Eq. (10.27) for all $j > 1$. For $j = 2$, we find the equation for W_1

$$\frac{\partial W_1}{\partial x} - \frac{\partial}{\partial a}(W_0 W_1) = 0 \,,$$

and for $j = 3$, we obtain

$$\frac{\partial W_2}{\partial x} - \frac{\partial\left(2W_0 W_2 + W_1^2\right)}{\partial a} - \frac{1}{2}\frac{\partial^2 W_0 W_1}{\partial a^2} + \frac{2}{3}\frac{\partial^3 W_0}{\partial a^2 \partial x} = 0 \,.$$

These two coupled equations are solved by $W_1 = 0$ and $W_2 = (4x\omega)/(2a + x^2\omega)^2$. The next order equations are solved by $W_3 = 0$ and $W_4 = (4x\omega)/(2a + x^2\omega)^4$. Generalizing these, we get

$$W_0 = \frac{1}{2}\omega x - \frac{a}{x}; \; W_{2j-1} = 0; \; W_{2j} = (4x\omega)/(2a + x^2\omega)^{2j}, \quad \text{for } j = 1, 2, 3, \cdots,$$

yielding a sum that converges to

$$W(x, a, \hbar) = \frac{1}{2}\omega x - \frac{a}{x} + \left(\frac{2\omega x\hbar}{\omega x^2 + 2a - \hbar} - \frac{2\omega x\hbar}{\omega x^2 + 2a + \hbar}\right) \,.$$

With the identification $a = (\ell + 1)\hbar$, and $\hbar = 1$, we get

$$W = \frac{\omega x}{2} - \frac{\ell+1}{x} + \left(\frac{2\omega x}{\omega x^2 + 2\ell + 1} - \frac{2\omega x}{\omega x^2 + 2\ell + 3}\right) \,. \tag{10.28}$$

10.3   A New Potential Derived Using the Above Procedure

We have so far shown that (a) the set of known $\hbar$ independent superpotentials is complete, and (b) the known extended shape invariant models indeed satisfy the partial differential equations (10.25)–(10.27). In this section we show that the formalism given above can also be used to generate new potentials. As an example, we demonstrate that starting from the Morse superpotential (see Table 10.1), we arrive at a new extended potential that has the same energy spectrum as the Morse potential.[5] This new potential is shape invariant and exhibits properties that differ from other known extended shape invariant potentials: though generated from the Morse superpotential, the asymptotic limits of this new potential are different from those of Morse. Thus, this new potential cannot be generated using the isospectral deformation techniques of Chapter 9.

Taking a cue from our derivation of the extended potential in the last section, Eq. (10.28), with $A = -a$, we begin with $W_0 = -a - e^{-x}$ and $W_1 = 0$. The equation for W_2 then reads

$$\frac{\partial W_2}{\partial x} - \frac{\partial W_0 W_2}{\partial a} = 0 .$$

This equation is solved by

$$W_2(x,a) = -Qe^{-3x}\left(2P + Qe^{-2x} + 2a\,Qe^{-x}\right) ,$$

where P and Q are constant parameters. Setting W_3 to be zero, the equation for W_4 is

$$2\frac{\partial W_4}{\partial x} - 2\frac{\partial\left(W_0 W_4 + \frac{1}{2}W_2^2\right)}{\partial a} - \frac{\partial}{\partial a}\left(\frac{\partial W_2}{\partial a}\frac{\partial W_0}{\partial a}\right)$$
$$- \frac{1}{3}\left(W_2\frac{\partial^3 W_0}{\partial a^3} + W_0\frac{\partial^3 W_2}{\partial a^3}\right) + \frac{\partial^3 W_2}{\partial x\partial a^2} = 0 .$$

The above equation is solved by

$$W_4(x,a) = -Q^3 e^{-7x}\left(2P + Qe^{-2x} + 2a\,Qe^{-x}\right) .$$

Generalizing this process yields $W_{2k-1} = 0$ and

$$W_{2k} = -Q^{2k-1}e^{-(4k-1)x}\left(2P + Qe^{-2x} + 2a\,Qe^{-x}\right) .$$

Computing the infinite sum $\sum_{j=0}^{\infty}\hbar^j W_j(x,a)$, we obtain

$$W(x,a,\hbar) = -a - e^{-x} + \frac{\hbar^2\left(2Pe^x + 2a\,Q + Qe^{-x}\right)}{e^{2x} + Q\,\hbar^2} . \qquad (10.29)$$

[5]J. Bougie, A. Gangopadhyaya, J.V. Mallow, C. Rasinariu, "Generation of a Novel Exactly Solvable Potential", *Phys. Lett.* A **379**, 2180–2183 (2015).

Substituting the above expression into Eq. (10.2) yields

$$W^2(x,a) - W^2(x, a+\hbar) + \hbar\frac{d}{dx}\left(W(x,a) + W(x, a+\hbar)\right) = \hbar(-2a - \hbar) ,$$

$$(10.30)$$

i.e., $g(a) = -a^2$. This leads to the energy eigenvalues $E_n^{(-)} = g(a + n\hbar) -$ $g(a) = a^2 - (a + n\hbar)^2$. As expected, these values are exactly the same as those of the Morse potential. Figures 10.2 and 10.3 show W and the partner potentials V_- and V_+, where we have chosen the specific values $P = 3, Q = 5, a = -3$, and $\hbar = 1$, as an example. The asymptotic limits for this extended superpotential are finite and their absolute values are equal. Consequently, the resulting potential has identical limits at $\pm\infty$.

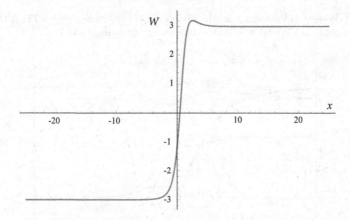

Fig. 10.2   Superpotential W for $P = 3, Q = 5, a = -3$, and $\hbar = 1$. Its asymptotic values imply an unbroken supersymmetry.

The ground state eigenfunction can be obtained from the first order differential equation $\mathcal{A}^-(x,a)\psi_0(x,a) = 0$, yielding

$$\psi_0(x,a) = N\left(\hbar^2 Q + e^{2x}\right)^a \exp\left[-ax + \frac{(1 - 2\hbar^2 P)\tan^{-1}\left(\frac{e^x}{\hbar\sqrt{Q}}\right)}{\hbar\sqrt{Q}}\right],$$

$$(10.31)$$

where N is the normalization constant. The excited states can be obtained recursively by applying the $\mathcal{A}^+$ operator to ψ_0. The first excited state is $\psi_1(x,a) = \mathcal{A}^+(x,a)\psi_0(x, a+\hbar)/\sqrt{E_1}$, and so on. For example, with our particular choice of parameters, the potential $V_-(x)$ holds three bound states, corresponding to the eigenenergies $0, 5$ and 8. The fourth energy level is at 9, which does not hold a bound state.

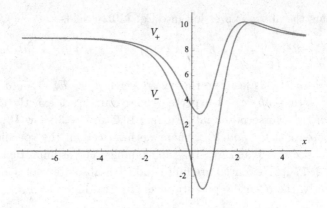

Fig. 10.3   Potentials V_- and V_+ for the same values of the parameters as Fig. 10.2. The deeper potential is V_-, and it holds the zero-energy groundstate.

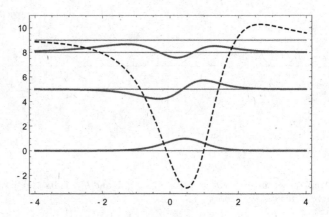

Fig. 10.4   The three bound-state eigenfunctions (shown in blue) for V_- (dashed line) with $P = 3, Q = 5, a = -3$, and $\hbar = 1$. Each eigenfunction is shown superposed on its corresponding eigenenergy (shown as a horizontal red line).

For these values of parameters we obtain for the ground state eigenfunction

$$\psi_0(x) = \frac{40}{(e^{2x} + 5)^3} \sqrt{\frac{105}{1 + e^{-\sqrt{5}\pi}}} \exp\left[3x - \sqrt{5}\tan^{-1}\left(\frac{e^x}{\sqrt{5}}\right)\right],$$

and for the first two excited states

$$\psi_1(x) = \frac{20\left(e^{2x} + 2e^x - 5\right)\sqrt{\frac{70}{1 + e^{\sqrt{5}\pi}}}}{(e^{2x} + 5)^3} \exp\left[2x - \sqrt{5}\tan^{-1}\left(\frac{e^x}{\sqrt{5}}\right) + \frac{\sqrt{5}\pi}{2}\right],$$

$$\psi_2(x) = \frac{4\left(3e^{4x} + 20e^{3x} - 20e^{2x} - 100e^x + 75\right)\sqrt{\frac{5}{1+e^{\sqrt{5}\pi}}}}{\left(e^{2x} + 5\right)^3}$$

$$\times \exp\left[x - \sqrt{5}\tan^{-1}\left(\frac{e^x}{\sqrt{5}}\right) + \frac{\sqrt{5}\pi}{2}\right].$$

As expected, the fourth energy level does not hold a bound state, as its corresponding eigenfunction is not square integrable.

Problem 10.1. *Show that ψ_3 is not renormalizable.*

Figure 10.4 superimposes on the potential $V_-(x)$ the three bound-state eigenfunctions together with their corresponding eigenenergies.

Chapter 11

Singular Potentials in SUSYQM

In this chapter we will study systems described by singular potentials, as they are abundant in quantum mechanics. For these systems, the value of the potential becomes unbounded either at a set of points known as the singular points or over a region. In some sense, the term "singular potential" is a misnomer: the actual issue is not *whether* the potential blows up, but *how* it does so. As we shall show, only potentials with a singular term of the form α/r^2 need concern us. Such terms often originate from the non-zero angular momentum of a system. It is the particular value of the coefficient α that allows these systems to be solved. Similar "fortuitous" coefficients appear in other potentials as well. In addition to the non-zero angular momentum type, potentials such as the Rosen-Morse, Eckart, and Pöschl-Teller also have this kind of singularity. We will find that in these cases $\alpha \geq -\frac{1}{4}$. Otherwise, the particle would be "sucked into" the singularity; i.e., in such cases there is no lower bound on the ground state energy as we approach the singularity. Compare this for example to the hydrogen atom, where despite the potential becoming unboundedly negative as $r \to 0$, the ground state energy is a well-defined $-13.6\ eV$. We will see that supersymmetric quantum mechanics provides natural barriers so that potentials with $\alpha < -\frac{1}{4}$ simply do not arise.

In a one-dimensional system, a term of the type α/x^2 has the potential of breaking the entire real line into two non-communicating halves. This, as we shall see, happens if $\alpha \geq \frac{3}{4}$. In such a case a particle situated on one side of the singularity never tunnels to the other side. We will also see that for $\alpha \geq \frac{3}{4}$, the eigenvalue problem is well defined and can be solved by conventional means. For these one- and three-dimensional cases we say that we have a "hard" singularity.

Problem 11.1. *Determine the value of α for these three superpotentials.*

$$\text{Rosen-Morse: } W(x, A, B) = -A \cot x - B/A$$
$$\text{Eckart : } W(x, A, B) = -A \coth r + B/A$$
$$\text{Pöschl-Teller: } W(x, A, B) = -A \coth r - B \operatorname{cosech} r.$$

For $-\frac{1}{4} \le \alpha < \frac{3}{4}$, the behavior of both solutions at the origin is such that they are square integrable, and, therefore acceptable. Such singularities are called "soft" singularities. Hence, the boundary conditions are not sufficient to determine eigenvalues and eigenfunctions, and there is no *a priori* mechanism to select any specific linear combination. Thus, in the context of conventional quantum mechanics, the spectrum of the system can be determined only by imposing additional ad hoc constraints.[1] They neither suck the particle in, nor do they break up the real axis. Potentials with this type of soft singularity are called "transition potentials." We will show that in some cases these transition potentials emerge naturally in supersymmetric quantum mechanics.

We thus have two problems to address: Why is the inverse square potential the only one we need consider, and why for this case alone are there constraints on α? The answer to both questions can be provided in a semi-quantitative way, as follows. The mean energy of a system is the sum of its average kinetic and potential energies: $E_{\text{mean}} = \langle T \rangle + \langle V \rangle$. Let us assume that the wave function is appreciable mostly within a distance r_0 of the origin. Then from uncertainty arguments we can make the following estimates: since the particle's position $\Delta x \sim r_0$, its momentum uncertainty is of order $\Delta p \sim \hbar/r_0$. Treating Δp as a standard deviation, $\Delta p^2 = \langle p^2 \rangle - \langle p \rangle^2$. But $\langle p \rangle = 0$ because it takes into account the opposite directions of p. Thus, the average kinetic energy $\langle T \rangle = \langle p^2 \rangle/2m \simeq \hbar^2/2mr_0^2$. Let us assume that the potential obeys an inverse power law: $V = -\alpha/r^s$. Now let us consider what happens for various values of s. For $s > 2$, V is the dominant term as $r_0 \to 0$. The mean energy is therefore unbounded, so there can be no finite ground state. This is the case of the particle being sucked into the singularity. Conversely, for $s < 2$ the energy is bounded at some finite negative value. Again, we mention as an example the hydrogen atom, with

---

[1]See L. D. Landau and E. M. Lifshitz, *Quantum Mechanics*, NY: Pergamon Press (1977);

W. M. Frank, D. J. Land and R. M. Spector, "Singular potentials," *Rev. Mod. Phys.* **43**, 36 (1971);

A. Gangopadhyaya, P. K. Panigrahi and U. P. Sukhatme, "Analysis of inverse-square potentials using supersymmetric quantum mechanics," *J. Phys. A* **27**, 4295 (1994).

well-defined ground state energy. It is only for the case $s = 2$ that the kinetic and potential energy terms are of the same order near r_0. Therefore the spectrum will depend on the magnitude of α as compared with $\hbar^2/2m$.

To summarize, we need only consider the case $s = 2$. For that case, the particle will be sucked into the singularity when $\alpha < -\frac{1}{4}$; i.e., there is no finite-energy ground state. For $\alpha > \frac{3}{4}$, the eigenvalues and eigenfunctions can be obtained from the normalization condition. However, for $-\frac{1}{4} \leq \alpha \leq \frac{3}{4}$, the problem cannot be solved without imposition of additional constraints. SUSYQM on the other hand will provide the eigenvalues and eigenfunctions without the need for additional constraints needed in ordinary quantum mechanics. The various situations are illustrated in the figure below.

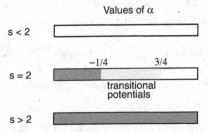

Fig. 11.1 The lightly shaded region depicts the range of α $\left(-\frac{1}{4} \leq \alpha \leq \frac{3}{4}\right)$ that allows for the ambiguity in determination of the spectrum. For $s < 2$ there is no ambiguity, and for the domain $s > 2$ there are no bound states.

Let us study the above statements through an example. We consider a system described by a potential with an α/r^2 singularity. The radial part $R(r)$ of the three-dimensional wave function obeys

$$-\frac{1}{r^2}\frac{dR}{dr}\left(r^2\frac{dR}{dr}\right) + V(r)R = ER . \tag{11.1}$$

We define a new function[2] $\psi = rR$. The equation for ψ then becomes an effective one-dimensional Schrödinger equation

$$-\frac{d^2\psi}{dr^2} + V(r)\psi = E\psi(r) . \tag{11.2}$$

The normalization of this function ψ is derived from the normalization of R:

$$\int_0^\infty |R|^2 r^2 dr = \int_0^\infty |\psi|^2 dr = 1 .$$

---

[2]We called these u and R in our discussion of hydrogen in Chapter 2, to stay with customary notation in most quantum mechanics texts.

For the last integral to be finite, it is necessary that the behavior of $|\psi|^2$ near the origin be less divergent than the function $\frac{1}{r}$; i.e., ψ should not blow up at the origin any faster than $\frac{1}{\sqrt{r}}$.

Problem 11.2. *Show that normalizability of $\psi(r)$ requires that the function be less singular than $\frac{1}{\sqrt{r}}$ near the origin.*

In the vicinity of the origin, finite terms such as $E\,\psi$ are much smaller than the α/r^2 term and can be ignored. Equation (11.2) then reduces to

$$-\frac{d^2\psi}{dr^2} + \frac{\alpha}{r^2}\,\psi \approx 0 \ . \tag{11.3}$$

Multiplying Eq. (11.3) by r^2 and solving the resulting equation, we find that all eigenfunctions near the origin reduce to the linear combination

$$\psi \approx c_1\, r^{\left(\frac{1}{2}-\sqrt{\alpha+\frac{1}{4}}\right)} + c_2\, r^{\left(\frac{1}{2}+\sqrt{\alpha+\frac{1}{4}}\right)} . \tag{11.4}$$

Problem 11.3. *Show that ψ near the origin reduces to the combination given in Eq. (11.4).*

For $\alpha < -\frac{1}{4}$, both terms of Eq. (11.4) become oscillatory and thus cannot generate the requisite nodeless ground state.

For $\alpha \geq \frac{3}{4}$, the first term is not normalizable and hence $c_1 = 0$. c_2 is then obtained from the normalization. In addition, the wave function must vanish at at the non-singular end of the domain. This quantizes the energy. Hence, the spectrum is determinable for $\alpha \geq \frac{3}{4}$.

Now, let us explore the transitional region $-\frac{1}{4} \leq \alpha < \frac{3}{4}$. For α within this range, both linearly independent solutions of the Schrödinger equation are less divergent that $\frac{1}{\sqrt{r}}$ near the origin, and thus normalizable. Hence we cannot say *a priori* which is the correct eigenfunction; in principle it could be a linear combination of the two. We must find a way to impose additional constraints. There are several ways to do so; we shall outline one here. These methods turn out to select the term with the higher power of r as the correct one. Conventionally, this higher power term is called the "less singular" function. For negative powers of r, when at least one term is singular, this is meaningful; the terminology is used even when neither of the terms is singular.

Problem 11.4. *Show that for $-\frac{1}{4} \leq \alpha < \frac{3}{4}$ both terms of Eq. (11.4) are less singular than $\frac{1}{\sqrt{r}}$.*

This approach of retaining the less singular wave function then leads to the determination of eigenvalues and eigenfunctions. But let us remark that, as we shall show later, SUSYQM makes this selection naturally; i.e., with no need for extra constraints.

Using the method of the first reference in Footnote 1, we first "regularize" the potential by taking it to be a constant within a small neighborhood of radius r_0 around the singular point. I.e., for $r < r_0$, the Schrödinger equation is

$$-\frac{d^2\psi}{dr^2} + \frac{\alpha}{r_0^2}\psi = 0 ,$$ (11.5)

and its solutions are given by

$$\psi = A\sinh\left(\frac{\sqrt{\alpha}}{r_0} r\right) + B\cosh\left(\frac{\sqrt{\alpha}}{r_0} r\right) .$$

Problem 11.5. *Derive the above solution for $r < r_0$.*

Since ψ vanishes at the origin, we must set $B = 0$. Hence the solution inside the regularized region is $A\sinh\left(\frac{\sqrt{\alpha}}{r_0} r\right)$. The solution for $r > r_0$, is given by the linear combination $\psi = c_1 r^{\left(\frac{1}{2} - \sqrt{\alpha + \frac{1}{4}}\right)} + c_2 r^{\left(\frac{1}{2} + \sqrt{\alpha + \frac{1}{4}}\right)}$.

Problem 11.6. *Derive the above solution for $r > r_0$.*

Matching the boundary conditions for the wave function and its derivative and dividing one by the other,[3] we obtain

$$\sqrt{\alpha}\coth\left(\sqrt{\alpha}\right) = \frac{\left(\frac{1}{2} - \sqrt{1+4\alpha}\right)\frac{c_1}{c_2} + \left(\frac{1}{2} + \sqrt{1+4\alpha}\right) r_0^{2\sqrt{1+4\alpha}}}{\frac{c_1}{c_2} + r_0^{2\sqrt{1+4\alpha}}} .$$ (11.6)

Solving for $\frac{c_1}{c_2}$ we obtain

$$\frac{c_1}{c_2} = \frac{r_0^{2\sqrt{1+4\alpha}}\left(\frac{1}{2} + \sqrt{1+4\alpha} - \sqrt{\alpha}\coth\left(\sqrt{\alpha}\right)\right)}{\left(\sqrt{\alpha}\coth\left(\sqrt{\alpha}\right) - \frac{1}{2} + \sqrt{1+4\alpha}\right)} .$$ (11.7)

Taking the limit $r_0 \to 0$, we find that $c_1 \to 0$, which means that the less singular wave function gets selected as the proper ground state eigenfunction.[4] This additional choice has allowed us to find this.

As we claimed above, SUSYQM provides an alternate way of arriving at the same prescription without the need for regularization. To see how

---

[3]This is the logarithmic derivative, familiar from the infinite square well.
[4]For more details, see Footnote 1.

this happens let us start with an appropriate superpotential near the origin:
$W(r, \alpha) \approx -\frac{(1+\sqrt{1+4\alpha})}{2r}$ such that the potential $V_-(r, \alpha) \approx \alpha/r^2$ falls in the transitional region $-\frac{1}{4} \leq \alpha < \frac{3}{4}$. We will then show that the eigenspectrum of the supersymmetric partner potential $V_+(r, \alpha)$ can be determined unambiguously.

Solving for V_+ first and then using the SUSYQM relations $E_{n+1}^{(-)} = E_n^{(+)}$, and $\psi_{n+1}^{(-)} = A^+ \psi_n^{(+)}$, we can solve the eigenvalue problem for the potential V_-. Thus, SUSYQM provides a straightforward prescription for choosing the "less singular" solution.

Problem 11.7. *Show that for* $W(r, \alpha) = -\frac{(1+\sqrt{1+4\alpha})}{2r}$, *the partner potentials are* $V_-(r, \alpha) = \frac{\alpha}{r^2}$ *and* $V_+(r, \alpha) = \frac{(1+\sqrt{1+4\alpha})(3+\sqrt{1+4\alpha})}{4r^2}$.

Since near the origin the superpotential $W(r, \alpha)$ is given by $W(r, \alpha) \approx -\frac{(1+\sqrt{1+4\alpha})}{2r}$, the resulting partner potentials are

$$V_-(r, \alpha) \approx \frac{\alpha}{r^2} \; ; \quad V_+(r, \alpha) \approx \frac{\left(1 + \sqrt{1 + 4\alpha}\right)\left(3 + \sqrt{1 + 4\alpha}\right)}{4r^2} .$$

Their eigenfunctions are

$$\psi^{(-)}(r, \alpha) \approx c_1 \, r^{\left(\frac{1}{2} - \sqrt{\alpha + \frac{1}{4}}\right)} + c_2 \, r^{\left(\frac{1}{2} + \sqrt{\alpha + \frac{1}{4}}\right)} \tag{11.8}$$

and

$$\psi^{(+)}(r, \alpha) \approx c_1' \, r^{-\left(\frac{1}{2} + \sqrt{\alpha + \frac{1}{4}}\right)} + c_2' \, r^{\left(\frac{3}{2} + \sqrt{\alpha + \frac{1}{4}}\right)} . \tag{11.9}$$

From Eq. (11.8), for $-\frac{1}{4} \leq \alpha < \frac{3}{4}$, both linearly independent solutions are less divergent than $\frac{1}{\sqrt{r}}$ near the origin. Hence, $\psi^{(-)}(r, \alpha)$ is square integrable. Thus, normalization does not place any constraints on values of c_1 and c_2.

However, with the value of α within this transitional region, one of the terms in Eq. (11.9) is normalizable. Thus, the requirement of square integrability of $\psi^{(+)}(r, \alpha)$ uniquely determines the proper linear combination near the origin, and hence unambiguously determines the eigenvalues and eigenfunctions for the partner potential V_+. Each eigenfunction $\psi^{(-)}(r, \alpha)$ can then be obtained by applying operator A^+ to the corresponding eigenfunction $\psi^{(+)}(r, \alpha)$ of V_+. Eigenvalues of V_- will be identical to those of V_+ with the exception of an additional zero energy ground state for V_- in the case of unbroken SUSY.

We will now work out a full fledged example with the potential given over the entire domain and not just near the singularity.

Let us consider the Pöschl-Teller II superpotential

$$W(r, A, B) = A \tanh r - B \coth r , \quad 0 \leq r < \infty . \tag{11.10}$$

For $A > B$ and $B > 0$ the above superpotential corresponds to a case of unbroken SUSY as discussed in Chapter 4. The corresponding supersymmetric partner potentials are

$$V_-(r, A, B) = -A(A+1)\,\mathrm{sech}^2 r + B(B-1)\,\mathrm{cosech}^2 r + (A-B)^2 ,$$

$$V_+(r, A, B) = -A(A-1)\,\mathrm{sech}^2 r + B(B+1)\,\mathrm{cosech}^2 r + (A-B)^2 . \tag{11.11}$$

The above set of potentials is shape invariant under the change of parameters $A \to A - 1$ and $B \to B + 1$. I.e., $V_+(r, A, B) = V_+(r, A-1, B+1) +$ constant. Without loss of generality we will assume $\frac{1}{2} < A < \infty$, and $\frac{1}{2} \leq B < \infty$. To clearly see the ambiguity in the eigenvalue problem, we proceed with the analysis of the Schrödinger equation for $V_-(r)$. The general solution is

$$\psi^{(-)}(r) = \cosh^{-A} r \left[c_1 \sinh^B r F\left(a', b', c'; -\sinh^2 r\right) \right.$$

$$\left. + c_2 \sinh^{(1-B)} r F\left(a' + 1 - c', b' + 1 - c', 2 - c'; -\sinh^2 r\right) \right] , \tag{11.12}$$

where the constants a', b', and c' are given by

$$a' = \frac{1}{2}(B - A + Q') ; \quad b' = \frac{1}{2}(B - A - Q') ; \quad c' = B + \frac{1}{2} ; \quad \text{and}$$

$$Q' = \sqrt{(B-A)^2 - E} . \tag{11.13}$$

Near the point $r = 0$, the solution reduces to

$$\psi^{(-)}(r) \approx c_1 r^B + c_2 r^{(1-B)} . \tag{11.14}$$

Thus for $B \geq \frac{3}{2}$ the wave function becomes non-normalizable unless $c_2 = 0$. With $c_2 = 0$, a subsequent constraint: the vanishing of the wave function at infinity as demanded by normalizability suffices to determine the eigenvalues E in terms of the parameters A and B. However, if $\frac{1}{2} \leq B < \frac{3}{2}$, both terms of Eq. (11.14) are well defined. Hence, normalizability places no constraints on their coefficients. In such cases, we solve the eigenvalue problem for V_+ instead. Its eigenfunctions near the origin are of the form:

$$\psi^{(+)}(r) \approx \tilde{c}_1 r^{(B+1)} + \tilde{c}_2 r^{(-B)} . \tag{11.15}$$

The normalizability of $\psi^{(+)}$ for $B < \frac{3}{2}$ requires that we set $\tilde{c}_2 = 0$. To determine the eigenvalues of V_+, we have to study the behavior of $\psi^{(+)}(r)$ at infinity. Using an alternate asymptotic form of the hypergeometric function:

$$F(a, b, c; z) \approx \frac{\Gamma(c)\Gamma(b-a)}{\Gamma(b)\Gamma(c-a)}(-z)^{-a} + \frac{\Gamma(c)\Gamma(a-b)}{\Gamma(a)\Gamma(c-b)}(-z)^{-b}, \quad (11.16)$$

we find

$$\psi^{(+)}(r) \approx \frac{\Gamma(c)\Gamma(b-a)}{\Gamma(b)\Gamma(c-a)}e^{-Qr} + \frac{\Gamma(c)\Gamma(a-b)}{\Gamma(a)\Gamma(c-b)}e^{Qr}, \quad (11.17)$$

where the constants a, b, c and Q are obtained from Eq. (11.13) by replacing A by $A - 1$ and B by $B + 1$. They are given by

$$a = \frac{1}{2}(B - A + 2 + Q) \ ; \quad b = \frac{1}{2}(B - A + 2 - Q) \ ; \quad c = B + \frac{3}{2} \ ;$$

$$Q = \sqrt{(B - A + 2)^2 - \bar{E}} \ ; \quad \bar{E} = E + (A - B - 2)^2 - (A - B)^2 . \quad (11.18)$$

The second term on the right hand side of Eq. (11.17) must vanish to have a well defined bound state. This can be achieved if either a or $(c - b)$ is equal to a negative number; say $-k$.[5] If $a = -k$, then the eigenvalues of V_+ are given by[6]

$$E_k^{(+)} = (A - B)^2 - [A - B - 2k - 2]^2 \quad k = 0, 1, \ldots, n , \quad (11.19)$$

where n is the number of bound states that the potential will hold. It is the largest integer such that $Q > 0$ i.e. $A - B - 2 > 2n$.

The eigenvalues for V_- will be the same as for V_+, except that V_- will have an additional state (its ground state) with zero energy. The eigenfunctions of V_+ are given by

$$\psi^{(+)}(r) = (\sinh r)^{1+B} (\cosh r)^{A+1}$$

$$\times F\left(-k, \frac{1}{2}(B - A + 2 - Q), B + \frac{3}{2} \ ; -\sinh^2 r\right) . \quad (11.20)$$

It is customary to express these eigenfunctions in terms of the Jacobi polynomials P_k, which are related to hypergeometric functions F. In terms of Jacobi polynomials, the eigenfunctions are given by

$$\psi_k^{(+)}(r) = (\sinh r)^{(1+B)} (\cosh r)^{-(A-1)} P_k^{(B+\frac{1}{2}, -A+\frac{1}{2})} (\cosh 2r) ,$$

$$k = 0, 1, \ldots, n . \quad (11.21)$$

---

[5]The inverse of the Gamma function Γ is zero at negative integers.

[6]If instead, the second condition holds; i.e., $c - b = -k$, the eigenvalues are given by $E_k^{(-)} = (A - B)^2 - [A + B + 2k + 1]^2$, $k = 0, 1, \ldots, n$. The condition on A for n-bound states in the second case, obtained by requiring that $Q > 0$, is given by $A < -B - 2n - 1$, which cannot be satisfied for any n, as we have assumed $-\frac{1}{2} \le A < \infty$ and $\frac{1}{2} \le B < \infty$.

The eigenfunctions of the system with potential V_- are now obtained by applying the operator A^+ to the function $\psi^{(+)}$. Near the origin, $\psi^{(+)}(r) \approx r^{B+1}$. Operating with A^+ on $\psi^{(+)}$ lowers the power of r by one, and hence $\psi^{(-)}(r) \approx r^B$. Comparing this result with Eq. (11.14), we see that SUSYQM automatically chooses the term with higher power of r. This is also the "less singular" of the two terms, and is consistent with the prescription used in the references given in Footnote 1. Thus, SUSYQM naturally selects the less divergent term without any need for additional assumptions such as regularization. We find that the SUSYQM-based formalism gives a clear-cut way of finding the eigenvalues and eigenfunctions of transition potentials in the region of ambiguity.

Now we will show that there are cases where α/x^2-type interactions appear naturally. There is an even more interesting fact about these potentials: they are necessarily of transitional nature; i.e., the value of α automatically falls within the region $-\frac{1}{4} \le \alpha < \frac{3}{4}$.

As we have seen in Chapter 5, in any shape invariant system the eigenvalues can be generated by

$$E_0^{(-)} = 0 \ , \ E_n^{(-)} = g(a_n) - g(a_0) \ , \tag{11.22}$$

and their corresponding eigenfunctions are given by

$$\psi_0^{(-)} \propto e^{-\int_{x_0}^x W(y,a_0)dy} \ , \ \psi_n^{(-)}(x, a_0) = \left[-\frac{d}{dx} + W(x, a_0) \right] \psi_{n-1}^{(-)}(x, a_1) \ ,$$

$$n = 1, 2, 3, \ldots \tag{11.23}$$

where $a_1 = f(a_0), \ldots a_k = f^k(a_0)$. If f^2 is the identity function; i.e., $f^2(a_0) = a_0$, we have a case of a cyclic potential of order 2. In this case we obtain the equally spaced spectrum of the harmonic oscillator. We now consider a function such that $f(a_0) = a_1 \ne a_0$, but $f(a_1) = a_0$. We can derive a potential that is generated by such a function and determine its eigenvalues and eigenfunctions. Since $f^2(a_0) = a_0$, we have

$$a_0 = a_2 = a_4 = \ldots \ , \ \text{and} \ f(a_0) = a_1 = a_3 = a_5 = \ldots .$$

There are many functions that meet the above requirement. One such function is $f(a_0) = C/a_0$, where C is an arbitrary constant. Another simple possibility is $f(a_0) = -a_0$. A less trivial transformation is $f(a_0) = \frac{1}{2C} \log[\coth(Ca_0)]$. From Eq. (11.22), the eigenvalues will be spaced alternately by $R(a_0) \equiv \omega_0$ and $R(a_1) \equiv \omega_1$. This spectrum corresponds to two shifted sets of equally spaced eigenvalues.

To explicitly compute the superpotential associated with these eigenvalues, we need to solve the shape invariance equations,

$$W^2(x, a_0) + W'(x, a_0) = W^2(x, a_1) - W'(x, a_1) + \omega_0 \ , \tag{11.24}$$

$$W^2(x, a_1) + W'(x, a_1) = W^2(x, a_0) - W'(x, a_0) + \omega_1 . \tag{11.25}$$

After some algebra we find that the superpotential $W(x, a_0)$ is given by

$$W(x, a_0) = \frac{1}{2} \omega x + \frac{1}{2} \frac{\Omega}{\omega} \frac{1}{x} , \tag{11.26}$$

where $\omega = \frac{1}{2}(\omega_0 + \omega_1)$ and $\Omega = \frac{1}{2}(\omega_0 - \omega_1)$. This yields $\omega_0 = (\Omega + \omega)$ and $\omega_1 = (\omega - \Omega)$.

Problem 11.8. *Derive the above superpotential.*

The corresponding potential $V_-(x)$ is

$$V_-(x, a_0) = \frac{1}{4} \frac{\Omega(\Omega + 2\omega)}{\omega^2} \frac{1}{x^2} + \frac{1}{4} \omega^2 x^2 - \frac{(\omega - \Omega)}{2} . \tag{11.27}$$

Problem 11.9. *Show that for positive values of ω_0 and ω_1, $\frac{1}{4} \frac{\Omega(\Omega + 2\omega)}{\omega^2}$ is always between $-\frac{1}{4}$ and $\frac{3}{4}$.*

Since for all positive values of ω_0 and ω_1, the coefficient of the $1/x^2$-term is constrained by $-\frac{1}{4} \leq \left(\frac{1}{4} \frac{\Omega(\Omega + 2\omega)}{\omega^2} \right) < \frac{3}{4}$, the potential of Eq. (11.27) is necessarily of the transitional type; i.e., a particle on the left can tunnel through the singularity and vice versa.

However, there is a problem. The superpotential $W(x, a_0)$ has an infinite discontinuity at the origin. Such a discontinuity is not acceptable, since while writing Eqs. (11.24) and (11.25), we implicitly assumed the existence of a continuous superpotential with a well-defined derivative. To remedy this we regularize the superpotential. We construct a continuous superpotential $\widetilde{W}(x, a_0, \epsilon)$ which reduces to $W(x, a_0)$ in the limit $\epsilon \to 0$. One such choice is

$$\widetilde{W}(x, a_0, \epsilon) = W(x, a_0) \tanh^2 \frac{x}{\epsilon} . \tag{11.28}$$

The moderating factor $\tanh^2 \frac{x}{\epsilon}$ provides a smooth interpolation through the discontinuity, since it is unity everywhere except in a small region of order ϵ around $x = 0$. In this region, $\widetilde{W}(x, a_0, \epsilon)$ is linear with a slope $\frac{1}{2\epsilon^2} \frac{\omega}{\Omega}$, which becomes larger as ϵ gets smaller. The potential $\widetilde{V}_-(x, a_0, \epsilon)$ corresponding to the superpotential $\widetilde{W}(x, a_0, \epsilon)$ is

$$\widetilde{V}_-(x, a_0, \epsilon) = \widetilde{W}^2(x, a_0, \epsilon) - \widetilde{W}'(x, a_0, \epsilon) . \tag{11.29}$$

In the limit $\epsilon \to 0$, $\frac{1}{2\epsilon} \operatorname{sech}^2 \frac{x}{\epsilon} = \delta(x)$ and $\tanh \frac{x}{\epsilon} = [2\theta(x) - 1]$ where $\theta(x)$ is the unit step (Heaviside) function. Thus, $\widetilde{V}_-(x, a_0, \epsilon)$ reduces to

$$\widetilde{V}_-(x, a_0) = V_-(x, a_0) - 4 W(x, a_0) [2\theta(x) - 1] \delta(x) . \tag{11.30}$$

Note that the potential $\widetilde{V}_-(x, a_0)$ has an additional singularity at the origin, given by $-\frac{2}{\omega}\frac{\Omega}{x}\frac{1}{x}[2\theta(x) - 1]\,\delta(x)$. The sign of this singularity depends upon the values of ω_0 and ω_1. Let us consider the two cases $\omega_0 > \omega_1$ and $\omega_0 < \omega_1$ separately.

Case 1: $\omega_0 > \omega_1$. An example of a superpotential for this case is shown in Fig. 11.2. The corresponding potential is drawn in Fig. 11.3, along with a few low lying energy levels. Note the sharp attractive shape of the potential near $x = 0$. In the limit $\epsilon \to 0$, this attractive δ-function singularity produces a bound state at $E_0 = 0$.

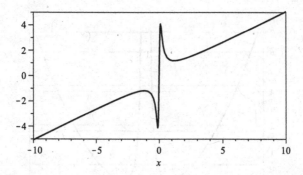

Fig. 11.2 The superpotential $\widetilde{W}(x, a_0, \epsilon)$ given by Eq. (11.28) for $\epsilon = 0.1$ and parameters $\omega_0 = 1.7, \omega_1 = 0.3$.

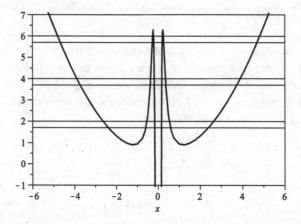

Fig. 11.3 The potential $\widetilde{V}_-(x, a_0)$ given by Eq. (11.30) for $\epsilon = 0.1$ and parameters $\omega_0 = 1.7, \omega_1 = 0.3$.

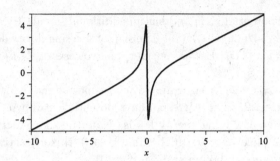

Fig. 11.4   The superpotential $\widetilde{W}(x, a_0, \epsilon)$ given by Eq. (11.28) for $\epsilon = 0.1$ and parameters $\omega_0 = 0.3, \omega_1 = 1.7$.

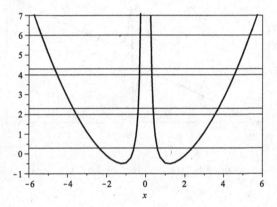

Fig. 11.5   The potential $\widetilde{V}_-(x, a_0)$ given by Eq. (11.30) for $\epsilon = 0.1$ and parameters $\omega_0 = 0.3, \omega_1 = 1.7$.

Case 2: $\omega_0 < \omega_1$. An example of a superpotential for this case is shown in Fig. 11.4. Again, proceeding as in case 1, we find a repulsive singularity in the potential at the origin. This is shown in Fig. 11.5.

We obtain eigenvalues and eigenfunctions for the potential $\widetilde{V}_-(x, a_0)$ from Eq. (11.23). The lowest four are:

$$E_0 = 0; \qquad \psi_0 \propto x^{-\frac{\omega_0 - \omega_1}{2(\omega_0 + \omega_1)}} \ e^{-\frac{1}{8}(\omega_0 + \omega_1)x^2},$$

$$E_1 = \omega_0; \qquad \psi_1 \propto x^{1 + \frac{\omega_0 - \omega_1}{2(\omega_0 + \omega_1)}} \ e^{-\frac{1}{8}(\omega_0 + \omega_1)x^2},$$

$$E_2 = \omega_0 + \omega_1;$$

$$\psi_2 \propto \left(\frac{\omega_0 - \omega_1}{\omega_0 + \omega_1} - 1 + \frac{\omega_0 + \omega_1}{2}x^2 \right) x^{-\frac{\omega_0 - \omega_1}{2(\omega_0 + \omega_1)}} \ e^{-\frac{1}{8}(\omega_0 + \omega_1)x^2},$$

$$E_3 = 2\omega_0 + \omega_1;$$

$$\psi_3 \propto \left(-\frac{\omega_0 - \omega_1}{\omega_0 + \omega_1} - 3 + \frac{\omega_0 + \omega_1}{2}x^2 \right) x^{1 + \frac{\omega_0 - \omega_1}{2(\omega_0 + \omega_1)}} \ e^{-\frac{1}{8}(\omega_0 + \omega_1)x^2}.$$

In Figs. 11.6 and 11.7, we have drawn the lowest few eigenfunctions for two choices of the parameters ω_0 and ω_1. Note that for $\omega_0 > \omega_1$, the ground state eigenfunction diverges at the origin, whereas for $\omega_0 < \omega_1$ it vanishes. For the intermediate situation $\omega_0 = \omega_1$, the ground state is finite at the origin and is just the standard Gaussian solution of a one-dimensional oscillator.

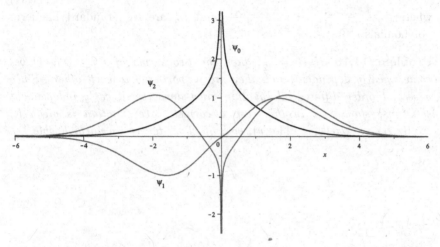

Fig. 11.6   The first three eigenfunctions corresponding to the potential of Eq. (11.30) for $\omega_0 = 1.7, \omega_1 = 0.3$.

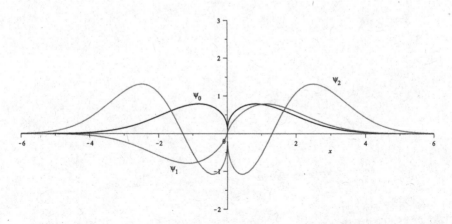

Fig. 11.7   The first three eigenfunctions corresponding to the potential of Eq. (11.30) for $\omega_0 = 0.3, \omega_1 = 1.7$. Note that the ground state wave function does not cross the x-axis, hence nodeless.

General expressions for these eigenfunctions and corresponding eigenenergies are given by

$$E_{2n} = n(\omega_0 + \omega_1) , \qquad \psi_{2n} \propto x^{-\frac{d}{2}} e^{-\frac{1}{4}\omega x^2} L_n^{-\frac{d}{2}-\frac{1}{2}} \left[\frac{\omega x^2}{2}\right] ,$$

$$E_{2n+1} = n(\omega_0 + \omega_1) + \omega_0 , \quad \psi_{2n+1} \propto x^{1+\frac{d}{2}} e^{-\frac{1}{4}\omega x^2} L_n^{\frac{d}{2}+\frac{1}{2}} \left[\frac{\omega x^2}{2}\right] ,$$

$$(11.31)$$

where $d \equiv \frac{\omega_0 - \omega_1}{(\omega_0 + \omega_1)}$, $\omega \equiv \frac{(\omega_0 + \omega_1)}{2}$, and L_n are the standard Laguerre polynomials.

Problem 11.10. *If $B = \frac{1}{2}$, then the two terms of Eq. (11.14) become linearly dependent; i.e., they are proportional to each other. Since a second order differential equation must have two linearly independent solutions, show that in this case a correct wave function is given by $\psi^{(-)}(r, \alpha) \approx \sqrt{r}\,(c_1 + c_2 \log r)$, both of whose terms are normalizable.*

Chapter 12

WKB and Supersymmetric WKB

In this chapter, we will study the WKB (Wentzel-Kramers-Brillouin)[1] method, which is one way to obtain approximate solutions for which exact solutions do not exist. We will then study the supersymmetric variant of the method, SWKB. We will discover the peculiar fact that the SWKB method yields exact solutions for all of the known translational shape invariant potentials.

12.1 WKB

This is a semi-classical approximation method; that is, it combines principles from both classical and quantum mechanics. In this case, approximate solutions to the Schrödinger equation are linked with the classical turning points of a particle; viz., the region in which a classical particle is constrained, where a quantum mechanical wave function is not. An example is the harmonic oscillator, for which a classical particle trajectory cannot exceed the amplitude of the oscillator, but the quantum mechanical wave function can and does.

We begin with the behavior of a particle in regions where the potential V is constant, such as a bound state of a finite square well. The time independent wave function is $\psi(x) \sim Ae^{\pm ikx}, E > V$; viz., inside the well, or $\psi(x) \sim Ae^{\pm \kappa x}, E < V$; viz., in the walls, where $k = \sqrt{E - V}$ and $\kappa = \sqrt{V - E}$. Let us confine ourselves to the former. This means that we are in the so-called classical region, whose limits on x are given by the solutions of $E = V(x)$. We now modify V to allow it to vary slowly in comparison to the variation in the wave function, characterized by k. (To

---

[1]Harold Jeffreys actually discovered this method before them, and is often credited by including his initial: JWKB.

put it another way, the scale of the change in the magnitude of $V(x)$ is much longer than the wavelength $\lambda = 2\pi/k$ of the wave function.)

From the definition of the classical momentum, $p(x) \equiv \sqrt{E - V(x)}$ and the quantum mechanical relationship $p \equiv -id/dx$, the Schrödinger equation becomes

$$\frac{d^2\psi}{dx^2} = -p(x)^2\psi .$$

The wave function may then be written in terms of amplitude and phase:

$$\psi(x) = A(x)e^{i\phi(x)} .$$

Our goal is to find $A(x)$ and $\phi(x)$.

The Schrödinger equation now becomes:

$$\frac{d^2\psi}{dx^2} = [A'' + 2iA'\phi' + iA\phi'' - A(\phi')^2]e^{i\phi} = -p^2 A\, e^{i\phi} .$$

We have suppressed the x-dependence of p and, as usual, set $\hbar$ and $2m$ equal to 1. Canceling $e^{i\phi}$,

$$A'' + 2iA'\phi' + iA\phi'' - A(\phi')^2 = -p^2 A . \tag{12.1}$$

Separating real and imaginary parts,

$$A'' - A(\phi')^2 = -p^2 A , \tag{12.2}$$

$$2A'\phi' + A\phi'' = 0 . \tag{12.3}$$

The first of these can be written as

$$A'' = A[(\phi')^2 - p^2] . \tag{12.4}$$

The second can be written as

$$(A^2\phi')' = 0 . \tag{12.5}$$

Integrating Eq. (12.5) yields the exact solution for $A(\phi)$:

$$A = \frac{C}{\sqrt{\phi'}} .$$

Equation (12.4) has no exact solution, so it is here that the approximation must be made. It is reasonable to assume that the amplitude of the wave function, $A(x)$, like $V(x)$, varies slowly compared to the oscillations of the wave function. Hence its second derivative is small: $A''/A \simeq 0$. Then $(\phi')^2 \simeq p^2$. Treating this as an equality, $\phi' = \pm p$. This is known as the zeroth order approximation.[2] Let us integrate this, taking as our limits x_0 and x

$$\phi(x) - \phi(x_0) = \pm \int_{x_0}^{x} p(x')dx'.$$

We can set $\phi(x_0) = 0$ and choose $x_0 = 0$, whence $\phi(x) = \pm \int_0^x p(x')dx'$. So the approximate wave function is

$$\psi(x) \simeq Ae^{i\phi} \simeq \frac{C}{\sqrt{p(x)}}e^{\pm i \int_0^x p(x')dx'}. \tag{12.6}$$

This is the general WKB result. Note that we have used the classical momentum $p = \sqrt{E - V}$, but the wave function is non-zero beyond the classical turning points, where $E < V$ and p is thus imaginary. In those regions, $p(x) \to -i|p(x)|$ and $e^{\pm i \int_0^x p(x')dx'} \to e^{\pm \int_0^x |p(x')|dx'}$.

Let us apply the WKB method to an arbitrarily shaped potential $V(x)$ added to the bottom of an infinite square well whose potential is taken as 0, in the range $0 \leq x \leq a$ as illustrated in Fig. 12.1.

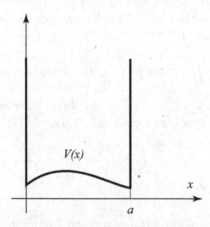

Fig. 12.1   Arbitrarily shaped potential $V(x)$ added to the bottom of an infinite square well.

For the time being, let us consider those regions where $E > V(x)$. Then ψ is in general a linear combination of terms involving $\exp\left[\pm i \int_0^x p(x')dx'\right]$.

---

[2]In general, one writes ϕ' as an infinite sum expanded in powers of $\hbar$ and then solves equations for each order to determine ϕ'. For more details, see J.L. Dunham, "The Wentzel-Brillouin-Kramers Method of Solving the Wave Equation", *Phys. Rev.* **41**, 713 (1932).

We can rewrite it as

$$\psi(x) \simeq \frac{1}{\sqrt{p(x)}} \left[C_1 \sin \left(\int_0^x p(x') dx' \right) + C_2 \cos \left(\int_0^x p(x') dx' \right) \right] .$$

Using the boundary conditions for infinite vertical walls, $\psi(0) = 0$ implies $C_2 = 0$ and $\psi(a) = 0$ implies

$$\int_0^a p(x) dx = n\pi . \tag{12.7}$$

In the regions where $E < V(x)$, $\psi(x) \sim \exp \left[\pm \int_0^x | p(x') | \, dx' \right]$. Then

$$\psi(x) \simeq \frac{1}{\sqrt{| p(x) |}} \left[C_1 \sinh \left(\int_0^x | p(x') | \, dx' \right) + C_2 \cosh \left(\int_0^x | p(x') | \, dx' \right) \right] .$$

Let us note some interesting and important features of this result.

- First, the form of Eq. (12.7) is reminiscent of the Bohr-Sommerfeld condition from pre-Schrödinger quantum theory: this is what makes the WKB method semi-classical.
- Second, knowing the form of $V(x)$ immediately gives us the approximate energy levels E_n^{WKB}:

$$\int_0^a p(x) dx = \int_0^a \sqrt{E - V(x)} dx = n\pi .$$

- Third, if we set $V(x) = 0$, we should obtain the exact solution for the infinite well: $p(x) = \sqrt{E}$, a constant. Then

$$n\pi = \int_0^a p(x) dx = pa = \sqrt{E} a ,$$

giving $E = n^2 \pi^2 / a^2$.

Problem 12.1. *An infinite well has a square barrier of height V_0 and width $a/3$ in its middle (see Fig. 12.2):*

$$V(x) = \begin{cases} 0, & 0 < x < a/3 \\ V_0, & a/3 < x < 2a/3 \\ 0, & 2a/3 < x < a \\ \infty & elsewhere \end{cases}$$

For $E > V_0$, use the WKB method to find the approximate value of the energy level E_n in terms of V_0 and the infinite well energy $E_n^0 = (n\pi/a)^2$.

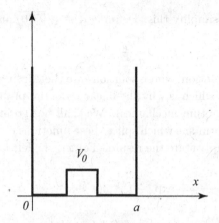

Fig. 12.2   The potential for problem (12.1).

For $E < V_0$; viz., within the barrier, the WKB method can be used to obtain the tunneling probability. We shall not do this here.[3]

The condition $\int_0^a p(x)dx = n\pi$ is specific to cases with two infinite vertical walls. The factor on the right hand side will change for cases with one vertical wall, such as a half harmonic oscillator with bumps, or no vertical walls, such as a full harmonic oscillator with bumps. For these cases, the analysis gets more complicated. The reason is that for any finite potential, $E \simeq V$ at the classical turning points. At these points, p goes to zero, so the WKB wave function $\psi \sim \frac{1}{\sqrt{p}}$ diverges.

To resolve this dilemma, we need to make another approximation. Since the WKB wave function is already approximate, we can modify it in some small way to avoid the turning-point catastrophe. This is usually done by a "patching" technique. Instead of the WKB wave functions to the right and left of the well boundary matching exactly (the source of the problem) they are spliced to a simpler patching function ψ_p which traverses the boundary. To be certain that this is not severely changing the WKB results, we insist that the patching function only be valid in a region very close to the turning points. The usual way to do this is to linearize the potential in those regions. Choosing for the moment $x = 0$ at the right hand turning point, $V(x) \simeq E + V'(0)\,x$. The Schrödinger equation is then

$$-\frac{d^2\psi_p}{dx^2} + [E + V'(0)\,x]\,\psi_p = E\psi_p\,.$$

---

[3]See A. Messiah, *Quantum Mechanics*, v.1, Paris: Dunod Publ. (1964).

Two substitutions simplify this. Defining $\alpha \equiv \sqrt[3]{V'(0)}$ and $z \equiv \alpha x$ yields

$$\frac{d^2\psi_p}{dz^2} = z\psi_p. \tag{12.8}$$

This is the Airy Equation, whose solutions are the Airy functions. They are quite complicated (which is why the linear potential problem is rarely seen in introductory quantum mechanics). We shall not go into any derivation of the connection formulas which splice these functions onto the WKB wave functions.[4] We simply state the results. For x_1, the left hand turning point of a potential well,

$$\psi(x) \simeq \begin{cases} \frac{C_3}{\sqrt{|p(x)|}} \exp\left[-\int_x^{x_1} |p(x')|dx'\right], & x < x_1 \\ \frac{2C_3}{\sqrt{p(x)}} \sin\left[\int_{x_1}^x p(x')dx' + \pi/4\right], & x > x_1 \end{cases} \tag{12.9}$$

and for x_2, the right hand turning point

$$\psi(x) \simeq \begin{cases} \frac{C_4}{\sqrt{|p(x)|}} \exp\left[-\int_{x_2}^x |p(x')|dx'\right], & x > x_2 \\ \frac{2C_4}{\sqrt{p(x)}} \sin\left[\int_x^{x_2} p(x')dx' + \pi/4\right], & x < x_2 \end{cases} \tag{12.10}$$

(We could find C_3 and C_4 by normalizing the wave functions, but we don't need them.)

We are now ready to find the quantization condition for $\int p(x)dx$. For the case of one vertical wall at $x_1 \equiv 0$, $\psi(0) = 0$. Since we are in the region within the turning points $x < x_2$, the second line of Eq. (12.10) with $x = 0$ makes the argument of the sine a multiple of π: $\int_0^{x_2} p(x)dx + \pi/4 = n\pi$. Then the quantization condition for one vertical wall is

$$\int_0^{x_2} p(x)dx = (n - 1/4)\pi, \quad n = 1, 2, 3, \dots \tag{12.11}$$

For the case of no vertical walls, we compare the solutions of Eq. (12.9) and Eq. (12.10) for $x_1 < x < x_2$. They must agree because they cover the same region. I.e.;

$$C_3 \sin\left[\int_{x_1}^x p(x')dx' + \pi/4\right] = C_4 \sin\left[\int_x^{x_2} p(x')dx' + \pi/4\right].$$

This is only possible if the constants C_3 and C_4 are the same magnitude and the sines are the same. The latter are non-trivially the same if their arguments differ by some multiple of π.

---

[4] For these details, see e.g. D. J. Griffiths, *Quantum Mechanics*, 2nd ed., CA: Benjamin-Cummings (2004).

Since any minus sign in the argument of the sine can be absorbed into the constant, that multiple is $n\pi$, and not, as we might have thought, $2n\pi$. So in fact the constants are related by $C_4 = \pm C_3$. Thus

$$\int_x^{x_2} p(x')dx' + \pi/4 = \pm \left[\int_{x_1}^x p(x')dx' + \pi/4 \right] + n\pi \ .$$

For WKB quantization we need $\int_{x_1}^{x_2} p(x')dx'$. For the plus sign, we obtain

$$\int_x^{x_2} p(x')dx' + \pi/4 = + \left[\int_{x_1}^x p(x')dx' + \pi/4 \right] + n\pi \ .$$

Then

$$\int_x^{x_2} p(x')dx' - \int_{x_1}^x p(x')dx' = n\pi \ .$$

But this does not give the right integral for quantization. On the other hand, for the minus sign, we obtain

$$\int_x^{x_2} p(x')dx' + \pi/4 = - \left[\int_{x_1}^x p(x')dx' + \pi/4 \right] + n\pi \ .$$

This gives us the appropriate integration:

$$\int_{x_1}^x p(x')dx' + \int_x^{x_2} p(x')dx' = \int_{x_1}^{x_2} p(x')dx' \tag{12.12}$$

$$= (n - 1/2)\pi \ , \qquad n = 1, 2, 3, \ldots$$

We have now obtained the WKB conditions for potentials with two, one, and zero vertical walls. For the first, we have done the example of the infinite square well and obtained the exact energies, despite the fact that the WKB method is an approximate one. There are several other potentials for which the WKB method give exact results. Let us now solve two of them: the half-harmonic oscillator (one vertical wall) and the full harmonic oscillator (no vertical walls).

Problem 12.2. *The half-harmonic oscillator potential is*

$$V(x) = \begin{cases} \infty, & x = 0 \\ \frac{1}{4}\omega^2 x^2, & x > 0 \end{cases}$$

Show that the WKB approximation gives the exact result for the energies.[5]

Problem 12.3. *The harmonic oscillator potential is $V(x) = \frac{\omega^2 x^2}{4}, -\infty < x < \infty$. Show that the WKB approximation gives the exact result for the energies.*

---

[5]Keep in mind that the usual notation starts with $n = 0$, while the WKB notation starts with $n = 1$.

12.2 SWKB

In Chapter 5 we introduced shape invariance and described the method by which we could obtain the eigenvalues and eigenfunctions of translational shape invariant potentials. Surprisingly, the exact energies for shape invariant potentials can be obtained from the SWKB method. Since any supersymmetric potential is given in terms of the superpotential W by $V = W^2 - W'$, the ordinary WKB approximation is

$$\int_{x_1}^{x_2} p(x)dx = \int_{x_1}^{x_2} \sqrt{E - V(x)}dx$$

$$= \int_{x_1}^{x_2} \sqrt{E - (W^2 - W')}dx = \text{N}\pi \ , \qquad (12.13)$$

where $\text{N} = n, n-1/4, n-1/2$ for two, one and no vertical walls respectively, where n is an integer. This formulation cannot in general give exact results, since it is the WKB approximation itself. But if W' is removed, then

$$\int_{\chi_1}^{\chi_2} \sqrt{E - W^2} \, dx = n\pi \ , \qquad (12.14)$$

where χ_1 and χ_2 are the solutions of $E - W^2(x) = 0$. This is the SWKB method which for integer n gives the exact energies for all of the conventional translational shape invariant potentials.[6] Let us emphasize that to our knowledge,

- No one has yet explained why this is so, although much work has been done and progress has been made;
- The method does not necessarily work for potentials other than those that have translational shape invariance.

Let us describe the SWKB method using two examples.

12.2.1 *Infinite Square Well: SWKB*

As an example, let us solve the infinite square well problem. As we saw in Chapter 4 we began with the superpotential $W(x) = -b \cot x$ with the parameter $b > 0$, where we chose the domain $0 < x < \pi$. We found the

---

[6]By this we mean those listed in Section 5.3. They are independent of $\hbar$.

supersymmetric partner potentials to be:

$$V_-(x, b) = W^2(x) - \frac{dW}{dx} = b(b - 1)\, \mathrm{cosec}^2 x - b^2$$

and $\qquad\qquad\qquad\qquad\qquad\qquad\qquad\qquad$ (12.15)

$$V_+(x, b) = W^2(x) + \frac{dW}{dx} = b(b + 1)\, \mathrm{cosec}^2 x - b^2 \ .$$

For $b = 1$, the potential $V_-(x, 1)$ was an infinite one-dimensional square well potential with bottom at -1. Then the SWKB equation is

$$J(E) = \int_{-\cot^{-1}\sqrt{E}}^{\cot^{-1}\sqrt{E}} \sqrt{E - \cot^2 x}\, dx \ = \ n\pi \ . \qquad (12.16)$$

As we integrate this expression, we will demonstrate techniques that are often used in physics to solve complicated integrals. We first make a substitution $x = \frac{\pi}{2} + y$. This leads to

$$J(E) = \int_{-\tan^{-1}\sqrt{E}}^{\tan^{-1}\sqrt{E}} \sqrt{E - \tan^2 y}\, dy \ . \qquad (12.17)$$

To solve it further, we introduce a parameter α which we will later set equal to one. We define

$$I(\alpha, E) = \int_{-\tan^{-1}\sqrt{E}}^{\tan^{-1}\sqrt{E}} \sqrt{E - \alpha \cdot \tan^2 y}\, dy \ . \qquad (12.18)$$

Thus, $J(E) = I(1, E)$. Now, differentiating $I(\alpha, E)$ with respect to α, we obtain

$$\frac{\partial I(\alpha)}{\partial \alpha} = -\frac{1}{2} \int_{-\tan^{-1}\sqrt{E}}^{\tan^{-1}\sqrt{E}} \left(\frac{\tan^2 y}{\sqrt{E - \alpha \cdot \tan^2 y}} \right) dy$$

$$= -\frac{1}{2\sqrt{\alpha}} \sin^{-1}\left(\sqrt{\frac{\alpha}{E}} \tan y \right)$$

$$+ \frac{1}{2\sqrt{E + \alpha}} \sin^{-1}\left(\sqrt{\frac{E + \alpha}{E}} \tan y \right) . \qquad (12.19)$$

Problem 12.4. *Derive the above expression for* $\frac{\partial I(\alpha)}{\partial \alpha}$.

This implies that

$$I(\alpha, E) = \int d\alpha \left[-\frac{1}{2\sqrt{\alpha}} \sin^{-1}\left(\sqrt{\frac{\alpha}{E}} \tan y \right) \right. \qquad (12.20)$$

$$\left. + \frac{1}{2\sqrt{E + \alpha}} \sin^{-1}\left(\sqrt{\frac{E + \alpha}{E}} \sin y \right) \right] .$$

In the first integral, setting $\sqrt{\alpha} = \beta$, we obtain

$$\int d\alpha \left[\frac{1}{\sqrt{\alpha}} \sin^{-1}\left(\sqrt{\frac{\alpha}{E}} \tan y \right) \right] = \int 2 \, d\beta \, \sin^{-1}\left(\frac{\beta}{\sqrt{E}} \tan y \right) \equiv T(\beta) \, .$$

$$(12.21)$$

Now using $\int dz \, \sin^{-1} z = z \sin^{-1} z + \sqrt{1 - z^2}$, we obtain

$$T(\beta) = \int 2\frac{\sqrt{E}}{\beta} \, d\left(\frac{\beta}{\sqrt{E}} \tan y \right) \sin^{-1}\left(\frac{\beta}{\sqrt{E}} \tan y \right)$$

$$= 2\beta \, \sin^{-1}\left(\frac{\beta}{\sqrt{E}} \tan y \right) + 2 \sqrt{1 - \beta^2 \frac{\tan^2 y}{E}} \frac{\sqrt{E}}{\tan y} \, .$$

Similarly integrating $K(\alpha) = \int d\alpha \, \dfrac{\sin^{-1}\left(\sqrt{\frac{E+\alpha}{E}} \sin y \right)}{\sqrt{E+\alpha}}$, we obtain

$$K(\alpha) = 2\sqrt{E + \alpha} \sin^{-1}\left(\frac{\sqrt{E + \alpha}}{\sqrt{E}} \sin y \right) + 2\frac{\sqrt{E}}{\sin y} \sqrt{1 - (E + \alpha) \frac{\sin^2 y}{E}} \, ,$$

where $\gamma = \sqrt{E + \alpha}$. Now, collecting terms, we obtain

$$I(\alpha, E) = \sqrt{E + \alpha} \, \sin^{-1}\left(\frac{\sqrt{E + \alpha}}{\sqrt{E}} \sin y \right) - \sqrt{\alpha} \sin^{-1}\left(\sqrt{\frac{\alpha}{E}} \tan y \right) \, .$$

Thus,

$$J(E) = I(1, E) = 2 \int_0^{\tan^{-1} \sqrt{E}} \sqrt{E - \tan^2 y} \, dy$$

$$= 2\left[\sqrt{E + 1} \, \frac{\pi}{2} - \frac{\pi}{2} \right] = \left[\sqrt{E + 1} - 1 \right] \pi \, .$$

$$(12.22)$$

Now, setting $\left[\sqrt{E + 1} - 1 \right] \pi = n\pi$, we obtain $E_n = (n + 1)^2 - 1$. For $n = 0$, we obtain $E = 0$, as expected for unbroken SUSY.

Problem 12.5. *Obtain the energies for the harmonic oscillator from the SWKB method.*

12.2.2 Morse Potential: WKB and SWKB

We will solve for the energies using the WKB method, for which it is exact, then leave the SWKB method for a problem. The Morse superpotential is

$$W(x, A, B) = A\left(1 - e^{-z}\right) . \tag{12.23}$$

The WKB condition for the Morse potential is given by

$$\int \left[E - A^2\left(1 - e^{-z}\right)^2 + Ae^{-z}\right]^{\frac{1}{2}} dz = n\pi . \tag{12.24}$$

Let us do a change of variables $e^{-z} = y$; the left hand side of Eq. (12.24) becomes

$$-A \int \frac{dy}{y} \left[\underbrace{\left(\frac{E + A + \frac{1}{4}}{A^2}\right)}_{\beta^2} - \left\{(1 - y) + \frac{1}{2A}\right\}^2\right]^{\frac{1}{2}} .$$

With two more changes of variables: $(1 - y) + \frac{1}{2A} = u$ and $u = \beta \sin\theta$, the above integral becomes

$$A\beta^2 \int_{-\pi/2}^{\pi/2} \left(\frac{\cos^2\theta}{1 - \beta\sin\theta + \frac{1}{2A}}\right) d\theta .$$

An integration then leads to

$$\frac{1}{2A\beta^2} \left[2\sqrt{-4A^2\left(\beta^2 - 1\right) + 4A + 1}\, \tan^{-1}\left(\frac{2A\beta - (2A + 1)\tan\left(\frac{\theta}{2}\right)}{\sqrt{-4A^2\left(\beta^2 - 1\right) + 4A + 1}}\right)\right]$$

$$+ \left[\frac{-2A\beta\cos(\theta) + 2A\theta + \theta}{2A\beta^2}\right] . \tag{12.25}$$

Setting $-4A^2\left(\beta^2 - 1\right) + 4A + 1 = 4A^2(1 - \alpha^2)$, where $\alpha = \frac{\sqrt{E}}{A}$, we obtain

$$\frac{4A\sqrt{1 - \alpha^2}\, \tan^{-1}\left(\frac{2A\beta - (2A + 1)\tan\left(\frac{\theta}{2}\right)}{2A\sqrt{1 - \alpha^2}}\right) - 2A\beta\cos(\theta) + 2A\theta + \theta}{2A\beta^2} . \tag{12.26}$$

Substituting the limits for θ; i.e., $\pm\pi/2$,

$$\frac{1}{2}\left(4A\sqrt{1 - \alpha^2}(-\pi/2) + (2A + 1)\pi\right) = \left(n + \frac{1}{2}\right)\pi . \tag{12.27}$$

Solving the above equation for energy, we obtain

$$E_n = A^2 - (A - n)^2 , \tag{12.28}$$

the exact eigenvalues for the Morse potential.

Problem 12.6. *Fill in the steps from Eq. (12.24) to Eq. (12.28).*

Problem 12.7. *Obtain the energy spectrum for the Morse potential using the SWKB method.*

Let us make some concluding remarks.

- First, there is no evidence that SWKB is exact for solvable potentials other than those that are translationally shape invariant.[7]
- Second, as per Footnote 2, Eqs. (12.7) and (12.14) are the zeroth order WKB and SWKB aproximations respectively. For the potentials listed in Table 10.1, all higher order contributions to the quantization are purportedly zero, thus rendering Eq. (12.14) exact.[8] However, it has only been shown that up to order $\hbar^6$ all contributions to the SWKB quantization formula[9] vanish.
- Third, while WKB yields the correct spectrum for the infinite square well, harmonic oscillator, and Morse potential, it is also known to produce exact results for the Coulomb potential, provided the centrifugal term $\frac{\ell(\ell+1)}{r^2}$ is replaced by $\frac{(\ell+1/2)^2}{r^2}$. This is called the Langer correction,[10] and is not yet understood. It appears that the Langer correction arises naturally in SUSYQM formalism.[11]

There is much work to be done to account for the exactness of the SWKB method for all known shape invariant potentials listed in Table 10.1.

---

[7]D. Barclay, A. Khare and U. Sukhatme, "Is the Lowest Order Supersymmetric WKB Approximation Exact for All Shape Invariant Potentials?", *Phys. Lett. A* **183**, 263–266 (1993).

[8]K. Raghunathan, M. Seetharaman and S.S. Vasan, "On The Exactness of the SUSY Semiclassical Quantization Rule", *Phys. Lett.* **188B**, 351–352 (1987).

D. Barclay and C.J. Maxwell, "Shape invariance and the SWKB series", *Phys. Lett. A* **157**, 351 (1991).

[9]R. Adhikari, R. Dutt, A. Khare, and U. Sukhatme, "Higher order WKB approximations in supersymmetric quantum mechanics", *Phys. Rev. A* **38**, 1679 (1986).

[10]R. E. Langer, "On the connection formulas and solutions of the wave equation." *Phys. Rev.* **51**, 669 (1937).

[11]R. De, R. Dutt, R. Adhikari and A. Comtet, "Supersymmetric WKB approach to scattering problems", *Phys. Lett. A* **152**, 381–387 (1991).

Chapter 13

Dirac Theory and SUSYQM

The Dirac equation is the relativistic counterpart — more precisely, generalization — of the Schrödinger equation. We shall focus on calculation of the hydrogen atom fine structure, in order to understand the connection between non-relativistic and relativistic theory, and then relate the latter to SUSYQM. The calculations will comprise

- A perturbation calculation starting from the Schroedinger equation;
- An exact calculation from the Dirac equation;
- An exact calculation from SUSYQM.

13.1 Introduction

Let us first obtain the fine structure as a perturbation of the Schrödinger equation. This is the form that undergraduate physics students have seen. It introduces spin, based on experimental spectroscopic evidence, but not accounted for by the Schrödinger equation. The time-dependent Schrödinger equation is

$$H\Psi(\mathbf{r},t) \equiv -\frac{1}{2m}\nabla^2\Psi(\mathbf{r},t) + V(\mathbf{r},t)\ \Psi(\mathbf{r},t) = i\partial\Psi(\mathbf{r},t)/\partial t \qquad (13.1)$$

with H a function of both $\mathbf{r}$ and t. With the $V = V(r)$, we obtain the time-independent Schrödinger equation:

$$H\psi(\mathbf{r}) \equiv -\frac{1}{2m}\nabla^2\psi(\mathbf{r}) + V(\mathbf{r})\ \psi(\mathbf{r}) = E\psi(\mathbf{r})\ . \qquad (13.2)$$

Now a spin function $\chi(s)$ is "tacked onto" the wave function:

$$\psi(r,\theta,\phi,s) = R_{n\ell}(r)Y_\ell^m(\theta,\phi)\chi(s)\ . \qquad (13.3)$$

As we saw in Chapter 2, the angular functions for any central potential are the spherical harmonics. For the Coulomb potential, $V = -e^2/r$ and the radial functions are the associated Laguerre polynomials.

$\chi(s)$ plays no direct role in the Schrödinger equation's energy eigenvalues. Its only contribution vis-a-vis energy is to add on, *ad hoc*, a contribution to the fine structure, which itself is not accounted for by the Schrödinger equation. About half of the fine structure is accounted for by the first-order relativistic correction to the classical kinetic energy. The rest is accounted for by treating the "spinning" electron as a magnetic dipole in the field created by the proton orbiting the electron, as seen by the electron. Since the proton's angular momentum in the electron's frame is proportional to the orbital angular momentum **L** of the electron in the proton's frame, the so-called spin-orbit coupling is a quantity proportional to $\mathbf{L} \cdot \mathbf{S}$.

The result is

$$E_{fs}/E_n = \frac{\alpha^2}{n^4} \left[\frac{n}{j + 1/2} - \frac{3}{4} \right],$$

where $\alpha = e^2/\hbar c$ is the fine structure constant. In our units, $\alpha = e^2$.

Note that the fine structure energy depends only on the quantum numbers j and n. Instead of calling it E_{fs} we could just as well call it E_{nj}, and we will. In other words, it is degenerate with respect to the quantum numbers ℓ and s. Figure 13.1 shows the first few levels for hydrogen.[1] At the bottom are the values of ℓ. Above each are the Bohr energy levels with the fine structure corrections (vastly magnified). Those levels of different ℓ but the same j have the same energy. For example, the level $\ell = 2, j = \ell - 1/2 = 3/2$ has the same energy as $\ell = 1, j = \ell + 1/2 = 3/2$. Is this true of the exact result, or is it an artifact of the perturbation calculation? As we shall see, the former is the case.

13.2 Spin

As we noted above, the appearance of unexpected spectral lines was accounted for by postulating an intrinsic angular momentum called spin. Its magnitude was $1/2$, and it had only two possible orientations, designated "up" and "down". It resided in its own two-dimensional space, independent of the ordinary three-dimensional space in which the orbital angular momentum resides.

---

[1] The standard notation for spectra is $^{2S+1}L_j$, where L is represented by its letter designation. Thus, $s = 1/2, \ell = 1, j = 3/2$ is $^2P_{3/2}$.

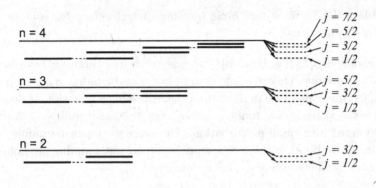

Fig. 13.1 The first few levels for the hydrogen spectrum.

We can construct a two-dimensional spin space with spin orientation chosen along the z-axis and eigenvalues $\pm 1/2$.

$$\chi_z^+ = \begin{pmatrix} 1 \\ 0 \end{pmatrix}, \quad \chi_z^- = \begin{pmatrix} 0 \\ 1 \end{pmatrix}.$$

Since we have chosen this basis set, the spin operator that yields $\pm 1/2$ must be

$$s_z = \frac{1}{2} \begin{pmatrix} 1 & 0 \\ 0 & -1 \end{pmatrix}.$$

The matrix itself is called σ_z, the "Pauli spin matrix in the z direction":

$$\sigma_z \equiv \begin{pmatrix} 1 & 0 \\ 0 & -1 \end{pmatrix}.$$

We now claim that

$$s_x = \frac{1}{2} \begin{pmatrix} 0 & 1 \\ 1 & 0 \end{pmatrix}$$

and

$$s_y = \frac{1}{2} \begin{pmatrix} 0 & -i \\ i & 0 \end{pmatrix}.$$

Problem 13.1. *Show that these give the correct values for spin in their respective directions.*

The choice of which axis to call "z" was arbitrary. However, once having chosen z, we must choose x and y to preserve cyclic order: $\sigma_x \sigma_y = i\sigma_z$.

Since spin cannot be derived from the Schrödinger equation, it does not change the spatial wave functions. Any contribution it makes to the total energy must be a small perturbation. The magnetic quantum numbers m_ℓ and m_s, introduced in the vector model of spin-orbit coupling are not good quantum numbers except in this approximation.

13.3 The Dirac Equation

Let us now turn to the development of the Dirac equation and use it to calculate the exact fine structure. We will begin with the consideration of a free particle, and then move on to a sketch of the solution for the hydrogen atom.

In special relativity, position and time play equivalent roles. Therefore, a relativistic equation must be symmetric in these. But the time dependent Schrödinger equation is linear in time and quadratic in position. Thus, it is necessary to make it either linear or quadratic in both.

The first attempts were to make the time dependence quadratic. Some led to interesting results. One was the Klein-Gordon equation. A serious shortcoming was that it predicted a particle probability density that was not positive definite. However, it turned out to be useful for particles of spin zero.

Dirac went the other way, seeking a first-order equation in position and time; i.e., one which would explicitly contain $\partial/\partial t$ and $\mathbf{p} = \nabla/i$. It should give the correct solution for a particle of any charge or none, as well as for a charged particle in electric and/or magnetic fields; viz., for both scalar and vector potentials. He looked for a linear form of the relativistic hamiltonian to match the linearity of the time-dependence. He accomplished this by writing the hamiltonian as

$$H = \boldsymbol{\alpha} \cdot \mathbf{p} + \beta m$$
$$= \alpha_1 p_1 + \alpha_2 p_2 + \alpha_3 p_3 + \beta m \ .$$

This is manifestly symmetric in the four-vector (p_1, p_2, p_3, E_0) where $E_0 = m$, thus yielding an equation which is first order in both ∇ and $\partial/\partial t$.

What then are the properties of α_i and β? They should be space- and time-independent. They must also be dimensionless.

We can now recapture $H\psi = E\psi$ by setting the relativistic form of the square of the energy, $\mathbf{p}^2 + m^2$, equal to the square of the Dirac hamiltonian, taking care not to disturb the order of operations, since we do not know if the components α_i commute with each other and/or with β. Writing

$$\mathbf{p}^2 + m^2 = (\boldsymbol{\alpha} \cdot \mathbf{p} + \beta m) \cdot (\boldsymbol{\alpha} \cdot \mathbf{p} + \beta m) \qquad (13.4)$$

we obtain

$$\mathbf{p}^2 + m^2 = \left[\sum_{i=1}^{3} \alpha_i^2 p_i^2 + \sum_{j \neq i} \alpha_i \alpha_j p_i p_j \right] + m^2 \beta^2 + m \left[\sum_{i=1}^{3} p_i (\alpha_i \beta + \beta \alpha_i) \right] .$$

Since p_i are arbitrary, matching left and right sides, we get

$$\mathbf{p}^2 = \sum_{i=1}^{3} \alpha_i^2 p_i^2 \Rightarrow \alpha_i^2 = 1 \quad \text{and} \quad m^2 = \beta^2 m^2 \Rightarrow \beta^2 = 1 . \qquad (13.5)$$

The cross terms must separately vanish for each i, since β and each of the α_is were assumed to be independent of each other

$$0 = \alpha_i \alpha_j + \alpha_j \alpha_i, \ j \neq i \qquad (13.6)$$

and

$$0 = \alpha_i \beta + \beta \alpha_i, \ i = 1, 2, 3 . \qquad (13.7)$$

The coefficients α_i and β can be written in matrix form. A conventional choice is

$$\alpha_1 = \begin{pmatrix} 0\,0\,0\,1 \\ 0\,0\,1\,0 \\ 0\,1\,0\,0 \\ 1\,0\,0\,0 \end{pmatrix} \qquad (13.8)$$

$$\alpha_2 = \begin{pmatrix} 0 & 0 & 0 & -i \\ 0 & 0 & i & 0 \\ 0 & -i & 0 & 0 \\ i & 0 & 0 & 0 \end{pmatrix} \qquad (13.9)$$

$$\alpha_3 = \begin{pmatrix} 0 & 0 & 1 & 0 \\ 0 & 0 & 0 & -1 \\ 1 & 0 & 0 & 0 \\ 0 & -1 & 0 & 0 \end{pmatrix} \qquad (13.10)$$

$$\beta = \begin{pmatrix} 1 & 0 & 0 & 0 \\ 0 & 1 & 0 & 0 \\ 0 & 0 & -1 & 0 \\ 0 & 0 & 0 & -1 \end{pmatrix} . \tag{13.11}$$

Writing the matrices as

$$\alpha_i = \begin{pmatrix} 0 & \sigma_i \\ \sigma_i & 0 \end{pmatrix} \tag{13.12}$$

$$\beta = \begin{pmatrix} 1 & 0 \\ 0 & -1 \end{pmatrix}, \tag{13.13}$$

where the σ_is are the Pauli spin matrices and $\mathbf{1} = \begin{pmatrix} 1 & 0 \\ 0 & 1 \end{pmatrix}$ simplifies calculation of the anti-commutation relations.

Problem 13.2. *Verify the anti-commutation relations Eqs. (13.6) and (13.7).*

Putting all of this together, the Dirac equation for a free particle is

$$H\psi = i\frac{\partial \psi}{\partial t} = E\psi = (\boldsymbol{\alpha} \cdot \mathbf{p} + \beta m)\psi = \left(\boldsymbol{\alpha} \cdot \frac{\nabla}{i} + \beta m \right) \psi . \tag{13.14}$$

But since the α_i and β_i are 4×4 matrices, ψ must be a 4-component column vector:

$$\psi = \begin{pmatrix} \psi_1 \\ \psi_2 \\ \psi_3 \\ \psi_4 \end{pmatrix} .$$

For the Coulomb potential the Dirac equation becomes

$$H\psi = [\boldsymbol{\alpha} \cdot \mathbf{p} + \beta m + V]\,\psi = E\psi . \tag{13.15}$$

where $V = -e^2/r$ for the hydrogen atom.

13.4 Constants of the Motion

As in the non-relativistic case, we look for an angular momentum operator that commutes with H. This is the criterion for conservation of that momentum. In Chapter 3 we showed that any operator that commutes with H is conserved. Its expectation values are time-independent; consequently, its eigenvalues are good quantum numbers. We note that, just as in the case of the Schrödinger equation, the results we will obtain are the same for the free particle and the hydrogen atom, since the hydrogen potential has no angular dependence. This is the case because **L** commutes[2] with r^2. From this we can conclude that **L** commutes with any scalar central potential $V(|\mathbf{r}|)$.

We first try the orbital angular momentum $\mathbf{L} = \mathbf{r} \times \mathbf{p}$. Since it operates in ordinary 3-space, it commutes with the α_is but not the p_is. The commutator of **L** with **p** was given in Problem 7.1. Using these results, we find that $[\mathbf{L}, H]$ does not vanish; in fact, it equals $i(\boldsymbol{\alpha} \times \mathbf{p})$.

Problem 13.3. *Prove this. (Show for one component, and generalize.)*

We must then look for an additional angular-momentum-like operator whose commutator with H is $-i(\boldsymbol{\alpha} \times \mathbf{p})$, which we can add to **L** to obtain an operator that does commute with H.

We define a 4-dimensional spin matrix as

$$\mathbf{S} = \begin{pmatrix} \mathbf{s} & 0 \\ 0 & \mathbf{s} \end{pmatrix}. \tag{13.16}$$

Its commutator with H is given by

$$[\mathbf{S}, H] = -i(\boldsymbol{\alpha} \times \mathbf{p}).$$

Problem 13.4. *Show this.*

Therefore, $\mathbf{L} + \mathbf{S}$ commutes with H. We define the total angular momentum $\mathbf{J} \equiv \mathbf{L} + \mathbf{S}$, which is conserved. We thus see how spin emerges naturally from the Dirac equation.

Of what operators in addition to the hamiltonian is ψ an eigenfunction? That is, which of their quantum numbers are good? The square of the angular momentum vector is not: $[\mathbf{L}^2, H] \neq 0$. So $\mathbf{L}^2$ is not conserved. $\mathbf{S}^2$ does commute with the hamiltonian. To confirm this, we observe that the square of each of the Pauli matrices is the identity operator, which commutes with everything.

We have seen that **L**, **S**, and $\mathbf{L}^2$ do not commute with H. Hence, m_ℓ, m_s and ℓ are not good quantum numbers. This leaves only s, j, and m_j as good quantum numbers. We therefore expect them to appear as indices in the Dirac wave function.

---

[2]The proof is tedious but straightforward, and may be found in many introductory quantum mechanics textbooks.

13.5   The Fine Structure

Returning to consideration of the wave function, we redefine

$$\psi \equiv \begin{pmatrix} \Phi \\ \Omega \end{pmatrix} \quad \text{where } \Phi \equiv \begin{pmatrix} \psi_1 \\ \psi_2 \end{pmatrix} \text{ and } \Omega \equiv \begin{pmatrix} \psi_3 \\ \psi_4 \end{pmatrix}. \tag{13.17}$$

After some algebra, we obtain

$$\alpha_1 \begin{pmatrix} \psi_1 \\ \psi_2 \\ \psi_3 \\ \psi_4 \end{pmatrix} = \begin{pmatrix} \psi_4 \\ \psi_3 \\ \psi_2 \\ \psi_1 \end{pmatrix}$$

$$\alpha_2 \begin{pmatrix} \psi_1 \\ \psi_2 \\ \psi_3 \\ \psi_4 \end{pmatrix} = \begin{pmatrix} -i\psi_4 \\ i\psi_3 \\ -i\psi_2 \\ i\psi_1 \end{pmatrix}$$

$$\alpha_3 \begin{pmatrix} \psi_1 \\ \psi_2 \\ \psi_3 \\ \psi_4 \end{pmatrix} = \begin{pmatrix} \psi_3 \\ -\psi_4 \\ \psi_1 \\ -\psi_2 \end{pmatrix}$$

$$\beta \begin{pmatrix} \psi_1 \\ \psi_2 \\ \psi_3 \\ \psi_4 \end{pmatrix} = \begin{pmatrix} \psi_1 \\ \psi_2 \\ -\psi_3 \\ -\psi_4 \end{pmatrix}.$$

Placing these results into the Dirac equation, with $p_i = -i\,\partial/\partial x_i$, we obtain a set of four coupled differential equations.

$$(-i\,\partial\psi_4/\partial x - \partial\psi_4/\partial y) - i\,\partial\psi_3/\partial z + m\psi_1 = E\psi_1$$

$$(-i\,\partial\psi_3/\partial x + \partial\psi_3/\partial y) + i\,\partial\psi_4/\partial z + m\psi_2 = E\psi_2$$

$$(-i\,\partial\psi_2/\partial x - \partial\psi_2/\partial y) - i\,\partial\psi_1/\partial z - m\psi_3 = E\psi_3 \tag{13.18}$$

$$(-i\,\partial\psi_1/\partial x + \partial\psi_1/\partial y) + i\,\partial\psi_2/\partial z - m\psi_4 = E\psi_4.$$

Choosing the momentum in the z-direction, we find their solutions, whose mutual couplings are given by

$$p\psi_3 - (E-m)\psi_1 = 0$$
$$-p\psi_4 - (E-m)\psi_2 = 0$$
$$p\psi_1 - (E+m)\psi_3 = 0 \qquad (13.19)$$
$$-p\psi_2 - (E+m)\psi_4 = 0 .$$

The equations are decoupled into two pairs: ψ_1, ψ_3 and ψ_2, ψ_4. The two equations involving the former produce the determinantal equation

$$\begin{vmatrix} E-m & -p \\ -p & E-m \end{vmatrix} = 0$$

whence $E = \pm\sqrt{p^2 + m^2}$. (The same result is obtained using ψ_2, ψ_4.)
We find

$$\psi_4 = (E-m)\psi_2/p$$

and

$$\psi_3 = p\psi_1/(E+m) .$$

Substituting $p = \pm\sqrt{E^2 - m^2}$,

$$\psi_3 = \sqrt{(E-m)/(E+m)}\,\psi_1$$

and

$$\psi_4 = \sqrt{(E-m)/(E+m)}\,\psi_2 .$$

In the highly relativistic case $E \gg m$, ψ_3 and ψ_4 are comparable in size to ψ_1 and ψ_2. In less extreme situations, ψ_3 and ψ_4 are smaller than ψ_1 and ψ_2 (e.g., for $v = c/2$, $\Omega = 0.27\Phi$.) The former have come to be known as the small parts of the wave function and the latter, the large parts. Equation (13.17) divided ψ neatly into these components.

At this point we have learned all we need from studying the free particle, and we move on to explicit consideration of the hydrogen atom. As $v \to 0$, the Dirac wave function must reduce to the Schrödinger wave function with the spin tacked on. We can separate the radial part of the wave function, since only it depends on the potential, from the rest, which is now analogous to the spherical harmonics, but with spin explicitly included.

$$\psi \equiv \begin{pmatrix} \Phi \\ \Omega \end{pmatrix} \sim \frac{1}{r} \begin{pmatrix} G_n(r)\mathcal{Y}^m_{\tilde{\omega},\,\ell\,j}(\theta,\phi,s) \\ i\,F_n(r)\mathcal{Y}^m_{\tilde{\omega},\,\ell'\,j}(\theta,\phi,s) \end{pmatrix} \qquad (13.20)$$

$\tilde{\omega}$, is the parity of the wave function. G is the large part of the radial wave function; F is the small part.[3]

If we take all possibilities into account, we obtain two sets of 4-component wave functions, one for $j = \ell + 1/2$ and one for $j = \ell - 1/2$, with the proviso that $\ell' \neq \ell$. All of this is tedious but straightforward, and true for any central potential. Thus, for example, if $j = 3/2$, then ℓ and ℓ' can be either 1 or 2. We saw that the perturbation result for the fine structure, a combination of the lowest order relativistic effect and the spin-orbit coupling, each of which had an ℓ dependence, itself depended only on j. This is the cause. Thus, for $j = 3/2$, the states $^2P_{3/2}$ and $^2D_{3/2}$ have the same energy.[4]

As speed decreases, G should reduce to R, the radial part of the Schrödinger equation, F should vanish, and the $\mathcal{Y}$s should reduce to the non-relativistic products of the spherical harmonics and the spin functions. Schematically,

$$\mathcal{Y}^{m_\ell = m_j \pm \frac{1}{2}}_{\tilde{\omega},\, \ell = j \pm 1,\, j}\left(\theta, \phi, \frac{1}{2}\right) \longrightarrow Y_\ell^m(\theta, \phi)\ \chi\left(\frac{1}{2}, \pm\frac{1}{2}\right)$$

with positive or negative parity.

Let us now obtain the radial functions for hydrogen. Since the Dirac equation is a first order differential equation, the result of a good deal of algebra[5] is a pair of coupled first order equations in G and F:

$$\left(-\frac{d}{d\rho} + \frac{\tau}{\rho}\right) F = \left(-\nu + \frac{\gamma}{\rho}\right) G \qquad (13.21)$$

$$\left(\frac{d}{d\rho} + \frac{\tau}{\rho}\right) G = \left(\frac{1}{\nu} + \frac{\gamma}{\rho}\right) F . \qquad (13.22)$$

The constants are defined as follows.

$$\kappa \equiv \sqrt{m^2 - E^2}$$

$$\rho \equiv \kappa r$$

$$\nu \equiv \left(\sqrt{\frac{m - E}{m + E}}\right)$$

---

[3]The i multiplying F traces its origin to the fact that the components of α_2 are imaginary.

[4]Technically, these are levels, not states. The latter include the magnetic quantum number m_j, and are degenerate in the absence of an external magnetic field.

[5]H. Bethe and E.E. Salpeter, *Quantum Mechanics of One- and Two-Electron Atoms.* NY: Springer (1977).

$$\tau \equiv \left[j + \frac{1}{2}\right]\tilde{\omega}$$

$$\gamma \equiv Ze^2.$$

γ is just $Z\alpha$, allowing for calculation of the fine structure for hydrogenic atoms.

We now carry out the same "peel-off + series expansion + truncation" procedure as we did for the Schrödinger equation for hydrogen.

$$G \sim \rho^s\, e^{-\rho}\, \Sigma_{i=0}^{\infty}\, a_i \rho^i$$

$$F \sim \rho^s\, e^{-\rho}\, \Sigma_{i=1}^{\infty}\, b_i \rho^i.$$

We substitute these into the pair of coupled differential equations, and match terms of the same order, giving us recursion relations between successive terms. In order for the wave functions to be normalizable, we must truncate the series for G and F. This gives us the allowed energies.

$$E_{nj} = m\left[1 + \left(\frac{\alpha}{n - (j + \frac{1}{2}) + \sqrt{(j + \frac{1}{2})^2 - \alpha^2}}\right)^2\right]^{-\frac{1}{2}}. \tag{13.23}$$

This expression includes the rest energy m.

Problem 13.5. *Prove that the exact result exclusive of the rest energy, expanded to order α^4 produces the Bohr result plus the fine structure*

$$E_{nj} = E_n\left\{\frac{1}{n^2} + \frac{\alpha^2}{n^4}\left[\frac{n}{j + 1/2} - \frac{3}{4}\right]\right\}.$$

E_n is the Bohr energy.

Note that this approximation does not reintroduce the "bad" quantum number ℓ. So the ℓ-degeneracy is not an artifact of the approximate fine structure calculation, but is true for the exact case. The equal energy of states of the same j in Fig. 13.1 comes from the Dirac equation.

13.6 SUSYQM

Finally, we shall obtain the fine structure using SUSYQM. This is also exact, since it is directly related to the hydrogenic radial functions for the Dirac equation. We begin by connecting the pair of coupled equations for F and G in Eqs. (13.21) and (13.22). The operation $\left(-\frac{d}{d\rho} + \frac{\tau}{\rho}\right) F$ yields $\left(-\nu + \frac{\gamma}{\rho}\right) G$ while the operation $\left(\frac{d}{d\rho} + \frac{\tau}{\rho}\right) G$ yields $\left(\frac{1}{\nu} + \frac{\gamma}{\rho}\right) F$. This looks suspiciously like raising and lowering operations associated with a superpotential of form τ/ρ.

But here we encounter a problem. The eigenfunctions of partner hamiltonians should be of the form $A^+ F \sim G$, $A^- G \sim F$ where all of the ρ (that is, r) dependence resides on the left side, in $A^\pm$. This is not the case in Eqs. (13.21) and (13.22). There is a dependence on the right hand sides of form F/ρ and G/ρ.

We can rewrite Eqs. (13.21) and (13.22) in a more compact matrix form:

$$\begin{pmatrix} dG/dr \\ dF/dr \end{pmatrix} + \frac{1}{r} \begin{pmatrix} k & -\gamma \\ \gamma & -k \end{pmatrix} \begin{pmatrix} G \\ F \end{pmatrix} = \begin{pmatrix} (m+E)F \\ (m-E)G \end{pmatrix}.$$

Here we see the problem: the second term on the left is mixing F and G. If it can be diagonalized, then the equations can be recast in SUSYQM form. A solution[6] is to apply a linear transformation of the functions G, F into $\mathcal{G}, \mathcal{F}$:

$$D \equiv \begin{pmatrix} k+s & -\gamma \\ \gamma & k+s \end{pmatrix}$$

$$\begin{pmatrix} \mathcal{G} \\ \mathcal{F} \end{pmatrix} = D \begin{pmatrix} G \\ F \end{pmatrix}.$$

Substituting $\gamma \equiv \sqrt{(k^2 - s^2)}$ yields the appropriate supersymmetric form of the coupled equations:

$$\left(\frac{d}{d\mu} + \frac{s}{\mu} - \frac{\gamma}{s}\right) \mathcal{G} = \left(\frac{k}{s} + \frac{m}{E}\right) \mathcal{F} \tag{13.24}$$

$$\left(\frac{-d}{d\mu} + \frac{s}{\mu} - \frac{\gamma}{s}\right) \mathcal{F} = \left(\frac{k}{s} - \frac{m}{E}\right) \mathcal{G}, \tag{13.25}$$

where $\mu \equiv Er$.

This is precisely the form needed for supersymmetric partner eigenfunctions, related by $A^\pm = \left(\pm\frac{d}{d\mu} + \frac{s}{\mu} - \frac{\gamma}{s}\right)$. We can then decouple the equations, yielding the eigenvalue equations for $\mathcal{G}$ and $\mathcal{F}$:

$$A^- A^+ \mathcal{F} \equiv H_+ \mathcal{F} = \left(\frac{\gamma^2}{s^2} + 1 - \frac{m^2}{E^2}\right) \mathcal{F} = \left(\frac{k^2}{s^2} - \frac{m^2}{E^2}\right) \mathcal{F} \tag{13.26}$$

---

[6] A. Sukumar, "Supersymmetry and the Dirac equation for a central Coulomb field," *J. Phys. A* **18**, L697 (1985).

$$A^+ A^- \mathcal{G} \equiv H_- \mathcal{G} = \left(\frac{\gamma^2}{s^2} + 1 - \frac{m^2}{E^2} \right) \mathcal{G} = \left(\frac{k^2}{s^2} - \frac{m^2}{E^2} \right) \mathcal{G} . \tag{13.27}$$

Thus, the hamiltonians H_- and H_+ share common eigenvalues, except for the extra case where $A^- \mathcal{G} = 0$, which yields the ground state wave function $\mathcal{G}(0) \sim \mu^s e^{-\mu/s}$ with eigenvalue $E^{(0)} = \frac{m^2}{k^2/s^2}$.

Note that neither H_- nor H_+ is the Dirac hamiltonian, so their eigenvalues are not the hydrogen energy levels. Those will be obtained from the terms $\frac{m^2}{E^2}$.

The application of the differential forms of $A^\pm$ in Eqs. (13.26) and (13.27) yields

$$\left[-\frac{d^2}{d\mu^2} - 2\frac{\gamma}{\mu} - \frac{s(s-1)}{\mu^2} + 1 - \frac{m^2}{E^2} \right] \mathcal{F} = 0 \tag{13.28}$$

$$\left[\frac{d^2}{d\mu^2} + 2\frac{\gamma}{\mu} - \frac{s(s+1)}{\mu^2} + 1 - \frac{m^2}{E^2} \right] \mathcal{G} = 0 . \tag{13.29}$$

Problem 13.6. *Obtain Eqs. (13.28) and (13.29) from Eqs. (13.26) and (13.27).*

From $A^\pm = \left(\pm\frac{d}{d\mu} + \frac{s}{\mu} - \frac{\gamma}{s} \right)$ we can obtain the superpotential from either $A^- A^+$ or $A^+ A^-$. The former gives us $A^- A^+ = H_+ = W^2 + dW/d\mu$. We compare this with

$$A^- A^+ = -\frac{d^2}{d\mu^2} + \frac{s^2}{\mu^2} - \frac{s}{\mu^2} + \frac{\gamma^2}{s^2} - 2\frac{\gamma}{s} .$$

By inspection, $s^2/\mu^2 + \gamma^2/s^2 - 2\gamma/s = (s/\mu - \gamma/s)^2$. We try $W = s/\mu - \gamma/s$. Then $dW/d\mu = -s/\mu^2$. So indeed $W = s/\mu - \gamma/s$.

Now we see that we actually have the equivalent of a system of partner potentials $W^2 \pm \frac{dW}{d\mu}$.

But we can do even better. From Eqs. (13.28) and (13.29) we see that the simple replacement of s by $s+1$ in the first yields the second. In other words, we·have shape invariance:

$$H_+(\mu, s, \gamma) = H_-(\mu, s+1, \gamma) + \frac{\gamma^2}{s^2} - \frac{\gamma^2}{(s+1)^2} . \tag{13.30}$$

The eigenvalues are

$$E_n^{(-)} = E_{n-1}^{(+)} = \gamma^2 \left[\frac{1}{s^2} - \frac{1}{(s+1)^2} \right] . \tag{13.31}$$

Finally, we relate the actual Dirac fine structure energy spectrum of hydrogen E_n to the spectrum $E_n^{(-)}$ from Eqs. (13.26), (13.27) (where it is called simply E) and (13.31).

$$\left(\frac{k^2}{s^2} - \frac{m^2}{E_n^2} \right) \equiv E_n^{(-)} .$$

This yields

$$E_n = \frac{m}{\sqrt{1 + \frac{\gamma^2}{(s+n)^2}}}, \quad n = 0, 1, 2, \ldots \tag{13.32}$$

We thus obtain the expression for the total energy, including the fine structure:

$$E_{nj} = m \left\{ \left[1 + \left(\frac{\alpha}{n - (j + \frac{1}{2}) + \sqrt{(j + \frac{1}{2})^2 - \alpha^2}} \right)^2 \right]^{-\frac{1}{2}} \right\} . \tag{13.33}$$

This is identical to Eq. (13.23).

Chapter 14

Natanzon Potentials

In 1971 Russian physicist G. A. Natanzon connected the one dimensional Schrödinger equation to the hypergeometric equation.[1] In the course of this he discovered all of the conventional translational shape invariant potentials.

The Natanzon potentials are not the most general ones that for which the Schrödinger equation reduced to the hypergeometric equation,[2] but they suffice to generate the complete set of conventional superpotentials: Pöschl-Teller, Scarf, Rosen-Morse, and Eckart from the full hypergeometric equation and Morse, 3D-Oscillator, and Coulomb from the confluent hypergeometric equation.

The equation for a Natanzon potential $U(r)$ is

$$U[z(r)] = \frac{-fz(1-z) + h_0(1-z) + h_1 z}{R(z)} - \frac{1}{2}\{z, r\} , \qquad (14.1)$$

where f, h_0, and h_1 are constants. z is an as-yet-to-be-determined function of r. As we shall soon see, this functional relationship is determined from two constraints: one stemming from the transformation of the hypergeometric equation into a Schrödinger equation, and the other to preserve the shape of SUSYQM operators $A^\pm$ under coordinate transformations. The variable r is itself either the radial or linear coordinate for each shape-invariant superpotential. $R(z)$ is a general quadratic function $az^2 + bz + c$.

---

[1] G. A. Natanzon, "Study of the one-dimensional Schroedinger equation generated from the hypergeometric equation", *Vestnik Leningradskogo Universiteta*, **10**, 22–28 (1971) – translated by H. C. Rosu, http://lanl.arxiv.org/abs/physics/9907032v1.

[2] F. Cooper, J. N. Ginocchio and A. Khare, "Relationship between supersymmetry and solvable potentials", *Phys. Rev. D* **36**, 2458–2473 (1987).

The Schwartzian derivative $\{z, r\}$ is defined by

$$\{z, r\} \equiv \frac{d^3z/dr^3}{dz/dr} - \frac{3}{2}\left[\frac{d^2z/dr^2}{dz/dr}\right]^2 . \tag{14.2}$$

Thus, if we can obtain expressions for z, that will give us the functions of r which yield the superpotentials. Once having done so, we follow a procedure similar to that in Chapter 8: finding a second order differential equation involving the Natanzon potential and obtaining the condition for removing the first order term, thus rendering it in the form of a Schrödinger equation.

From this, the general form of the Natanzon potential subjected, as we mentioned above, to two constraints on z, reduces to the set of conventional translational shape invariant potentials. The first is

$$\frac{dz}{dr} = \frac{2z(1-z)}{\sqrt{R(z)}} . \tag{14.3}$$

The second is

$$\frac{dz}{dr} = z^{1+\beta}(1-z)^{-\alpha-\beta} . \tag{14.4}$$

α and β are as-yet-undetermined constants. Once we have found them, we will be able to generate $z(r)$ and thence the superpotentials. Combining Eqs. (14.3) and (14.3), we obtain

$$z^{1+\beta}(1-z)^{-\alpha-\beta} = \frac{2z(1-z)}{\sqrt{R(z)}} . \tag{14.5}$$

After some computation, we find that there are only a finite number of values of α and β consistent with $R(z)$, a quadratic polynomial[3] in z.

Problem 14.1. *Find the allowed values of the sets* (α, β).

Integrating Eq. (14.4), we obtain $z(r)$. From this, we can use Eq. (14.3) to determine $R(z)$. Putting that back into the Natanzon potential, Eq. (14.1) will give us the superpotential.

Let us demonstrate the procedure for $\alpha = \beta = 0$. $dz/dr = z \rightarrow z = e^r$. From Eq. (14.5),

$$z = \frac{2z(1-z)}{\sqrt{R(z)}} ,$$

---

[3]For the details of this procedure, see A. Gangopadhyaya, J.V. Mallow, and U.P. Sukhatme, "Translational shape invariance and the inherent potential algebra", *Phys. Rev. A* **58**, 4287–4292 (1998).

whence $R(z) = 4(1 - z)^2$. The Natanzon potential is then

$$U[z(r)] = \frac{-fz(1 - z) + h_0(1 - z) + h_1 z}{4(1 - z)^2} - \frac{1}{2}\{z, r\} \qquad (14.6)$$

$$-\frac{1}{2}\{z, r\} = \frac{1}{4} .$$

Defining $A \equiv -f/4, B \equiv h_0/4, C \equiv h_1/4$, and substituting $z = e^r$, we get

$$U = A\frac{e^r}{1 - e^r} + B\frac{1}{1 - e^r} + C\frac{e^r}{(1 - e^r)^2} + \frac{1}{4} .$$

Some manipulation yields

$$U = -A\frac{\cosh\left(\frac{r}{2}\right) + \sinh\left(\frac{r}{2}\right)}{2\sinh\left(\frac{r}{2}\right)} - B\frac{\cosh\left(\frac{r}{2}\right) - \sinh\left(\frac{r}{2}\right)}{2\sinh\left(\frac{r}{2}\right)} + \frac{C}{4\sinh^2\left(\frac{r}{2}\right)} + \frac{1}{4}$$

$$= -\frac{(A + B)}{2}\coth\left(\frac{r}{2}\right) - \frac{(A + B)}{2} + \frac{C}{4}\operatorname{cosech}^2\left(\frac{r}{2}\right) + \frac{1}{4} .$$

Since this must have the form $W^2 - \frac{dW}{dr}$, we can examine various possibilities, and eventually choose

$$W = \tilde{m}_1 \coth\left(\frac{r}{2}\right) + \tilde{m}_2 .$$

This generates U with the identifications

$$\tilde{m}_1\tilde{m}_2 = -\frac{A + B}{4}, \tilde{m}_1^2 + \tilde{m}_1 = \frac{C}{4}, \tilde{m}_1^2 + \tilde{m}_2^2 = -\frac{A + B}{2} + \frac{1}{4} .$$

This is the Eckart superpotential.

Problem 14.2. *Without doing any calculation, show that the case $\alpha = 0$, $\beta = -1$ also yields the Eckart superpotential.*

Problem 14.3. *Obtain the Rosen-Morse superpotential from $\alpha = -1$, $\beta = 0$.*

Table 14.1 gives the values α, β, the functions $z(r)$ which they generate, and the concomitant superpotentials. It shows all allowed values of α, β and the superpotentials that they generate. Constants $\tilde{m}_1$ and $\tilde{m}_2$ are linear functions of the Natanzon constants f, h_0, h_1. Replacing the hyperbolic functions by their trigonometric analogs gives us the full set of two-term translational shape invariant potentials.

Table 14.1   Superpotentials associated with various values of α and β.

α	β	$z(r)$	**Superpotential**	**Name**
0	0	$z = e^r$	$\tilde{m}_1 \coth \frac{r}{2} + \tilde{m}_2$	Eckart
0	$-\frac{1}{2}$	$z = \sin^2 \frac{r}{2}$	$\tilde{m}_1 \operatorname{cosec} r + \tilde{m}_2 \cot r$	Gen. Pöschl-Teller trigonometric
0	-1	$z = 1 - e^r$	$\tilde{m}_1 \coth \frac{r}{2} + \tilde{m}_2$	Eckart
$-\frac{1}{2}$	0	$z = -\operatorname{sech}^2 \frac{r}{2}$	$\tilde{m}_1 \operatorname{cosech} r + \tilde{m}_2 \coth r$	Pöschl-Teller II
$-\frac{1}{2}$	$-\frac{1}{2}$	$z = \tanh^2 \frac{r}{2}$	$\tilde{m}_1 \tanh \frac{r}{2} + \tilde{m}_2 \coth \frac{r}{2}$	Gen. Pöschl-Teller
-1	0	$z = \frac{1 + \tanh \frac{r}{2}}{2}$	$\tilde{m}_1 \tanh \frac{r}{2} + \tilde{m}_2$	Rosen-Morse

Chapter 15

The Quantum Hamilton-Jacobi Formalism and SUSYQM

In this chapter we introduce another formulation of quantum mechanics, the Quantum Hamilton-Jacobi (QHJ) formalism[1] and its connection to SUSYQM. In this formalism, we work with the quantum momentum function $p(x)$, which is the quantum analog of the classical momentum function $p_c(x) = \sqrt{E - V(x)}$.

In QHJ formalism the spectrum of a quantum mechanical system is determined by the solution of the following non-linear differential equation:

$$p^2 - i\,p' = E - V(x,\alpha) \equiv p_c^2 \,. \tag{15.1}$$

This equation is related to the Schrödinger equation

$$-\psi'' + (V(x,\alpha) - E)\,\psi = 0 \tag{15.2}$$

via the correspondence

$$p(x) = -i\psi'(x)/\psi(x) \quad \text{whence} \quad \psi(x) \sim e^{-\int p(x)\,dx} \,. \tag{15.3}$$

The quantum momentum function (QMF) is clearly a function of x, E, and parameter α; i.e., $p \equiv p(x, E, \alpha)$, however we will suppress this dependence unless necessary. From Eq. (15.2), we see that function p for the zero-energy ground state equals $i\,W(x,\alpha)$, the superpotential of the system. For non-zero energies, the QMF has singularities at the zeroes of the wave function $\psi(x)$. The structure of these singularities plays an important role in determining the spectrum of the hamiltonian. In Eq. (15.1), note that if the potential is a non-singular function, the singularities of p^2, if any, would have to be canceled by those of the term p'. We will see that Eq. (15.1) helps us determine the singularities of the p-function.

---

[1] Even though this procedure has similarities with the Hamilton-Jacobi formalism of classical mechanics, the material covered in this chapter is self contained.

The Quantum Hamilton-Jacobi method is centered around analyzing the pole structure of $p = -i\left(\frac{\psi'}{\psi}\right)$. Since the n-th eigenfunction of the hamiltonian has n nodes within the turning points determined by the energy E_n and the potential, the corresponding function p will have n singular points. We will now show that these singularities are simple poles; i.e., near a singular point x_0, the function has the form $p \approx \frac{\delta}{x-x_0}$, where the constant δ is known as the residue.

Near a point x_0 where the wave function $\psi(x)$ is zero, we can write the wave function as $\psi(x) = h(x)(x - x_0)$. Its derivative is then given by $\psi'(x) = h'(x - x_0) + h(x)$. The singular term of the QMF is then given by

$$p(x) \equiv -i\,\psi'(x)/\psi(x) = \frac{-i}{x - x_0} . \qquad (15.4)$$

I.e., p has a pole at x_0 with a residue of $-i$. Since these poles are due to the zeroes of the wave function and their location depends on energy, they are defined as the "moving poles." In addition to these moving poles, QMF may also have a poles at the singular points of the *potential*. Since these are fixed for a given system described by the potential V, they are known as the "fixed poles". Now let us visualize the x-axis to be the real axis of a complex plane. These moving poles would then all fall on the real axis. If we consider a closed path that encircles these moving poles, and there are n of them, the Cauchy theorem would imply that the line integral of p carried out on this closed path traveling in a counter-clockwise direction would yield:

$$\frac{1}{2\pi} \oint p(x)\, dx = n , \qquad (15.5)$$

where the contour involved includes all n moving poles of the system and none of the fixed poles.

The strength of QHJ derives from the fact that the contour of the above integration, which was put in to encircle moving poles only, can be deformed to enclose the fixed poles instead (albeit traveling in the opposite direction.) To understand this, visualize the complex plane folded into a sphere where the boundary at infinity is pulled together to a point. Now, exactly as a rubber-band around Australia on a globe can be stretched to encircle all nations other than Australia, similarly a contour on a complex plane around moving poles can be deformed to enclose all fixed poles instead. Figures 15.1(a) through 15.1(f) show such a deformation. Thus,

$$\oint p(x)\, dx \bigg|_{\text{Fixed poles}} = -\oint p(x)\, dx \bigg|_{\text{Moving poles}} . \qquad (15.6)$$

The difference in sign is due to the change in direction in which we move along the contour as it is deformed. The counter-clockwise becomes clockwise as shown in Fig. 15.1. Thus, the condition of Eq. (15.5) becomes

$$\frac{1}{2\pi} \oint p\,dx \bigg|_{\text{Fixed poles}} = -n \ . \tag{15.7}$$

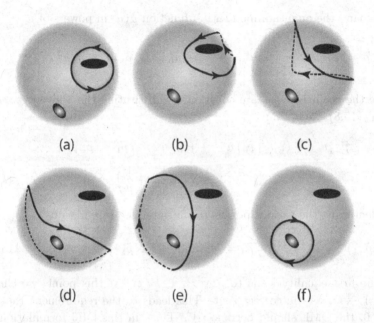

Fig. 15.1   Continuously moving a contour on the complex plane.

In QHJ, we first determine the singularity structure of p from Eq. (15.1) by explicitly inserting the potential V. Then Eq. (15.7) determines n as a function of the energy E_n and parameter α. Inverting this equation, we find the energy $E_n(\alpha)$.

Since Eq. (15.5) determines exact eigenvalues of the system, it is known as the quantization condition. It is important to note that unlike semi-classical quantization discussed in Chapter 12, the condition Eq. (15.5) is exact.

As an example, let us consider the problem of a harmonic oscillator described by the potential $V(x,\omega) = \frac{1}{4}\,\omega x^2$. The corresponding super-potential is given by $W(x,\omega) = \frac{1}{2}\,\omega x$. At first sight, this seems to be a potential with no singularity. However, it does diverge as $x \to \infty$ and will force the QMF to have a fixed pole at infinity. To see the form of p at infinity let us change the variable to $u = \frac{1}{x}$. Near $u = 0$, we write $p(x) = p\left(\frac{1}{u}\right)$ as $\widetilde{p}(u,)$. The form of Eq. (15.1) in the vicinity of $u = 0$ is

$$\widetilde{p}^2 + iu^2 \frac{d\widetilde{p}}{du} = E - V\left(\frac{1}{u}, \alpha\right) = E - \frac{1}{4}\,\omega\left(\frac{1}{u}\right)^2 = p_c^2. \qquad (15.8)$$

We expand the quantum momentum function $\widetilde{p}(u)$ in powers of u:

$$\widetilde{p}(u, \alpha_0) = \sum_{k=-1}^{\infty} \widetilde{p}_k\,u^k = \frac{\widetilde{p}_{-1}}{u} + \widetilde{p}_0\,u^0 + \widetilde{p}_1\,u^1 + \widetilde{p}_2\,u^2 + \cdots,$$

where the coefficients $\widetilde{p}_k$ are constants. Substituting the above expansion in Eq. (15.8), we obtain

$$\widetilde{p}_{-1}^2 u^{-2} + \left(2\,\widetilde{p}_{-1}\,\widetilde{p}_0\right)u^{-1} + \left(\widetilde{p}_0^2 + 2\,\widetilde{p}_{-1}\,\widetilde{p}_1 - i\,\widetilde{p}_{-1}\right) + \cdots$$

$$= E - \frac{1}{4}\,\omega\left(\frac{1}{u}\right)^2. \qquad (15.9)$$

By identifying the various powers of the variable u, we obtain

$$\left(\widetilde{p}_{-1}\right)^2 = -\frac{1}{4}\,\omega^2\,; \quad \widetilde{p}_0 = 0; \quad \text{and} \quad 2\,\widetilde{p}_{-1}\,\widetilde{p}_1 - i\widetilde{p}_{-1} = E\,. \qquad (15.10)$$

The first equality leads to $\widetilde{p}_{-1} = \pm\frac{1}{2}i\omega$. At this point, we choose $\widetilde{p}_{-1} = \frac{1}{2}i\omega$ as the correct root. This leads to the requirement that for $E \to 0$, the QMF should become iW. Thus, in this QHJ formulation we make our first connection to supersymmetric quantum mechanics.[2] This choice then leads to $i\omega\widetilde{p}_1 + \frac{1}{2}\omega = E$. From the last equality, we obtain $\widetilde{p}_1 = -i\left(\frac{E}{\omega} - \frac{1}{2}\right)$. As we will soon see, this last coefficient is the only one that contributes to the contour integration around the fixed pole for this system.

The quantization condition from Eq. (15.7) yields

$$-n = \frac{1}{2\pi}\oint p(x, E_n, \alpha)\,dx\Bigg|_{\text{Fixed poles}}.$$

---

[2] For an alternate explanation of this choice without resorting to SUSYQM, see R. A. Leacock and M. J. Padgett, "Hamilton-Jacobi/action-angle quantum mechanics," *Phys. Rev. D* **28**, 2491–2502 (1983). However, we find the stipulation $\lim_{E\to 0} p \to iW$ to be a much easier method to unambiguously select the correct root. In our case that implies $\widetilde{p}_{-1}/u = \frac{1}{2}i\omega/u$. Hence, $\widetilde{p}_{-1} = \frac{1}{2}i\omega$.

Since the fixed pole is at infinity, we change variables to $u = 1/x$ and replace dx by $- du/u^2$. This then gives

$$
n = \frac{1}{2\pi} \oint_{u=0} \frac{1}{u^2} \, \widetilde{p} \, du
$$

$$
= \frac{1}{2\pi} \oint_{u=0} \frac{1}{u^2} \left(\frac{\widetilde{p}_{-1}}{u} + \widetilde{p}_0 \, u^0 + \widetilde{p}_1 \, u^1 + \widetilde{p}_2 \, u^2 + \cdots \right) du
$$

$$
= \frac{1}{2\pi} \oint_{u=0} \left(\frac{\widetilde{p}_{-1}}{u^3} + \frac{\widetilde{p}_0}{u^2} + \frac{\widetilde{p}_1}{u} + \widetilde{p}_2 \, u^0 + \cdots \right) du
$$

$$
= i \widetilde{p}_1 \, .
$$

This then yields $n = i\widetilde{p}_1 = \left(\frac{E_n}{\omega} - \frac{1}{2} \right)$. I.e., $E_n = \left(n + \frac{1}{2} \right) \omega$, the familiar result for the harmonic oscillator.

15.1   Connection to SUSYQM and Shape Invariance

We have already seen how SUSYQM helped us choose the correct root in QHJ formalism with relative ease. In this section we will work out the general connection between SUSYQM and shape invariance and QHJ theory.[3] Since parameters play a very important role in shape invariance, we will now explicitly display all variables that the QMF p and potential V depend upon. Since the partner potentials $V_\pm(x, \alpha)$ are given by $W^2(x, \alpha) \pm W'(x, \alpha)$, the QHJ equation (15.1) for the hamiltonian $H_-(x, \alpha)$ becomes

$$
p^{(-)\,2}(x, E^{(-)}, \alpha) - i\, p^{(-)\,\prime}(x, E^{(-)}, \alpha) = E^{(-)} - \left[W^2(x, \alpha) - W'(x, \alpha) \right] .
\tag{15.11}
$$

Thus, in the limit[4] $E^{(-)} \to 0$, we have

$$
\lim_{E \to 0} p^{(-)}(x, E^{(-)}, \alpha) \to i\, W(x, \alpha) .
\tag{15.12}
$$

Considering now the supersymmetric partner hamiltonian $H_+(x, \alpha)$, there exists another analogous equation for a partner QMF, $p^{(+)}(x, \alpha)$ given by

$$
p^{(+)\,2}(x, E^{(+)}, \alpha) - i\, p^{(+)\,\prime}(x, E^{(+)}, \alpha) = E^{(+)} - \left[W^2(x, \alpha) + W'(x, \alpha) \right] .
\tag{15.13}
$$

---

[3]C. Rasinariu, J. Dykla, A. Gangopadhyaya, and J. V. Mallow, "Exactly solvable systems and the Quantum Hamilton-Jacobi formalism", *Phys. Lett. A* **338**, 197–202 (2005).

[4]We will assume that our superpotential is such that supersymmetry remains unbroken for a range of values of the parameters α.

Since supersymmetry ensures that these equations lead to the same set of eigenvalues, except for the ground state, let us write $E \equiv E^{(-)} = E^{(+)}$. The corresponding Schrödinger equations are

$$H_{\pm}(x, \alpha)\, \psi^{(\pm)} = -\psi^{(\pm)''} + \left[W^2(x, \alpha) \pm W'(x, \alpha) \right] \psi^{(\pm)} = E\, \psi^{(\pm)},$$
(15.14)

and the solutions are connected to the QHJ solutions by

$$p^{(\pm)} = -i \left(\frac{\psi^{(\pm)'}}{\psi^{(\pm)}} \right).$$
(15.15)

Defining operators $A^+ = -\frac{d}{dx} + W$ and $A^- = \frac{d}{dx} + W$, we can rewrite the partner hamiltonians as $H_+ = A^- A^+$ and $H_- = A^+ A^-$. Furthermore, we find that A^+ and A^- behave as raising and lowering operators between the eigenstates of the partner hamiltonians. In particular,

$$\psi^{(-)} = C^{(-)} A^+ \psi^{(+)} = C^{(-)} \left(-\psi^{(+)'} + W\, \psi^{(+)} \right)$$
$$\psi^{(+)} = C^{(+)} A^- \psi^{(-)} = C^{(+)} \left(\psi^{(-)'} + W\, \psi^{(-)} \right),$$
(15.16)

where $C^{(-)}$ and $C^{(+)}$ are normalization constants. To find a relationship between $p^{(-)}$ and $p^{(+)}$ we exploit the connection between $\psi^{(-)}$ and $\psi^{(+)}$ respectively. Using Eqs. (15.14)–(15.16), we obtain

$$\frac{\psi^{(+)'}}{\psi^{(-)}} = C^{(+)} \left(W^2 - E + i\, W p^{(-)} \right)$$
(15.17)

$$\frac{\psi^{(-)'}}{\psi^{(+)}} = C^{(-)} \left(-W^2 + E + i\, W p^{(+)} \right).$$
(15.18)

Multiplying Eqs. (15.17) and (15.18) we obtain

$$-p^{(+)} p^{(-)} = C^{(-)} C^{(+)} \left(W^2 - E + i\, W p^{(-)} \right) \left(-W^2 + E + i\, W p^{(+)} \right).$$

To solve the above equation for $p^{(+)}$ or $p^{(-)}$, we first evaluate the product of the normalization constants $C^{(-)} C^{(+)}$. Since $\psi^{(+)} = C^{(+)} A^- \psi^{(-)}$, and $\psi^{(-)} = C^{(-)} A^+ \psi^{(+)}$, we obtain successively

$$\langle \psi^{(+)} | \psi^{(+)} \rangle = C^{(+)} \langle \psi^{(+)} | A^- | \psi^{(-)} \rangle$$
$$= C^{(+)} \langle \psi^{(+)} | A^- C^{(-)} A^+ | \psi^{(+)} \rangle$$
$$= C^{(+)} C^{(-)} E \langle \psi^{(+)} | \psi^{(+)} \rangle,$$
(15.19)

where we have used $A^- A^+ = H_+$ as noted earlier. Thus, $C^{(+)} C^{(-)} = 1/E$. This leads to

$$p^{(+)} = \frac{i\, W p^{(-)} + W^2 - E}{-p^{(-)} + i\, W},$$
(15.20)

or,

$$p^{(-)} = \frac{-iWp^{(+)} + W^2 - E}{-p^{(+)} - iW} . \qquad (15.21)$$

It is important to note at this point that both sides of Eqs. (15.20) and (15.21) are related to the same superpotential $W(x, \alpha)$ and can be denoted by $p^{(\pm)}(x, \alpha)$. Also note that $p^{(+)}$ in Eq. (15.20) is not defined for the ground state, for which $-p^{(-)} + iW = 0$.[5]

We have only applied the conditions of supersymmetry so far. To render a hamiltonian solvable, the superpotential $W(x, \alpha)$ needs to satisfy the shape invariance condition, thus enabling us to find the solution of H_+ or H_- by algebraic means.

As we have seen in Chapter 5, for shape invariant potentials $V^{(\pm)}(x, \alpha_i)$ we have $V^{(+)}(x, \alpha_i) = V^{(-)}(x, \alpha_{i+1}) + R(\alpha_i)$, where $R(\alpha_i)$ is a constant. Consequently, the eigenstates $\psi_n^{(+)}(x, \alpha_i)$ are identical to $\psi_n^{(-)}(x, \alpha_{i+1})$. In addition, since SUSYQM relates $\psi_n^{(+)}(x, \alpha_i)$ to $\psi_{n+1}^{(-)}(x, \alpha_i)$ via operators A and A^+, the states $\psi_{n+1}^{(-)}(x, \alpha_i)$ can be derived from states $\psi_n^{(-)}(x, \alpha_{i+1})$, and so forth. That is, a simple parameter shift allows for construction of the entire ladder of eigenstates $\psi_n^{(-)}$ or $\psi_n^{(+)}$. Furthermore, $E_n = \sum_{i=0}^{n-1} R(\alpha_i)$. We shall employ a similar technique to construct the ladders of QMF's $p^{(-)}$. Let us now replace the subscript n, which was a label for energy E, with E itself. Subscript $n - 1$ is replaced by $E - R(\alpha_i)$. Thus, shape invariance identifies $\psi_E^{(+)}(x, \alpha_i)$ with $\psi_{E-R(\alpha_i)}^{(-)}(x, \alpha_{i+1})$, which in QHJ becomes a relationship between quantum momentum functions. From Eq. (15.15), we find $p_{E-R(\alpha_i)}^{(-)}(x, \alpha_{i+1}) = p_E^{(+)}(x, \alpha_i)$. Substituting this relation in Eq. (15.21), we obtain the following recursion relation for $p_E^{(-)}(x, \alpha_i)$:

$$p_E^{(-)}(x, \alpha_i) = \frac{iW(x, \alpha_i) p_{E-R(\alpha_i)}^{(-)}(x, \alpha_{i+1}) - W^2(x, \alpha_i) + E}{p_{E-R(\alpha_i)}^{(-)}(x, \alpha_{i+1}) + iW(x, \alpha_i)} . \qquad (15.22)$$

This recursion relation, along with the initial condition given by Eq. (15.12) determines all functions $p_E^{(-)}(x, \alpha_i)$.

To provide a concrete example, let us determine the quantum momentum function related to the first excited state of the system with energy $E = R(\alpha_1)$. In Eq. (15.22), let us substitute $E - R(\alpha_2) = 0$. Then

---

[5]This is due to the unbroken nature of supersymmetry which implies that there is no normalizable $\psi^{(+)}$ at zero energy.

$$p_{E-R(\alpha_2)}^{(-)} = p_0^{(-)} = iW$$

$$-i\, p_{R(\alpha_1)}^{(-)}(x,\alpha_1) = \frac{W(x,\alpha_2)\cdot W(x,\alpha_1) + W^2(x,\alpha_1) - R(\alpha_1)}{W(x,\alpha_1) + W(x,\alpha_2)} \ ,$$

$$= W(x,\alpha_1) - \frac{R(\alpha_1)}{W(x,\alpha_1) + W(x,\alpha_2)} \ . \tag{15.23}$$

Therefore, starting from zero-energy QMF's $p_0^{(-)}(x,\alpha_i) = iW(x,\alpha_i)$, we have derived the higher level QMF; viz., $p_{R(\alpha_1)}^{(-)}(x,\alpha_1)$. This procedure can be iterated to generate $p_E^{(-)}(x,\alpha_1)$ for any eigenvalue E. Using the recursion relation (15.22) we can write

$$p_E^{(-)}(x,\alpha_1) \frac{i\,W(x,\alpha_1)\,p_{E-R(\alpha_1)}^{(-)}(x,\alpha_2) - W^2(x,\alpha_1) + E}{p_{E-R(\alpha_1)}^{(-)}(x,\alpha_2) + i\,W(x,\alpha_1)} \tag{15.24}$$

$$p_{E-R(\alpha_1)}^{(-)}(x,\alpha_2) = \frac{i\,W(x,\alpha_2)\,p_{E-R(\alpha_1)-R(\alpha_2)}^{(-)}(x,\alpha_3)}{p_{E-R(\alpha_1)-R(\alpha_2)}^{(-)}(x,\alpha_3) + i\,W(x,\alpha_2)}$$
$$- \frac{W^2(x,\alpha_2) - E - R(\alpha_1)}{p_{E-R(\alpha_1)+R(\alpha_2)}^{(-)}(x,\alpha_3) + i\,W(x,\alpha_2)} \ .$$

$$\tag{15.25}$$

Substituting (15.25) into (15.24) we obtain

$$p_E^{(-)}(x,\alpha_1) = \frac{A_3\,p_{E-R(\alpha_2)-R(\alpha_1)}^{(-)}(x,\alpha_3) + B_3}{C_3\,p_{E-R(\alpha_2)-R(\alpha_1)}^{(-)}(x,\alpha_3) + D_3} \ , \tag{15.26}$$

where

$A_3 = E - R(\alpha_1) - W(x,\alpha_2)W(x,\alpha_1) - W^2(x,\alpha_2) \ ,$

$B_3 = iW(x,\alpha_2)\left(W^2(x,\alpha_1) - E\right) + iW(x,\alpha_1)\left(W^2(x,\alpha_2) - E + R(\alpha_1)\right) \ ,$

$C_3 = -i\,W(x,\alpha_2) - i\,W(x,\alpha_1) \ ,$

$D_3 = E - W(x,\alpha_2)W(x,\alpha_1) - W^2(x,\alpha_1) \ .$

The recursion relation given by Eq. (15.26) is of the type

$$z_{i+1} = \frac{az_i + b}{cz_i + d} \ . \tag{15.27}$$

Such relations are called fractional linear transformation in z. The composition of two fractional linear transformations is tedious to compute. A short-cut is provided by the map

$$\frac{az + b}{cz + d} \mapsto \begin{bmatrix} a & b \\ c & d \end{bmatrix} \ . \tag{15.28}$$

Note that the function composition corresponds to matrix multiplication. That is, if f_1 and f_2 are two transformations given by

$$f_1(z) = \frac{a_1 z + b_1}{c_1 z + d_1} \ , \quad f_2(z) = \frac{a_2 z + b_2}{c_2 z + d_2} \ , \tag{15.29}$$

then

$$(f_2 \circ f_1)(z) = \frac{az + b}{cz + d} \ , \tag{15.30}$$

where the coefficients a, b, c and d are given by

$$\begin{bmatrix} a & b \\ c & d \end{bmatrix} = \begin{bmatrix} a_2 & b_2 \\ c_2 & d_2 \end{bmatrix} \cdot \begin{bmatrix} a_1 & b_1 \\ c_1 & d_1 \end{bmatrix} \ . \tag{15.31}$$

For any transformation f of form (15.27) there exists an inverse transformation f^{-1} if and only if $ad - bc \neq 0$. In this case the matrix associated with f is nonsingular and its inverse gives the coefficients of f^{-1}.

Now, let us associate to transformation (15.24) the matrix

$$m_1 = \begin{bmatrix} i\,W(x, \alpha_1) & E - W^2(x, \alpha_1) \\ 1 & i\,W(x, \alpha_1) \end{bmatrix} \ . \tag{15.32}$$

Similarly, we associate to transformation (15.25) the matrix

$$m_2 = \begin{bmatrix} i\,W(x, \alpha_2) & E - R(\alpha_1) - W^2(x, \alpha_2) \\ 1 & i\,W(x, \alpha_2) \end{bmatrix} \ . \tag{15.33}$$

Then the coefficients A_3, B_3, C_3, D_3 of Eq. (15.26) are obtained from

$$\begin{bmatrix} A_3 & B_3 \\ C_3 & D_3 \end{bmatrix} = m_2 \cdot m_1 \ . \tag{15.34}$$

Using the fractional linear transformation property, we can generalize this result. We have

$$p_E^{(-)}(x, \alpha_1) = \frac{A_{n+1}\, p_{E - \sum_{i=1}^{n} R(\alpha_i)}^{(-)}(x, \alpha_{n+1}) + B_{n+1}}{C_{n+1}\, p_{E - \sum_{i=1}^{n} R(\alpha_i)}^{(-)}(x, \alpha_{n+1}) + D_{n+1}} \ , \tag{15.35}$$

where

$$\begin{bmatrix} A_{n+1} & B_{n+1} \\ C_{n+1} & D_{n+1} \end{bmatrix} = m_n \cdot m_{n-1} \cdots m_1 \tag{15.36}$$

and

$$m_k = \begin{bmatrix} i\,W(x, \alpha_k) & E - \sum_{j=1}^{k-1} R(\alpha_j) - W^2(x, \alpha_k) \\ 1 & i\,W(x, \alpha_k) \end{bmatrix} \ ,$$

$$k = 1, 2, \ldots, n \ . \tag{15.37}$$

Note that the determinant of m_k is given by

$$\det(m_k) = E - \sum_{j=1}^{k-1} R(\alpha_j) . \tag{15.38}$$

Therefore, for those values of energy where $E = \sum_{j=1}^{k-1} R(\alpha_j)$ the matrix m_k is singular.

We have now connected QHJ theory with supersymmetric quantum mechanics, and have shown that the quantum momenta of supersymmetric partner potentials are connected via linear fractional transformations. Then, by making use of the matrix representation of the linear fractional transformations, we have provided an algorithm to generate any quantum momentum function for a general shape invariant potential. In the next section we will study the constraints placed upon the singularity structure of QMF by the shape invariance condition.

15.2 Determination of Eigenvalues Using Shape Invariance

In the last section we showed that in QHJ, shape invariance enables us to recursively determine the quantum momenta, and hence the eigenfunctions, analogous to SUSYQM. However, this derivation assumed that the eigenvalues E_n were given by the sum $\sum_{i=1}^{n-1} R(\alpha_i)$. Let us examine how the shape invariance condition also provides sufficient information to determine the singularity structure of the quantum momentum functions, and thus helps in determining eigenvalues of the system via QHJ as well.

The shape invariance condition, when written in terms of the superpotential W with $R(\alpha_0) = g(\alpha_1) - g(\alpha_0)$, takes the form:

$$W^2(x,\alpha_0) + \frac{dW(x,\alpha_0)}{dx} = W^2(x,\alpha_1) - \frac{dW(x,\alpha_1)}{dx} + g(\alpha_1) - g(\alpha_0) . \tag{15.39}$$

How does the above condition affect the pole structure of the superpotential? We will see that its behavior near fixed poles is constrained. In our analysis, we will divide all superpotentials into two categories: those with infinite domain $(0,\infty)$ or $(-\infty,\infty)$, and those with finite domain, such as (x_1, x_2). In QHJ formalism we embed these domains into a complex plane and analyze the behavior of QMFs near singular points. The superpotentials with finite domain generate infinitely many copies of the same domain on the real axis. The pole structure of QMFs is also repeated infinitely many times on the real axis. One of the ways to solve these systems is to map these finite domains to the infinite domains and thus turn them into an infinite-domain problem. For this reason, we will only consider systems with infinite domains.

Our objective is to show that using shape invariance and QHJ formalism, we can derive spectra for all infinite domain systems considered in Chapter 5. We divide all superpotentials into two categories: a) those that are of algebraic form and b) those can be converted into an algebraic form using a transformation such as $y = e^x$. We will term these cases algebraic and exponential, respectively.

15.3 Algebraic Cases

Let us first consider the superpotentials that can be explicitly expressed in an algebraic form. We would like to determine their structure near the fixed singularities, using shape invariance. In particular, we will be exploring how the superpotential diverges, if at all, near the origin and infinity. Let us first analyze the superpotential in the vicinity of the origin. Expanding in powers of the coordinate r,

$$
\begin{aligned}
W(r, \alpha_0) &= \sum_{k=-1}^{\infty} b_k(\alpha_0) \, r^k \\
&= \frac{b_{-1}(\alpha_0)}{r} + b_0(\alpha_0) \, r^0 + b_1(\alpha_0) \, r^1 + b_2(\alpha_0) \, r^2 + \cdots
\end{aligned}
$$

We have not included a term more singular than the $1/r$ in the superpotential to avoid the particle's "falling into the center."[6] The derivative of the superpotential is then given by

$$
\frac{d\,W(r, \alpha_0)}{dr} = -\frac{b_{-1}(\alpha_0)}{r^2} + 0 + b_1(\alpha_0) + 2\,b_2(\alpha_0)\,r + \cdots \, ;
$$

and the square of the superpotential is given by

$$
\begin{aligned}
W^2(r, \alpha_0) &= (b_{-1}(\alpha_0))^2 \, r^{-2} + (2\,b_{-1}(\alpha_0)\,b_0(\alpha_0))\,r^{-1} \\
&\quad + \left((b_0(\alpha_0))^2 + 2\,b_{-1}(\alpha_0)\,b_1(\alpha_0) \right) + \cdots
\end{aligned}
$$

Substituting these expansions into the shape invariance condition (15.39), we arrive at the following constraints on coefficients of the superpotential $W(x, \alpha_0)$:

$$
b_{-1}(\alpha_0)^2 - b_{-1}(\alpha_0) = b_{-1}(\alpha_1)^2 + b_{-1}(\alpha_1) \, , \tag{15.40}
$$

$$
2\,b_{-1}(\alpha_0)\,b_0(\alpha_0) = 2\,b_{-1}(\alpha_1)\,b_0(\alpha_1) \, , \tag{15.41}
$$

---

[6]See the discussion of such potentials in Chapter 11.

$$b_0(\alpha_0)^2 + 2\,b_{-1}(\alpha_0)\,b_1(\alpha_0) + b_1(\alpha_0)$$

$$= b_0(\alpha_1)^2 + 2\,b_{-1}(\alpha_1)\,b_1(\alpha_1) - b_1(\alpha_1) + R(\alpha_0)\,.$$

$$(15.42)$$

The constraint (15.40) is a quadratic equation for the coefficient $b_{-1}(\alpha_1)$ in terms of $b_{-1}(\alpha_0)$. It has the following two solutions: $b_{-1}(\alpha_1) = -b_{-1}(\alpha_0)$ or $b_{-1}(\alpha_1) = b_{-1}(\alpha_0) - 1$. Since $b_{-1}(\alpha_0)$ is the coefficient of the dominant term near the origin, its behavior plays an essential role in deciding whether supersymmetry remains unbroken. If the supersymmetry is unbroken for a given value of the coefficient $b_{-1}(\alpha_0)$, changing the sign of this coefficient would change the asymptotic value of the superpotential as $r \to 0$, and supersymmetry will no longer remain unbroken. Hence, the first solution, $b_{-1}(\alpha_1) = -b_{-1}(\alpha_0)$, is not acceptable if both $b_{-1}(\alpha_0)$ and $b_{-1}(\alpha_1)$ were to keep the system in the parameter domain needed for unbroken supersymmetry. Thus, the constraint (15.40) implies the relationship $b_{-1}(\alpha_1) = b_{-1}(\alpha_0) - 1$. For $\alpha_1 = \alpha_0 + 1$, this gives the difference equation $b_{-1}(\alpha_0 + 1) = b_{-1}(\alpha_0) - 1$, whose solution is

$$b_{-1}(\alpha_0) = -\alpha_0 + \text{constant}\,. \qquad (15.43)$$

We choose the constant to be zero. Equation (15.41) then leads to the difference equation

$$\alpha_0\,b_0(\alpha_0) = \alpha_1\,b_0(\alpha_1)\,. \qquad (15.44)$$

In other words, the product $\alpha_0\,b_0(\alpha_0)$ does not depend on the parameter α_0; we denote this product by β. Hence, we have

$$b_0(\alpha_0) = \frac{\beta}{\alpha_0}\,.$$

Thus, near the origin, the structure of the superpotential $W(x, \alpha_0)$ is given by:

$$W(r, \alpha_0)\big|_{r \to 0} = -\frac{\alpha_0}{r} + \frac{\beta}{\alpha_0} + \cdots \qquad (15.45)$$

Now, let us explore the structure of the superpotential near infinity. To do this, as we have done before, we define $u = \frac{1}{r}$. The shape invariance condition of Eq. (15.39) in terms of the variable u transforms into

$$W^2\left(\frac{1}{u}, \alpha_0\right) - u^2 \frac{dW\left(\frac{1}{u}, \alpha_0\right)}{du} = W^2\left(\frac{1}{u}, \alpha_1\right) + u^2 \frac{dW\left(\frac{1}{u}, \alpha_1\right)}{du} + R(\alpha_0)\,.$$

$$(15.46)$$

Problem 15.1. *Derive Eq. (15.46).*

Analogous to the expansion near the origin, let us expand $W\left(\frac{1}{u}, \alpha_0\right)$ in powers of u as:

$$W\left(1/u, \alpha_0\right) = \sum_{k=-1}^{\infty} c_k(\alpha_0)\, u^k$$

$$= \frac{c_{-1}(\alpha_0)}{u} + c_0(\alpha_0)\, u^0 + c_1(\alpha_0)\, u^1 + c_2(\alpha_0)\, u^2 + \cdots .$$

Similarly expanding W^2 and $\frac{dW(r,\alpha_0)}{dr}$, we obtain

$$W^2\left(1/u, \alpha_0\right) = \frac{(c_{-1}(\alpha_0))^2}{u^2} + \frac{(2\,c_{-1}(\alpha_0)\, c_0(\alpha_0))}{u}$$

$$+ \left((c_0(\alpha_0))^2 + 2\,c_{-1}(\alpha_0)\, c_1(\alpha_0)\right) + \cdots$$

and

$$\frac{dW(r,\alpha_0)}{dr} = -u^2\, \frac{dW(1/u, \alpha_0)}{du}$$

$$= c_{-1}(\alpha_0) - c_1(\alpha_0)\, u^2 - 2\,c_2(\alpha_0)\, u^3 + \cdots .$$

Now we substitute these expressions into the transformed shape invariance condition of Eq. (15.46) to get

$$c_{-1}(\alpha_0)^2 = c_{-1}(\alpha_1)^2 \tag{15.47}$$

$$2\,c_{-1}(\alpha_0)\, c_0(\alpha_0) = 2\,c_{-1}(\alpha_1)\, c_0(\alpha_1) \tag{15.48}$$

$$c_0(\alpha_0)^2 + 2\,c_{-1}(\alpha_0)\, c_1(\alpha_0) + c_{-1}(\alpha_0)$$

$$= c_0(\alpha_1)^2 + 2\,c_{-1}(\alpha_1)\, c_1(\alpha_1) - c_{-1}(\alpha_1) + R(\alpha_0) .$$

$$\tag{15.49}$$

Problem 15.2. *Derive Eqs. (15.47)–(15.49).*

From Eq. (15.47), we obtain $c_{-1}(\alpha_1) = \pm c_{-1}(\alpha_0)$. As we argued before, since we assumed that supersymmetry remains unbroken for a range of values of the parameter, we choose $c_{-1}(\alpha_1) = c_{-1}(\alpha_0)$; i.e., this coefficient does not depend on the parameter α_0. We denote it by c_{-1}. This then leads to $c_0(\alpha_0) = c_0(\alpha_1) = c_0$, another constant. Equation (15.49) now yields $2\,c_{-1}\, c_1(\alpha_0) + c_{-1} = 2\,c_{-1}\, c_1(\alpha_1) - c_{-1} + g(\alpha_1) - g(\alpha_0)$, which gives

$$c_1(\alpha_1) + \frac{g(\alpha_1)}{2\,c_{-1}} = c_1(\alpha_0) + \frac{g(\alpha_0)}{2\,c_{-1}} + 1 . \tag{15.50}$$

This difference equation is of the form

$$f(\alpha_1) = f(\alpha_0) + 1 \;, \tag{15.51}$$

where $f(\alpha_0) = c_1(\alpha_0) + \frac{g(\alpha_0)}{2\,c_{-1}}$. Since $\alpha_1 = \alpha_0 + 1$, the solution of Eq. (15.51) is $f(\alpha_0) = \alpha_0 + \Delta$. Hence, we have $c_1(\alpha_0) = \alpha_0 - \frac{g(\alpha_0)}{2\,c_{-1}} + \Delta$, where Δ is a constant independent of α_0. Thus, near infinity the superpotential is given by

$$\widetilde{W}\left(\frac{1}{u}, \alpha_0\right)\Bigg|_{r\to\infty} = \frac{c_{-1}}{u} + c_0$$
$$+ \left(\alpha_0 - \frac{g(\alpha_0)}{2\,c_{-1}} + \Delta\right) u + \dots \,. \tag{15.52}$$

The result we have obtained depended crucially upon the assumption that c_{-1} is not zero. If $c_{-1} = 0$, the structure of the potential near infinity will be very different and the constraints given by Eqs. (15.47)–(15.49) would no longer be valid.

For $c_{-1} = 0$, the superpotential near infinity is given by $\widetilde{W}(u, \alpha_0) = c_0 + c_1 u + c_2 u^2 + \cdots$. This leads to $\widetilde{W}^2(u, \alpha_0) = c_0^2 + 2c_0 c_1 u + (c_1^2 + 2c_0 c_2)u^2 + \cdots$, and $u^2 \frac{d\widetilde{W}}{du} = c_1 u^2 + 2c_2 u^3 + \cdots$. Since $\widetilde{W}^2(u, \alpha_0)$ and $\widetilde{W}^2(u, \alpha_1)$ must satisfy the shape invariance condition (15.39), matching the first two powers of u we obtain

$$c_0^2(\alpha_1) = c_0^2(\alpha_1) + R(\alpha_0) \;, \tag{15.53}$$

$$c_0(\alpha_0)c_1(\alpha_0) = c_0(\alpha_1)c_1(\alpha_1) \;. \tag{15.54}$$

Equation (15.53) may be rewritten as $c_0^2(\alpha_0) + g(\alpha_0) = c_0^2(\alpha_0 + 1) + g(\alpha_0 + 1)$ which means that the quantity $c_0^2(\alpha_0) + g(\alpha_0)$ is independent of the argument α_0, and hence equal to a constant Λ. This leads to

$$c_0(\alpha_0) = \pm\sqrt{-g(\alpha_0) + \Lambda} \;. \tag{15.55}$$

Now from Eq. (15.54) we see that the product $c_0(\alpha_0)c_1(\alpha_0)$ is also independent of the argument α_0, and hence it must also be equal to a constant, which we denote by β. From $c_0(\alpha_0)c_1(\alpha_0) = \beta$, we have

$$c_1(\alpha_0) = \frac{\beta}{c_0(\alpha_0)} = \frac{\beta}{\pm\sqrt{-g(\alpha_0) + \Lambda}} \;. \tag{15.56}$$

Thus, near ∞,

$$\widetilde{W}(u, \alpha_0) = \pm\left(\sqrt{-g(\alpha_0) + \Lambda}\right) \pm \left(\frac{\beta}{\sqrt{-g(\alpha_0) + \Lambda}}\right) u + \cdots \,. \tag{15.57}$$

We have now acquired significant knowledge about the structure of the superpotential near boundary points, simply from the requirement of shape invariance. Let us write these findings again. Near the origin, as given in Eq. (15.45), the structure is

$$W(r, \alpha_0) = -\frac{\alpha_0}{r} + \frac{\beta}{\alpha_0} + \cdots . \tag{15.58}$$

Near infinity, as $u \to 0$, there are two possible structures that depend on whether the superpotential has a singularity; i.e., whether the value of the coefficient $c_{-1}(\alpha_0)$ is zero or non-zero. They are:

$$\widetilde{W}(u, \alpha_0) = \begin{cases} \frac{c_{-1}}{u} + c_0 + \left(\alpha_0 - \frac{g(\alpha_0)}{2\,c_{-1}} + D\right) u + \cdots & c_{-1}(\alpha_0) \neq 0 \\[2mm] \pm\sqrt{-g(\alpha_0) + \Lambda} \pm \frac{\beta}{\sqrt{-g(\alpha_0)+\Lambda}}\, u + \cdots & c_{-1}(\alpha_0) = 0 . \end{cases} \tag{15.59}$$

With this understanding about the structure of $W(x, \alpha_0)$ at the end points of the domain, we now proceed to substitute Eqs. (15.58) and (15.59) into QHJ Eq. (15.7). This substitution will help us learn about their implications for the QMF, and thence the eigenvalues of the hamiltonian. We first expand the momentum $p(r)$ near the origin as:

$$p(r, \alpha_0) = \sum_{k=-1}^{\infty} p_k\, r^k = \frac{p_{-1}}{r} + p_0\, r^0 + p_1\, r^1 + p_2\, r^2 + \cdots .$$

The derivative of p is then given by $\frac{dp(r,\alpha_0)}{dr} = -\frac{p_{-1}}{r^2} + 0 + p_1 + 2\,p_2\, r + \cdots$. and the square of p is given by $p^2(r, \alpha_0) = \frac{(p_{-1})^2}{r^2} + \frac{(2\,p_{-1}\,p_0)}{r} + (p_0^2 + 2\,p_{-1}\,p_1) + \cdots$. Collecting terms with various powers of r, we find that near the origin the combination $p^2 + ip'$ is given by

$$\left(p^2 - ip'\right)\big|_{\text{origin}} \sim \frac{p_{-1}^2 + ip_{-1}}{r^2} + \frac{(2\,p_{-1}\,p_0)}{r} + (p_0^2 + 2\,p_{-1}\,p_1 + ip_1) + \cdots .$$

We substitute this series in the QHJ Eq. (15.11) and equate the coefficients of powers of r on both sides. From the coefficient of r^{-2}, we obtain $p_{-1}^2 + ip_{-1} + \alpha_0^2 - \alpha_0 = 0$. It has two solutions:

$$p_{-1} = i(\alpha_0 - 1) \text{ and } p_{-1} = i\alpha_0 . \tag{15.60}$$

Using Eq. (15.12), we choose $p_{-1} = -i\alpha_0$. We will later see that only this term in the expansion of p will contribute when we carry out the contour integration around the origin in the complex r plane. So we do not list other coefficients.

To determine the leading order behavior of the momentum function at infinity, let us first consider the case that $c_1(\alpha_0) \neq 0$. We expand the momentum function $p\left(\frac{1}{u}, \alpha_0\right) \equiv \widetilde{p}(u, \alpha_0)$ in powers of u:

$$\widetilde{p}(u, \alpha_0) = \sum_{k=-1}^{\infty} \widetilde{p}_k u^k = \frac{\widetilde{p}_{-1}}{u} + \widetilde{p}_0 u^0 + \widetilde{p}_1 u^1 + \widetilde{p}_2 u^2 + \cdots .$$

This leads to the following expression for the combination $p^2 - ip'$:

$$(p^2 - ip\,')\big|_\infty \equiv \widetilde{p}^2 + iu^2 \frac{d\widetilde{p}}{du} = \frac{\widetilde{p}_{-1}^2}{u^2} + \frac{2\widetilde{p}_{-1}\widetilde{p}_0}{u}$$
$$+ \left(\widetilde{p}_0^2 + 2\widetilde{p}_{-1}\widetilde{p}_1 - i\widetilde{p}_{-1}\right) + \cdots .$$

$$(15.61)$$

Substituting it into Eq. (15.11); i.e., in

$$\widetilde{p}^2 + iu^2 \frac{d\widetilde{p}}{du} = E - \left(\widetilde{W}^2 + u^2 \frac{d\widetilde{W}}{du}\right), \qquad (15.62)$$

we obtain

$$\frac{\widetilde{p}_{-1}^2}{u^2} + \frac{2\widetilde{p}_0\widetilde{p}_{-1}}{u} + \left(\widetilde{p}_0^2 + 2\widetilde{p}_{-1}\widetilde{p}_1 - i\widetilde{p}_{-1}\right) + \cdots$$
$$= -\frac{c_{-1}^2}{u^2} - \frac{2c_0 c_{-1}}{u} + E - c_0^2 - 2c_{-1}\left(\alpha_0 - \frac{g(\alpha_0)}{2c_{-1}}\right) + c_{-1} .$$

By identifying the various powers of the variable u, we obtain $\widetilde{p}_{-1} = ic_{-1}$; $\widetilde{p}_0 = ic_0$; and $\widetilde{p}_1 = i\alpha_0 - i\,\frac{g(\alpha_0)+E}{2c_{-1}}$. At infinity, as we shall soon see, it is this last coefficient $\widetilde{p}_1$ that contributes towards the contour integration.

We are now ready to determine the eigenvalue for the system using the information we have generated solely from shape invariance. The contribution to the contour integration is given by

$$\left[\oint p\,dr\right]_{\text{Moving poles}} = -\oint_{r=0} p\,dr - \oint_{r=\infty} p\,dr$$

$$= -\oint_{r=0} p\,dr + \oint_{u=0} \frac{1}{u^2}\,\widetilde{p}\,du$$

$$= -(2\pi i p_{-1}) + (2\pi i \widetilde{p}_1)$$

$$= -2\pi i\,(-i\,\alpha_0) + 2\pi i\left[i\,\alpha_0 - i\,\frac{g(\alpha_0)+E}{2c_{-1}}\right]$$

$$= 2\pi\left[-2\alpha_0 + \frac{g(\alpha_0)+E}{2c_{-1}}\right]. \qquad (15.63)$$

Note that due to the $\frac{1}{u^2}$ factor in the term $\oint_{u=0} (\frac{1}{u^2}\, \widetilde{p})\, du$, only the linear term of $\widetilde{p}$, $2\pi\, i\widetilde{p}_1$, contributes towards the integral.

It is worth noting that we cannot get the entire global behavior of the potential from expansions at the origin and infinity. If the potential is a symmetric function of r, as is the case for e.g. the three-dimensional harmonic oscillator, we obtain an additional contribution. If we embed the domain $(0, r)$ in a two-dimensional complex plane, a mirror image of the potential function in the region $\mathcal{R}e\,(r) < 0$ would then have singularities for $p(r)$, precisely at the same points as $p(r)$ does in the region $\mathcal{R}e\,(r) < 0$. As a consequence,

$$\oint p\, dr\,\Big|_{\text{Moving poles}} \equiv \oint p\, dr\,\Big|_{\mathcal{R}e\,(r)>0}$$

$$= -\oint p\, dr\,\Big|_{\mathcal{R}e\,(r)<0} - \oint_{r=0} p\, dr + \oint_{u=0} \frac{1}{u^2}\, \widetilde{p}\, du\ .$$

However, $\oint p\, dr\big|_{\mathcal{R}e\,(r)>0}$ and $\oint p\, dr\big|_{\mathcal{R}e\,(r)<0}$ have identical value due to the symmetry of the potential. This leads to

$$\oint p\, dr\,\Big|_{\text{Moving poles}} = \frac{1}{2}\left(-\oint_{r=0} p\, dr + \oint_{u=0} \frac{1}{u^2}\, \widetilde{p}\, du\right) = 2\pi n\ . \tag{15.64}$$

From Eq. (15.63), we then get

$$\frac{1}{2}\left\{2\pi\left[-2\alpha_0 + \frac{g(\alpha_0) + E}{2\, c_{-1}}\right]\right\} = 2\pi n\ . \tag{15.65}$$

Solving for the energy, Eq. (15.63) gives

$$E_n = 4\, c_{-1}\,(n + \alpha_0) - g(\alpha_0)\ . \tag{15.66}$$

To determine the value of $g(\alpha_0)$, we demand as per SUSYQM that the ground state energy E_0 be zero. Substituting $n = 0$, we obtain $g(\alpha_0) = 4\, c_{-1}\alpha_0$. Since $\alpha_n = \alpha_0 + n$, we obtain $g(\alpha_n) = 4\, c_{-1}\alpha_n = 4\, c_{-1}\,(\alpha_0 + n)$. Substituting for $g(\alpha_0)$ and $g(\alpha_n)$ in E_n, we obtain

$$E_n = 4\, n\, c_{-1} = g(\alpha_n) - g(\alpha_0)\ . \tag{15.67}$$

Thus, QHJ formalism also generates the correct eigenvalues from shape invariance. For the three-dimensional harmonic oscillator, we identify the coefficient c_{-1} with $\frac{1}{2}\omega$. As expected, the energy is

$$E_n = 4\, n\, c_{-1} = 2n\omega\ . \tag{15.68}$$

Problem 15.3. *In the above derivation, we assumed that the origin was a singular point. Repeat the procedure without a singular point at the origin. Show that for a shape invariant potential with only one fixed singularity, at infinity, the eigenvalues are same as that of a one-dimensional harmonic oscillator; i.e., $E_n \propto n$.*

Now let us consider the case in which the superpotential has no divergence at infinity; i.e., $c_{-1} = 0$. For this case, we will need to determine the structure of the quantum momentum function $p(x)$ at infinity. (The structure at the origin remains the same as that for $c_{-1} \neq 0$.) We substitute the expansion of the superpotential $\widetilde{W}(u, \alpha_0)$ from Eq. (15.59) into the QHJ equation at infinity given by Eq. (15.62), and we obtain

$$\frac{\widetilde{p}_{-1}^2}{u^2} + \frac{2\widetilde{p}_0 \widetilde{p}_{-1}}{u} + \left(\widetilde{p}_0^2 + 2\widetilde{p}_{-1}\widetilde{p}_1 - i\widetilde{p}_{-1}\right) + \cdots = E + g(\alpha_0) - \Lambda - 2\beta u + \cdots .$$

Now equating the coefficients of powers of u, we obtain $\widetilde{p}_{-1}(\alpha_0) = 0$; $\widetilde{p}_0^2(\alpha_0) = E + g(\alpha_0) - \Lambda$; $\widetilde{p}_0(\alpha_0)\widetilde{p}_1(\alpha_0) = \beta$; etc. From these, we obtain the following coefficients:

$$\widetilde{p}_0 = \pm\sqrt{E + g(\alpha_0) - \Lambda} \; ; \quad \text{and} \quad \widetilde{p}_1 = \pm\frac{\beta}{\sqrt{E + g(\alpha_0) - \Lambda}} .$$

The quantization condition then yields

$$\oint p \, dr \bigg|_{\text{Moving poles}} = -\oint_{r=0} p \, dr + \oint_{u=0} \frac{1}{u^2} \, \widetilde{p} \, du \tag{15.69}$$

$$= -(2\pi i p_{-1}) + (2\pi i \widetilde{p}_1)$$

$$= -2\pi i \, (-i\alpha_0) \pm 2\pi i \left(\frac{\beta}{\sqrt{E + g(\alpha_0) - \Lambda}} \right)$$

$$= 2\pi n .$$

Thus, we have $-\alpha_0 \pm i\frac{\beta}{\sqrt{E + g(\alpha_0) - \Lambda}} = n$. Solving for E_n,

$$E_n = \Lambda - g(\alpha_0) - \frac{\beta}{(n + \alpha_0)^2} . \tag{15.70}$$

Since we are in the domain of unbroken supersymmetry, we must have $E_0 = 0$. Hence substituting $n = 0$ in Eq. (15.70), we obtain $g(\alpha_0) = -\frac{1}{\alpha_0^2} + \Lambda$; i.e., $g(\alpha_n) = -\frac{1}{(\alpha_0 + n)^2} + \Lambda$. The energy E_n is

$$E_n = \beta^2 \left[\frac{1}{(\alpha_0)^2} - \frac{1}{(n + \alpha_0)^2} \right] = g(\alpha_n) - g(\alpha_0) . \tag{15.71}$$

Thus, we have determined the energy of this system described by a Coulomb-like potential that is singular at the origin and has no singularity at infinity, simply by using the shape invariance condition. Once we identify β^2 with e^2 and α_0 with $l + 1$, we obtain the energy spectrum for the Coulomb potential.

15.4 Exponential Cases

Here we consider superpotentials that arise as a linear combination of various powers of the exponential function e^x. For example, the Morse and Rosen-Morse potentials fall into this category. We assume that our generic superpotential has the general form

$$W(x, \alpha_0) = \sum_j b_j \left[e^x \right]^j . \tag{15.72}$$

This superpotential has singularities at $\pm \infty$. To analyze the nature of these singularities we first convert it into an algebraic form with the transformation $r = e^x$. This maps $x \in (-\infty, \infty)$ to $r \in (0, \infty)$. We then subject the resulting superpotential to the shape invariance condition. As $x \to -\infty$, and thus $r \to 0$, we assume the potential has the following structure:

$$W(r, \alpha_0) = b_{-1} r^{-1} + b_0 + b_1 r + \cdots . \tag{15.73}$$

Similarly, as $r \to \infty$, we express the superpotential in terms of a variable $u \equiv 1/r$:

$$\widetilde{W}(u, \alpha_0) = c_{-1} u^{-1} + c_0 + c_1 u + \cdots . \tag{15.74}$$

Now we will determine how shape invariance constrains various terms in these two expansions at the end points of the domain of r.

As with the algebraic superpotential, the following procedure differs depending on whether b_{-1} and c_{-1} are each zero or nonzero, making for four possible cases. We want to stress that we are considering only those shape invariant superpotentials that have at least one singularity at either the origin or infinity. *If the superpotential has no poles at these extreme points, then it must have poles at "finite" fixed-points.* This type of potential would fall under what we call trigonometric potentials and will not be considered.

It is interesting to note that the case $b_{-1} = 0$, and $c_{-1} \neq 0$ is identical to the case $b_{-1} \neq 0$, and $c_{-1} = 0$. Hence, we need consider only two cases[7]: $(b_{-1} \neq 0, \ c_{-1} = 0)$ and $(b_{-1} \neq 0, \ c_{-1} \neq 0)$.

Problem 15.4. *Show that the above claim is true.*

---

[7]The case $(b_{-1} = 0, \ c_{-1} = 0)$ has no fixed pole on the real-axis.

However, it turns out that for $(b_{-1} \neq 0,\; c_{-1} \neq 0)$, the quantization condition $\oint p\, dx = 2\pi n$ does not yield an equation involving energy, and hence energy cannot be determined. There are no known exponential-type shape invariant potentials in quantum mechanics for which the superpotential has singularities for both positive and negative infinity. Thus, we will only consider the case where $b_{-1} \neq 0$ and $c_{-1} = 0$.

Near $r = 0$, the superpotential is given by

$$W(r, \alpha_0) = \frac{b_{-1}(\alpha_0)}{r} + b_0(\alpha_0) + b_1(\alpha_0)r + \cdots . \qquad (15.75)$$

From W, we obtain

$$\frac{dW}{dx} = \frac{dW}{dr}\frac{dr}{dx} = r\frac{dW}{dr} - -\frac{b_{-1}}{r} + b_1 r , \qquad (15.76)$$

and

$$W^2 = \frac{b_{-1}^2}{r^2} + \frac{2b_{-1}b_0}{r} + b_0^2 + 2b_{-1}b_1 + \cdots . \qquad (15.77)$$

Hence, due to the above change of variables, the shape invariance condition now takes the form

$$W^2(\alpha_0) + r\frac{dW(\alpha_0)}{dr} = W^2(\alpha_1) - r\frac{dW(\alpha_1)}{dr} + R(\alpha_0) . \qquad (15.78)$$

Substituting the expansion of W into this equation and equating powers of r, we obtain

$$b_{-1}^2(\alpha_0) = b_{-1}^2(\alpha_1) ,$$

$$2b_{-1}(\alpha_0)b_0(\alpha_0) - b_{-1}(\alpha_0) = 2b_{-1}(\alpha_1)b_0(\alpha_1) + b_{-1}(\alpha_1) , \qquad (15.79)$$

$$b_0^2(\alpha_0) + 2b_{-1}(\alpha_0)b_1(\alpha_0) = b_0^2(\alpha_1) + 2b_{-1}(\alpha_1)b_1(\alpha_1) + R(\alpha_0) .$$

From the first equation, we see that $b_{-1}(\alpha_0)$ does not depend upon its argument; i.e., it is a constant. We denote this constant by b_{-1}. The second equation then gives $b_0(\alpha_1) = b_0(\alpha_0) - 1$. Substituting $\alpha_1 = \alpha_0 + 1$, we obtain the difference equation

$$b_0(\alpha_0 + 1) = b_0(\alpha_0) - 1 ,$$

whose solution is $b_0(\alpha_0) = -\alpha_0 + C$, where C is a constant. Now, from the last equation of the set we have

$$(-\alpha_0 + C)^2 + 2b_{-1}b_1(\alpha_0) = (-\alpha_1 + C)^2 + 2b_{-1}b_1(\alpha_1) + g(\alpha_1) - g(\alpha_0) .$$

This equation can be written as

$$b_1(\alpha_0) + \frac{g(\alpha_0) + (-\alpha_0 + C)^2}{2b_{-1}} = b_1(\alpha_1) + \frac{g(\alpha_1) + (-\alpha_1 + C)^2}{2b_{-1}} ,$$

which shows that the expression $b_1(\alpha_0) + \frac{g(\alpha_0) + (-\alpha_0 + C)^2}{2b_{-1}}$ is independent of its argument α_0, hence we set it equal to a constant D. Thus,

$$b_1(\alpha_0) = D - \frac{g(\alpha_0) + (-\alpha_0 + C)^2}{2b_{-1}} . \tag{15.80}$$

We carry out a very similar analysis in the region $r \to \infty$, where the superpotential has the form $\widetilde{W}(u, \alpha_0) = c_0(\alpha_0) + c_1(\alpha_0) u + c_1(\alpha_0) u^2 \cdots$, and we obtain

$$c_0(\alpha_0)^2 = c_0(\alpha_1)^2 + R(\alpha_0) . \tag{15.81}$$

Writing this equation as $c_0(\alpha_0)^2 + g(\alpha_0) = c_0(\alpha_1)^2 + g(\alpha_1)$, we see that the expression $c_0(\alpha_0)^2 + g(\alpha_0)$ is independent of α_0. We set it equal to a constant Θ. Thus,

$$c_0(\alpha_0) = \pm\sqrt{\Theta - g(\alpha_0)} . \tag{15.82}$$

Therefore, asymptotic forms of the superpotential are:
Near the origin

$$W(r, \alpha_0) = \frac{b_{-1}}{r} + (-\alpha_0 + C) + \left[D - \left(\frac{g(\alpha_0) + (-\alpha_0 + C)^2}{2b_{-1}} \right) \right] r + \cdots ; \tag{15.83}$$

and as $r \to \infty$ $(u \to 0)$

$$\widetilde{W}(u, \alpha_0) = \pm\sqrt{\Theta - g(\alpha_0)} + c_1(\alpha_0) u + c_1(\alpha_0) u^2 \cdots . \tag{15.84}$$

We now substitute these two asymptotic forms of the superpotential into the QHJ equation to determine the analytic structure of p, and from it, the eigenvalues of the shape invariant system.

Near $r = 0$, the QHJ equation is given by

$$p^2 - i r \frac{dp}{dr} = E - W^2(r, \alpha_0) + r \frac{dW(r, \alpha_0)}{dr} , \tag{15.85}$$

where the superpotential $W(r, \alpha_0)$ is given by Eq. (15.83). Expanding the quantum momentum function as

$$p(r) = p_{-1} r^{-1} + p_0 + p_1 r + \cdots , \tag{15.86}$$

and substituting it into the above QHJ equation, we obtain

$$p_{-1}^2 = -b_{-1}^2; \quad 2p_{-1}p_0 + i\, p_{-1} = -b_{-1} - 2b_{-1}b_0(\alpha_0); \quad \text{etc.} \tag{15.87}$$

From the first equality in the above equation, we obtain $p_{-1} = \pm i b_{-1}$. However, in the limit $E \to 0$, we must have $p \to iW$. This implies $p_{-1} = i b_{-1}$. Substituting this value of p_{-1} in the second equality, we obtain $p_0 = i\,b_0(\alpha_0) = i\,(-\alpha_0 + C)$. As we will soon see, we need go no further. The coefficient p_0 is all we need to determine the contour integral around $r = 0$.

Near $r \to \infty$, we expand the QMF as $\widetilde{p}\,(u) = \widetilde{p}_{-1}u^{-1} + \widetilde{p}_0 + \widetilde{p}_1 u + \cdots$. We substitute this expansion for $\widetilde{p}(u)$ and the superpotential given by Eq. (15.84) into the QHJ equation

$$\widetilde{p}^2 + i\,u\frac{d\widetilde{p}}{du} = E - \left(\widetilde{W}^2(u,\alpha_0) + u\frac{d\widetilde{W}(u,\alpha_0)}{du}\right) . \tag{15.88}$$

Now comparing various powers of u, we obtain $\widetilde{p}_{-1} = 0$, $(\widetilde{p}_0)^2 = \left(\pm\sqrt{\Theta - g(\alpha_0)}\right)^2 - E$, etc. Thus, we obtain $\widetilde{p}_0 = \pm\sqrt{\Theta - g(\alpha_0) - E}$.

The quantization condition for this system then yields

$$\oint_{x=-\infty} p\,dx + \oint_{x=\infty} p\,dx = -2\pi\,n . \tag{15.89}$$

Since near $x = -\infty$, the variable $r = e^x \to 0$ and as $x \to \infty$, the variable $u \equiv 1/r = e^{-x} \to 0$, we write the above condition as

$$\oint_{r=0}\left(\frac{p_0}{r}\right)dr - \oint_{u=0}\left(\frac{\widetilde{p}_0}{u}\right)du = -2\pi\,n .$$

This then gives

$$(C - \alpha_0) \pm \sqrt{\Theta - g(\alpha_0) - E_n} = n .$$

Solving for E_n we obtain

$$E_n = \Theta - (C - \alpha_0 - n)^2 - g(\alpha_0) . \tag{15.90}$$

Since we assume that supersymmetry is unbroken, we must have $E_0 = 0$. This implies $g(\alpha_0) = \Theta - (C - \alpha_0)^2$, and hence $g(\alpha_n) = \Theta - (C - (\alpha_0 + n))^2 = \Theta - (C - \alpha_0 - n)^2$. Thus, we have

$$E_n = (C - \alpha_0)^2 - (C - \alpha_0 - n)^2 = g(\alpha_n) - g(\alpha_0) . \tag{15.91}$$

If we identify $C - \alpha_0$ with the parameter A of the Morse superpotential $W = A - e^{-x}$, we obtain the energy $E_n = A^2 - (A - n)^2$.

Thus, through several examples we have shown that in the QHJ formalism, shape invariance and unbroken SUSY completely determine the spectra of hamiltonians. The eigenvalues are given by

$$E_n = g(\alpha_n) - g(\alpha_0) ,$$

exactly the expression we have seen in Chapter 5. It is true that these spectra could have been gotten much more easily using SUSYQM. However, this formalism is analytic as opposed to the algebraic method we saw in Chapter 5. This QHJ quantization, which looks very similar to the JWKB semi-classical quantization procedure we saw in Chapter 12, provides us with insight into the analytical structure of the QMFs.

Chapter 16

The Phase Space Quantum Mechanics Formalism and SUSYQM

So far we have encountered two ways of quantizing physical systems: the canonical way and the Quantum Hamilton Jacobi formalism. Currently, there are four independent paths to quantization. In chronological order, the first, the canonical way, is the quantization formalism based on Schrödinger's equation, or equivalently, using operators acting on a Hilbert space. The second way is the path integral formalism, which we won't address in this book. The third way of doing quantum mechanics is the Quantum Hamilton Jacobi formalism, which we discussed in Chapter 15. A fourth path to quantization is the phase space quantization, also known as the deformation quantization formalism, which is the subject of this chapter.

16.1 Introduction to Phase Space Quantum Mechanics

The phase space quantization method was developed primarily by Wigner, Weyl, Groenewold, and Moyal,[1] as an alternate quantization approach stemming from statistical mechanics. Here, we will consider only the one-dimensional case, where the phase space is simply (x, p). For the general case, see Weyl, Footnote 1. For clarity, we will use the "hat" notation to symbolize quantum operators and the "non-hat" notation for the corresponding functions on the phase space.

---

[1] E. P. Wigner, "On the Quantum Correction For Thermodynamic Equilibrium", *Phys. Rev.* **40**, 749 (1932);

Perspectives in Quantum Theory, Dover, NY (1979);

H. Weyl, "Quantenmechanik und Gruppentheorie", **46**, *Z. Phys.* 1 (1927);

H. J. Groenewold, "On the Principles of Elementary Quantum Mechanics", *Physica* **12**, 405 (1946);

J. E. Moyal, "Quantum Mechanics as a Statistical Theory", *Proc. Camb. Phil. Soc.* **45**, 99 (1949).

As a prelude to deformation quantization, let us consider the method used in canonical quantization. To each classical quantity $A(x, p)$ we associate an operator $\hat{A}(\hat{x}, \hat{p})$ in such a way that the Poisson bracket of two classical quantities $\{A, B\}_P \equiv \frac{\partial A}{\partial x} \frac{\partial B}{\partial p} - \frac{\partial A}{\partial p} \frac{\partial B}{\partial x}$ and the commutator of the corresponding operators $[\hat{A}, \hat{B}] = \hat{A}\hat{B} - \hat{B}\hat{A}$ are connected via

$$\{A, B\}_P \mapsto \frac{1}{i\hbar}[\hat{A}, \hat{B}] . \tag{16.1}$$

In particular

$$\{x, p\}_P = 1 \mapsto \frac{1}{i\hbar}[\hat{x}, \hat{p}] = 1 . \tag{16.2}$$

While the above quantization rule is simple to implement, it does not provide guidance for products of observables like xp, where due to non-commutativity of the corresponding quantum counterparts, there is an ambiguity on how to interpret such a product: is it $\hat{x}\hat{p}$ or $\hat{p}\hat{x}$? In order to remove the ambiguity for the classical product of observables, we can adopt the symmetrical ordered polynomial rule (Weyl's ordering) where $xp \rightarrow (\hat{x}\hat{p} + \hat{p}\hat{x})/2$, etc. Then, we observe that the quantization rule of Eq. (16.1) works perfectly up to polynomials of second order in x and p. However, it fails for higher orders. Groenewold showed that (Footnote 1) the identically zero expression in Poisson brackets

$$\{x^3, p^3\}_P + 3\{xp^2, x^2p\}_P = 0 , \tag{16.3}$$

becomes

$$\frac{1}{i\hbar}[\hat{x}^3, \hat{p}^3] + \frac{3}{i\hbar}\left[\frac{\hat{x}\hat{p}^2 + \hat{p}^2\hat{x}}{2}, \frac{\hat{x}^2\hat{p} + \hat{p}\hat{x}^2}{2}\right] = -3\hbar^2 , \tag{16.4}$$

thus exhibiting a "deficiency" of $-3\hbar^2$.

Problem 16.1. *Check Eqs. (16.3) and (16.4).*

One may think that the deficiency for polynomials of order three or above is caused by the ordering convention, but that is not true. The example below poses no ordering dilemma, yet it produces the same quantization deficiency. Indeed

$$\{x^3, p^3\}_P + \frac{1}{12}\{\{p^2, x^3\}_P, \{x^2, p^3\}_P\}_P = 0 , \tag{16.5}$$

yields[2]:

$$\frac{1}{i\hbar}[\hat{x}^3, \hat{p}^3] + \frac{1}{12} \times \frac{1}{i\hbar}\left[\frac{1}{i\hbar}[\hat{p}^2, \hat{x}^3], \frac{1}{i\hbar}[\hat{x}^2, \hat{p}^3]\right] = -3\hbar^2 . \tag{16.6}$$

---

[2]C. K. Zachos, D. B. Fairlie, and T. L. Curtright, *Quantum Mechanics in Phase Space*, World Scientific (2005).

These two examples illustrate that the Poisson bracket alone does not provide enough guidance for a coherent quantization mechanism.

Via a novel mathematical formalism known as "deformation quantization", or "phase space quantization", we can obtain a deficiency-free quantization scheme, where not the Poisson bracket, but rather its "deformation" $\{A, B\}_P \mapsto \{A, B\}_P + \mathcal{O}(\hbar^2)$ gives the correct answer. Phase space quantum mechanics formalism uses complex functions acting on phase space instead of operators acting on a Hilbert space. These functions are endowed with a novel multiplication rule, the "star-product", which is non-commutative, associative, and hermitian. The formalism maps operators $\hat{A}$ of quantum mechanics to complex functions $A(x, p)$, and vice versa. As Dirac observed[3]

> Two points of view may be mathematically equivalent and you may think for that reason if you understand one of them you need not bother about the other and can neglect it. But it may be that one point of view may suggest a future development which another point does not suggest, and although in their present state the two points of view are equivalent they may lead to different possibilities for the future. Therefore, I think that we cannot afford to neglect any possible point of view for looking at Quantum Mechanics and in particular its relation to Classical Mechanics.

Therefore, in the following, we will present exactly such a new point of view. We begin by introducing the Groenewold-Moyal product between two functions, also known as the star-product.

16.1.1   The Star Product

To any quantum operator $\hat{A}$ acting on a Hilbert space, we associate a complex phase space function $A(x, p)$ using the Weyl transform $\mathscr{W}(\hat{A})$ (Footnote 1), which in the coordinate $|x\rangle$ basis reads

$$\mathscr{W}(\hat{A}) \equiv A(x, p) = \hbar \int dy \, e^{-ipy} \left\langle x + \frac{\hbar y}{2} \left| \hat{A} \right| x - \frac{\hbar y}{2} \right\rangle. \qquad (16.7)$$

All integrals run from $-\infty$ to $+\infty$, unless specifically restricted. The complex function $A(x, p) = \mathscr{W}(\hat{A})$ is also known as the "Weyl's symbol" of $\hat{A}$.

Problem 16.2. *Prove that* $\mathscr{W}(\hat{x}) = x$ *and* $\mathscr{W}(\hat{p}) = p$, *respectively*.

---

[3]P. A. M. Dirac, "The Relation of Classical to Quantum Mechanics", 2nd *Can. Math. Congress, Vancover 1949*, U. Toronto Press (1951).

Without additional constraints, the reciprocal map $\mathscr{W}^{-1}$ is not unique. A sufficient condition for the uniqueness of $\mathscr{W}^{-1}$ is obtained by choosing an operator ordering.[4] We choose Weyl's ordering, which prescribes symmetrically ordered polynomials in $\hat{x}$ and $\hat{p}$. I.e., $\mathscr{W}^{-1}(xp) = (\hat{x}\hat{p} + \hat{p}\hat{x})/2$. Thus, we write the reciprocal map as the Fourier-like transform

$$\hat{A}(\hat{x}, \hat{p}) \equiv \mathscr{W}^{-1}(A(x,p)) \tag{16.8}$$
$$= \frac{1}{(2\pi)^2} \int du\, dv\, dx\, dp\, A(x,p)\, e^{i(\hat{p}-p)u + i(\hat{x}-x)v} \ .$$

With this definition, $\mathscr{W}^{-1}(x) = \hat{x}$ and $\mathscr{W}^{-1}(p) = \hat{p}$. More generally, we have $\mathscr{W}^{-1}((ax + bp)^n) = (a\hat{x} + b\hat{p})^n$.

Next, we define the phase space product between the Weyl symbols. This product should correspond to the product of the quantum operators in the Hilbert space. Groenewold[5] and Moyal[6] showed that using Weyl's ordering assumption, the product of functions on phase space can be written as

$$A(x,p) * B(x,p) = A(x,p)\, e^{\frac{i\hbar}{2}\left(\overleftarrow{\partial_x}\, \overrightarrow{\partial_p} - \overleftarrow{\partial_p}\, \overrightarrow{\partial_x}\right)} B(x,p) \ , \tag{16.9}$$

where the arrow indicates the direction in which the derivative acts. Using (16.7) one can prove that indeed $\mathscr{W}(\hat{A}\hat{B}) = A(x,p) * B(x,p)$. The Groenewold-Moyal product is also known as the "star-product" or $*$-product of $A(x,p)$ and $B(x,p)$.

Note that in Eq. (16.9) we have used the exponential notation for an operator, where

$$e^{\frac{i\hbar}{2}\left(\overleftarrow{\partial_x}\, \overrightarrow{\partial_p} - \overleftarrow{\partial_p}\, \overrightarrow{\partial_x}\right)} = 1 + \frac{i\hbar}{2}\left(\overleftarrow{\partial_x}\, \overrightarrow{\partial_p} - \overleftarrow{\partial_p}\, \overrightarrow{\partial_x}\right) + \cdots \ . \tag{16.10}$$

We see thus that the $*$-product can be written as a series in powers of $\hbar$

$$A(x,p) * B(x,p) = A(x,p)B(x,p)$$
$$+ \frac{i\hbar}{2}\left(\frac{\partial A(x,p)}{\partial x}\frac{\partial B(x,p)}{\partial p} - \frac{\partial A(x,p)}{\partial p}\frac{\partial B(x,p)}{\partial x}\right) \tag{16.11}$$
$$+ \mathcal{O}(\hbar^2) \ .$$

The $*$-product can be expressed in several equivalent forms, such as

$$A(x,p) * B(x,p) = e^{\frac{i\hbar}{2}\left(\partial_x\, \partial_{p'} - \partial_p\, \partial_{x'}\right)} A(x,p)B(x',p') \ , \tag{16.12}$$

---

[4]J. Tosiek and M. Przanowski, "Weyl-Wigner-Moyal Formalism. I. Operator Ordering", *Acta Phys. Pol.* **26**, 1703 (1995).

[5]H. J. Groenewold, "On the Principles of Elementary Quantum Mechanics", *Physica* **12**, 405 (1946).

[6]J. E. Moyal, "Quantum Mechanics as a Statistical Theory", *Proc. Camb. Phil. Soc.* **45**, 99 (1949).

calculated at $(x', p') = (x, p)$; or,

$$A(x,p) * B(x,p) = A(x + \frac{i\hbar}{2}\overrightarrow{\partial_p}, p - \frac{i\hbar}{2}\overrightarrow{\partial_x}) B(x,p) \qquad (16.13)$$

$$= A(x,p) B(x - \frac{i\hbar}{2}\overleftarrow{\partial_p}, p + \frac{i\hbar}{2}\overleftarrow{\partial_x}),$$

which are known also as Bopp shifts.[7] It can also be represented as an integral, such as the Fourier representation[8]

$$A(x,p) * B(x,p) = \frac{1}{(\pi\hbar)^2} \int dx_1 dp_1 dx_2 dp_2 \, A(x_1, p_1) B(x_2, p_2)$$

$$\times \exp\left\{ -\frac{2i}{\hbar} [p(x_1 - x_2) + x(p_2 - p_1) + (x_2 p_1 - x_1 p_2)] \right\}, \qquad (16.14)$$

or as the alternate integral[9]

$$A(x,p) * B(x,p) = \frac{1}{(\pi\hbar)^2} \int dx_1 dp_1 dx_2 dp_2 \, A(x+x_1, p+p_1) B(x+x_2, p+p_2)$$

$$\times \exp\left\{ \frac{2i}{\hbar} (x_1 p_2 - x_2 p_1) \right\}. \qquad (16.15)$$

We conclude this definition by emphasizing the cyclic, trace-like properties[10] of the *-product:

$$\int dx dp \, A(x,p) * B(x,p) = \int dx dp \, A(x,p) B(x,p) = \int dx dp \, B(x,p) * A(x,p). \qquad (16.16)$$

Thus, we have constructed a consistent way of moving back and forth from quantum operators acting on a Hilbert space, to complex functions acting on phase space:

$$\mathscr{W}(\hat{A}\hat{B}) = \mathscr{W}(\hat{A}) * \mathscr{W}(\hat{B}), \qquad (16.17)$$

$$\mathscr{W}^{-1}(A * B) = \mathscr{W}^{-1}(A) \, \mathscr{W}^{-1}(B). \qquad (16.18)$$

This novel product is in agreement with the product rules for the operators in quantum mechanics. It is non-commutative

$$A * B \neq B * A, \qquad (16.19)$$

---

[7]F. Bopp, *Werner Heisenberg und die Physik unserer Zeit*, Vieweg, Braunschweig (1961); M. de Gosson, *Symplectic Geometry and Quantum Mechanics*, Birkhauser, Basel (2006).

[8]J. von Neumann, "Die Eindeutigkeit der Schrödingerschen Operatoren", *Math. Ann.* **104**, 570 (1931);

G. Baker, "Formulation of Quantum Mechanics Based on the Quasi-Probability Distribution Induced on Phase Space", *Phys. Rev.* **109**, 2198 (1958).

[9]M. Hillery, R. F. O'Connell, M. O. Scully, and E. P. Wigner, "Distribution Functions in Physics: Fundamentals", *Phys. Rep.* **106**, 121 (1983).

[10]T. Curtright, T. Uematsu, and C. Zachos, "Generating all Wigner functions", *J. Math. Phys.* **42**, 2396 (2001).

associative

$$A * (B * C) = (A * B) * C ; \tag{16.20}$$

and hermitian

$$\overline{A * B} = \overline{B} * \overline{A} , \tag{16.21}$$

where the bar denotes complex conjugation.

For example, using Bopp shifts (16.13), we obtain:

$$x * p \equiv \left(x + \frac{i\hbar}{2}\overrightarrow{\partial_p} \right) p = x\,p + \frac{i\hbar}{2} , \tag{16.22}$$

$$p * x \equiv \left(p - \frac{i\hbar}{2}\overrightarrow{\partial_x} \right) x = p\,x - \frac{i\hbar}{2} . \tag{16.23}$$

Then the $*$-commutator, or the "Moyal bracket" of x and p is

$$[x, p]_* \equiv x * p - p * x = i\hbar , \tag{16.24}$$

which is consistent with the canonical commutation relation $[\hat{x}, \hat{p}] = i\hbar$. We thus see that using the $*$-product as multiplication rule for Weyl symbols on phase space, the Weyl transform provides a homomorphism[11] between the Moyal bracket and the commutator of operators.

16.1.2 The Moyal Bracket and the Correspondence Principle

We are now in a position to show that it is the Moyal bracket that provides a consistent path to quantization starting from classical mechanics, not the Poisson bracket.

The $*$-product and the Moyal bracket are $\hbar$ deformations[12] of the usual commutative product of functions $A(x,p)\,B(x,p)$, and of the Poisson bracket $\{A(x,p), B(x,p)\}_P = \partial_x A(x,p)\,\partial_p B(x,p) - \partial_p A(x,p)\,\partial_x B(x,p)$. Using Eq. (16.11) we can write

$$A(x,p) * B(x,p) = A(x,p)\,B(x,p) + \mathcal{O}(\hbar) , \tag{16.25}$$

$$\frac{1}{i\hbar}[A(x,p), B(x,p)]_* = \{A(x,p), B(x,p)\}_P + \mathcal{O}(\hbar^2) . \tag{16.26}$$

---

[11] A homomorphism is a map from one set to another that preserves in the second set the relations between the elements of the first.

[12] F. A. Berezin, "Feynman path integrals in a phase space", *Sov. Phys. Usp.* **23**, 763 (1980);

A. C. Hirshfeld and P. Henselder, "Deformation quantization in the teaching of quantum mechanics", *Am. J. Phys.* **70**, 537 (2002);

J. Hancock, M. A. Walton, and B. Wynder, "Quantum mechanics another way", *Eur. J. Phys.* **25**, 525–534 (2004).

Thus, this gives the name "deformation quantization." Property (16.26) implies that

$$\lim_{\hbar \to 0} \frac{1}{i\hbar}[A(x,p), B(x,p)]_* = \{A(x,p), B(x,p)\}_P , \qquad (16.27)$$

showing that in the classical limit one recovers the Poisson bracket.

The Moyal bracket obeys Jacobi's identity

$$[[A, B]_*, C]_* + [[C, A]_*, B]_* + [[B, C]_*, A]_* = 0 , \qquad (16.28)$$

as well as the Leibniz rule

$$[A, B * C]_* = [A, B]_* * C + B * [A, C]_* . \qquad (16.29)$$

Thus, one can endow the space of Weyl symbols not only with a Lie algebra structure with respect to the Moyal bracket, but also with an inner derivative.

An analogous statement holds true for the commutator algebra of quantum operators, but not for the Poisson-bracket algebra. One can ask whether this is an impasse due to the way in which the Weyl transform was defined, or whether this signals a deeper result. In other words, can one invent a different definition for the Weyl transform that preserves the Lie algebra structure induced by the Poisson bracket? Groenewold and van Hove showed[13] that there is no invertible linear map from all functions on phase space $A(x,p), B(x,p), \ldots$ to hermitian operators in Hilbert space $\hat{A}, \hat{B}, \ldots$ that will preserve the Poisson bracket structure; i.e., $\mathscr{W}^{-1}(\{A, B\}_P) = \frac{1}{i\hbar}[\mathscr{W}^{-1}(A), \mathscr{W}^{-1}(B)]$.

As we showed at the beginning of this chapter, using Poisson brackets as a starting point for quantization works only in the special cases where the series (16.25) terminates after the first order in $\hbar$. This is true for functions at most quadratic in p and x.

Through the Weyl invertible correspondence map, Eq. (16.7), the Hilbert space of Hermitian operators, with the operator commutator, has as a counterpart the algebra of Weyl symbols with the Moyal bracket. The correspondence between the quantum operators acting on a Hilbert space and the complex functions acting on phase space is illustrated in Fig. 16.1.

Up to an isomorphism, the Lie algebra generated by the Moyal bracket is the unique associative one-parameter ($\hbar$) deformation of the Poisson

[13]H. J. Groenewold, "On the Principles of Elementary Quantum Mechanics", *Physica* **12**, 405 (1946);

L. van Hove, "Sur certaines représentations unitaires", *Proc. R. Acad. Sci. Belgium* **26**, 1 (1951).

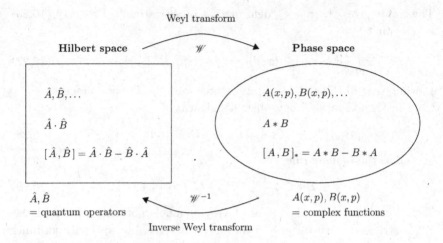

Fig. 16.1 The correspondence between the quantum operators acting on a Hilbert space and the complex functions acting on phase space.

bracket.[14] This uniqueness is extended (up to an isomorphism) to the $*$-product.

Thus, it is the Moyal bracket that gives the correct correspondence principle for the quantization scheme.[15]

---

[14] J. Vey, *Comment. Math. Helv.*, "Déformation du crochet de poisson sur une variété symplectique", **50**, 421 (1975);

M. Flato, A. Lichnerowicz, and A. Sternheimer, "Deformations of Poisson brackets, Dirac brackets and applications", *J. Math. Phys.* **17**, 1754 (1976);

F. Bayen, M. Flato, C. Fronsdal, A. Lichnerowicz, and D. Sternheimer, "Deformation Theory and Quantization. I. Deformations of Symplectic Structures", *Ann. of Phys.* **111**, 61 (1978);

W. Arveson, "Quantization and the uniqueness of invariant structures", *Comm. Math. Phys.* **89**, 77 (1983);

M. de Wilde and P. Lecomte, "Existence of star-products and of formal deformations of the Poisson Lie algebra of arbitrary symplectic manifolds", *Lett. Math. Phys.* **7**, 487 (1983);

E. Gozzi and M. Reuter, "A proposal for a differential calculus in quantum mechanics", *Int. J. Mod. Phys. A* **9**, 2191 (1994).

[15] H. J. Groenewold, "On the Principles of Elementary Quantum Mechanics", *Physica* **12**, 405 (1946);

C. Fronsdal, "Some Ideas About Quantization", *Rep. Math. Phys.* **15**, 111 (1978);

C. K. Zachos, D. B. Fairlie, and T. L. Curtright, *Quantum Mechanics in Phase Space*, World Scientific (2005).

16.1.3 *The Wigner Function*

Now that we have a coherent way of modeling quantum operators and their products, we turn our attention to modeling the quantum state. A pure state is described by the ket vector $|\psi\rangle$, or equivalently, by the corresponding density operator $\hat{\rho} = |\psi\rangle\langle\psi|$. For simplicity, we will consider here only pure states, because the generalization to mixed states is straightforward. Let us denote by $P(x,p)$ the normalized Weyl symbol of the density operator $\hat{\rho}$. We have

$$P(x,p) \equiv \frac{1}{2\pi\hbar}\, \mathscr{W}(\hat{\rho}) = \frac{1}{2\pi}\int dy\, e^{-ipy}\left\langle x + \frac{\hbar y}{2}\,\Big|\,\psi\right\rangle\langle\psi\,|\, x - \frac{\hbar y}{2}\right\rangle, \quad (16.30)$$

or

$$P(x,p) = \frac{1}{2\pi}\int dy\, e^{-ipy}\, \psi\left(x - \frac{\hbar y}{2}\right)\overline{\psi}\left(x + \frac{\hbar y}{2}\right). \quad (16.31)$$

$P(x,p)$ is the Wigner function,[16] and will play a central role in the deformation quantization technique. Here we list some of the properties of the Wigner function[17] (not all of which we shall use).

(i) $P(x,p)$ is real.

(ii)

$$\int dp\, P(x,p) = |\psi(x)|^2 = \langle x\,|\,\hat{\rho}\,|x\rangle \quad (16.32)$$

$$\int dx\, P(x,p) = |\psi(p)|^2 = \langle p\,|\,\hat{\rho}\,|p\rangle \quad (16.33)$$

$$\int dx dp\, P(x,p) = 1\,. \quad (16.34)$$

---

[16] E. P. Wigner, "On the Quantum Correction For Thermodynamic Equilibrium", *Phys. Rev.* **40**, 749 (1932).

[17] G. J. Iafrate, H. L. Grubin, and D. K. Ferry, "The Wigner Distribution Function", *Phys. Lett. A* **87**, 145 (1982);

V. I. Tatarskii, "The Wigner representation of quantum mechanics", *Sov. Phys. Usp.* **26**, 311 (1983);

M. Hillery, R. F. O'Connell, M. O. Scully, and E. P. Wigner, "Distribution Functions in Physics: Fundamentals", *Phys. Rep.* **106**, 121 (1983);

F. J. Narcowich and R. F. O'Connell, "Necessary and sufficient conditions for a phase-space function to be a Wigner distribution", *Phys. Rev. A* **34**, 1 (1986);

F. J. Narcowich, "Conditions for the convolution of two Wigner distributions to be itself a Wigner distribution", *J. Math. Phys.* **29**, 2036 (1988);

R. L. Hudson, "When is the Wigner quasi-probability density non-negative?" *Rep. Math. Phys.* **6**, 249 (1974);

N. C. Dias and J. N. Prata, "Wigner functions with boudaries", *J. Math. Phys.* **43**, 4602 (2002).

(iii) If $P_\psi(x,p)$ and $P_\phi(x,p)$ correspond to the states $\psi(x)$ and $\phi(x)$ respectively, then

$$\left| \int dx\, \overline{\psi}(x)\phi(x) \right|^2 = 2\pi\hbar \int dx\, dp P_\psi(x,p) P_\phi(x,p) \ . \tag{16.35}$$

The last property has two interesting consequences. If $\psi(x) = \phi(x)$ then

$$\int dx\, dp P_\psi^2(x,p) = \frac{1}{2\pi\hbar} \ , \tag{16.36}$$

and, if we choose $\psi(x)$ orthogonal to $\phi(x)$, we get

$$\int dx\, dp\, P_\psi(x,p) P_\phi(x,p) = 0 \ . \tag{16.37}$$

Equation (16.37) implies that $P(x,p)$ cannot be positive everywhere. Because it can also take negative values, the Wigner function is also known as a "pseudo-distribution".

(iv) $P(x,p)$ is the only pseudo-distribution for which each Galilean transformation corresponds to the same Galilean transformation of the quantum mechanical wave function.[18] I.e., if $\psi(x) \mapsto \psi(x+a)$, then $P(x,p) \mapsto P(x+a,p)$, and if $\psi(x) \mapsto \exp(ip'x/\hbar)\,\psi(x)$, then $P(x,p) \mapsto P(x,p-p')$.

(v) If $\psi(x) \mapsto \psi(-x)$, then $P(x,p) \mapsto P(-x,-p)$, and if $\psi(x) \mapsto \overline{\psi}(x)$ then $P(x,p) \mapsto P(x,-p)$.

To describe the time evolution of the system in phase space, let us take the time derivative of (16.30), and use Schrödinger's equation together with its conjugate. We get

$$\frac{\partial}{\partial t} P(x,p) = \frac{1}{2\pi i\hbar} \int dy\, e^{-ipy} \left\langle x + \frac{\hbar y}{2} \left| (\hat{H}\hat{\rho} - \hat{\rho}\hat{H}) \right| x - \frac{\hbar y}{2} \right\rangle$$

$$= \frac{1}{i\hbar} \mathscr{W}([\hat{H}, \hat{\rho}/2\pi\hbar]) \ ,$$

or

$$i\hbar \frac{\partial}{\partial t} P(x,p) = [H(x,p), P(x,p)]_* \ , \tag{16.38}$$

where $H(x,p)$ is the Weyl symbol of the hamiltonian $\hat{H}$. Equation (16.38) is mirroring the time evolution of the density operator $i\hbar \frac{\partial}{\partial t}\hat{\rho} = [\hat{H}, \hat{\rho}]$. The time evolution of the Wigner function can also be symbolically written using

---

[18]E. P. Wigner, *Perspectives in Quantum Theory*, Dover, NY (1979).

the sine notation,[19] which emphasizes the non-linear deformation involved by the Moyal bracket

$$\frac{\partial}{\partial t} P(x,p) = \frac{2}{\hbar} \sin \left\{ \frac{\hbar}{2} \left(\frac{\partial}{\partial x} \frac{\partial}{\partial p_1} - \frac{\partial}{\partial x_1} \frac{\partial}{\partial p} \right) \right\} H(x,p) \, P(x_1,p_1)$$

calculated at $(x_1, p_1) = (x, p)$.

For stationary states, $\frac{\partial}{\partial t} P(x,p) = 0$, hence the hamiltonian $*$-commutes with the Wigner function

$$[H(x,p), P(x,p)]_* = 0 \ . \tag{16.39}$$

In the stationary case we get more constraints on the Wigner function. If we apply the Weyl transform to $\hat{H}\hat{\rho} = E\hat{\rho}$, we obtain the $*$-eigenvalue equation[20]

$$H(x,p) * P(x,p) = E \, P(x,p) \ , \tag{16.40}$$

where E is the energy of the system. Hermiticity implies that $\hat{\rho}\hat{H} = E\hat{\rho}$, hence the symmetrical relation holds true as well

$$P(x,p) * H(x,p) = E \, P(x,p) \ . \tag{16.41}$$

Note that if P_E and $P_{E'}$ correspond to the eigenenergies E and E', then due to the associativity of the $*$-product, we have

$$P_E * H * P_{E'} = E \, P_E * P_{E'} = E' \, P_E * P_{E'} \ . \tag{16.42}$$

If $E \neq E'$, it follows that $P_E * P_{E'} = 0$. In the general case we have $P_E * P_{E'} = \mathscr{W}(\hat{\rho}_E \, \hat{\rho}_{E'})/(2\pi\hbar)^2$ which yields the "orthogonality–idempotence" relation of Wigner functions[21]

$$P_E * P_{E'} = \frac{\delta_{E,E'}}{2\pi\hbar} \, P_E \ . \tag{16.43}$$

---

[19]M. S. Bartlett and J. E. Moyal, "The Exact Transition Probabilities of Quantum-Mechanical Oscillators Calculated by the Phase-Space Method", *Proc. Camb. Phil. Soc.* **45**, 545 (1949).

[20]D. Fairlie, "The formulation of quantum mechanics in terms of phase space functions", *Proc. Camb. Phil. Soc.* **60**, 581 (1964);

T. Curtright, D. Fairlie, and C. Zachos, "Features of time-independent Wigner functions", *Phys. Rev. D* **58**, 025002 (1998);

N. C. Dias and J. N. Prata, "Formal solutions of stargenvalue equations", *Ann. of Phys.* **311**, 120 (2004);

M. de Gosson and F. Luef, "A New Approach to the *-Genvalue Equation", *Lett. Math. Phys.* **85**, 173 (2008).

[21]N. C. Dias and J. N. Prata, "Admissible states in quantum phase space", *Ann. of Phys.* **313**, 110 (2004).

In deformation quantization, $P(x,p)$ plays an analogous role to the probability density function in classical statistical mechanics. Namely, the average of $\hat{A}$ in state $|\psi\rangle$ is given by[22]

$$\langle \psi | \hat{A} | \psi \rangle \equiv \langle A(x,p) \rangle = \int dx\, dp\, P(x,p) * A(x,p) \ . \tag{16.44}$$

Note the resemblance to statistical mechanics. However there is a major difference in the case of the deformation quantization formalism: the function $P(x,p)$ is not a probability distribution in the statistical sense, because it can take negative values.

16.1.4 *Example: The Harmonic Oscillator*

As an example of the concepts we have introduced so far, let us consider the case of the simple harmonic oscillator. For convenience we take $2m = 1, \omega = 2$. Then the hamiltonian operator of the harmonic oscillator reads

$$\hat{H} = \hat{p}^2 + \hat{x}^2 \ , \tag{16.45}$$

with the corresponding Weyl symbol

$$H(x,p) = p^2 + x^2 \ . \tag{16.46}$$

Let us write the $*$-eigenvalue Eq. (16.40) for this case. We obtain

$$\left(p^2 + x^2 \right) * P(x,p) = E\, P(x,p) \ , \tag{16.47}$$

which, using Bopp shifts representation (16.13), becomes

$$\left[\left(p - \frac{i\hbar}{2}\overrightarrow{\partial_x} \right)^2 + \left(x + \frac{i\hbar}{2}\overrightarrow{\partial_p} \right)^2 \right] P(x,p) = EP(x,p) \ . \tag{16.48}$$

After some algebra we arrive at

$$\left[p^2 + x^2 + i\hbar(x\partial_p - p\partial_x) - \frac{\hbar^2}{4}(\partial_p^2 + \partial_x^2) - E \right] P(x,p) = 0 \ . \tag{16.49}$$

Problem 16.3. *Prove Eq. (16.49).*

If we separate the imaginary and the real parts, we obtain

$$(x\partial_p - p\partial_x)\, P(x,p) = 0 \ , \tag{16.50}$$

$$\left[p^2 + x^2 - \frac{\hbar^2}{4}(\partial_p^2 + \partial_x^2) - E \right] P(x,p) = 0 \ . \tag{16.51}$$

---

[22] J. E. Moyal, "Quantum Mechanics as a Statistical Theory", *Proc. Camb. Phil. Soc.* **45**, 99 (1949).

The symmetry of Eq. (16.50) indicates that $P(x, p)$ depends on only one variable, which is a symmetric combination of x and p. Thus, in (16.51) we make the change of variables $t = 2(p^2 + x^2)/\hbar$ and write $P(t) = e^{-t/2} L(t)$. We get

$$\left[t \partial_t^2 + (1 - t) \partial_t + \left(\frac{E}{2\hbar} - \frac{1}{2} \right) \right] L(t) = 0 \ . \tag{16.52}$$

For $L(t)$ to be normalizable, the zero derivative term of (16.52) must be a positive integer

$$\left(\frac{E}{2\hbar} - \frac{1}{2} \right) = n \ , \quad n = 0, 1, 2, \dots \tag{16.53}$$

This assures that the series solution of $L(t)$ terminates at a given rank, hence $L(t)$ is finite and $e^{-t/2} L(t)$ is normalizable. But condition (16.53) yields exactly the quantization formula for the energy of the harmonic oscillator ($\omega = 2$)

$$E_n = 2\hbar \left(n + \frac{1}{2} \right) \ , \quad n = 0, 1, 2, \dots \tag{16.54}$$

Consequently, Eq. (16.52) subject to the constraint (16.53), becomes the differential equation of Laguerre polynomials[23] $L_n(t)$. Thus, we obtain the analytic expression for the Wigner functions $P_n(x, p)$ of the harmonic oscillator

$$P_n(x, p) = \frac{(-1)^n}{\pi} e^{-\frac{p^2 + x^2}{\hbar}} L_n \left(\frac{p^2 + x^2}{\hbar/2} \right) \ , \quad n = 0, 1, 2, \dots \tag{16.55}$$

where L_n is the n-th Laguerre polynomial. The harmonic oscillator is one of the few systems where the Wigner functions are completely known analytically.[24] In Fig. 16.2 we illustrate the Wigner function for $n = 5$. Note the circular $x \leftrightarrow p$ symmetry of the solution, as reflected in the Eq. (16.50).

Finally, let us recall (Chapter 3) the algebraic solution induced by the factorization of the hamiltonian in terms of creation and annihilation operators $\hat{a}^+$ and $\hat{a}^-$. This method will naturally segue into the factorization of a general hamiltonian, which is the crux of the SUSYQM techniques. We write the hamiltonian as

$$\hat{H} = 2\hbar \left(\hat{a}^+ \hat{a}^- + 1/2 \right) \ , \tag{16.56}$$

---

[23]T. Chow, *Mathematical Methods for Physicists -A Concise Introduction*, Cambridge Univ. Press (2000).
[24]E. A. Akhundova, V. V. Dodonov, V. I. Man'ko, "Wigner Functions of Quadratic Systems", *Physica* **115 A**, 215 (1982).

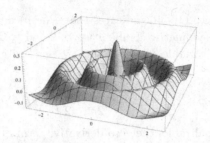

Fig. 16.2 The Wigner function $P_5(x,p)$ for the simple harmonic oscillator.

where $\hat{a}^- = (i\hat{p} + \hat{x})/\sqrt{2\hbar}$ and $\hat{a}^+ \equiv (\hat{a}^-)^\dagger = (-i\hat{p} + \hat{x})/\sqrt{2\hbar}$. Then $[\hat{a}^-, \hat{a}^+] = 1$, and after algebraic manipulations we obtain

$$\hat{H}|n\rangle = 2\hbar(n+1/2)|n\rangle , \quad n = 0, 1, 2, \ldots \tag{16.57}$$

$$\hat{a}^+|n\rangle = \sqrt{n+1}|n+1\rangle \tag{16.58}$$

$$\hat{a}^-|n\rangle = \sqrt{n}|n-1\rangle . \tag{16.59}$$

The harmonic oscillator wave functions are obtained by projecting the recursion relations

$$|n\rangle = \frac{(\hat{a}^+)^n}{\sqrt{n!}}|0\rangle , \tag{16.60}$$

onto the $|x\rangle$ basis, where the starting point is given by $\hat{a}^-|0\rangle = 0$.

In the phase space picture these relations become

$$H(x,p) = p^2 + x^2 = 2\hbar(a^+ * a^- + 1/2) , \tag{16.61}$$

where the annihilation and creation functions $a^-(x,p)$ and $a^+(x,p)$ are

$$a^- = (ip+x)/\sqrt{2\hbar} , \quad a^+ \equiv \overline{(a^-)} = (-ip+x)/\sqrt{2\hbar} . \tag{16.62}$$

Their Moyal bracket is

$$[a^-, a^+]_* = 1 . \tag{16.63}$$

Then, the $*$-eignevalue problem becomes

$$H(x,p) * P_n(x,p) = 2\hbar(n+1/2)P_n(x,p) , \tag{16.64}$$

where the Wigner function $P_n(x,p)$ is determined from the recursion relations[25]

$$P_n(x,p) = \frac{1}{n!}(a^+*)^n P_0(x,p)(*a^-)^n , \tag{16.65}$$

[25]T. Curtright, T. Uematsu, and C. Zachos, "Generating all Wigner functions", *J. Math. Phys.* **42**, 2396 (2001).

using as starting point

$$a^- * P_0(x, p) = 0 . \tag{16.66}$$

A direct calculation (Footnote 25) shows that the algebraic method recovers the known form for the Wigner function (16.55).

This algebraic technique will be extended to a general hamiltonian, by applying SUSYQM concepts to phase space formalism.

16.2 Supersymmetric Quantum Mechanics in Operator Form

We briefly rewrite the main ideas of SUSYQM in operator form, using the position $\hat{x}$ and momentum $\hat{p}$ operators. That would allow us to seamlessly express the supersymmetric ideas in the deformation quantization formalism, by applying the Weyl's transformation.

Given the hamiltonian $\hat{H}_- = \frac{\hat{p}^2}{2m} + \hat{V}_-(\hat{x})$ with the eigenenergies E_n (where $E_0 = 0$), let us denote its eigenvectors by $|n; -\rangle$. Then

$$\hat{H}_-|n; -\rangle = E_n^- |n; -\rangle , \quad E_0^- = 0 . \tag{16.67}$$

The hamiltonian $\hat{H}_-$ can be factorized[26] as

$$\hat{H}_- = \hat{A}^+ \hat{A}^- , \tag{16.68}$$

where

$$\hat{A}^- = \frac{i\hat{p}}{\sqrt{2m}} + \hat{W}(\hat{x}) , \quad \hat{A}^+ \equiv (\hat{A}^-)^\dagger = \frac{-i\hat{p}}{\sqrt{2m}} + \hat{W}(\hat{x}) . \tag{16.69}$$

The superpotential operator $\hat{W}$ is is related to $\hat{V}_-(\hat{x})$ by

$$\hat{V}_-(\hat{x}) = \hat{W}^2(\hat{x}) - \frac{\hbar}{\sqrt{2m}} \hat{W}'(\hat{x}) . \tag{16.70}$$

Equation (16.70) becomes a Riccati equation when projected on the $|x\rangle$ basis, thus allowing for finding $W(x)$. A sufficient condition for $E_0^- = 0$ is given by

$$\hat{A}|0; -\rangle = 0 , \tag{16.71}$$

which yields the explicit action of $\hat{W}(\hat{x})$ on the ground state eigenvector $|0; -\rangle$

$$\hat{W}(\hat{x})|0; -\rangle = -\frac{i\hat{p}}{\sqrt{2m}}|0; -\rangle . \tag{16.72}$$

---

[26]E. Schrödinger, "A Method of Determining Quantum-Mechanical Eigenvalues a Eigenfunctions", *Proc. R. Irish Acad.* **46 A**, 9 (1940).

If we project (16.72) on the $|x\rangle$ basis, we can express $W(x)$ in terms of the ground state eigenfunction $\psi_0^-(x) = \langle x|\, 0; -\rangle$ as[27]

$$W(x) = -\frac{\hbar}{\sqrt{2m}} \frac{\psi_0^{-\,\prime}(x)}{\psi_0^-(x)} \ . \tag{16.73}$$

By interchanging the operators $\hat{A}^+$ and $\hat{A}^-$ we generate the supersymmetric partner hamiltonian $\hat{H}_+ = \hat{A}^- \hat{A}^+$, which corresponds to a new potential operator $\hat{V}_+(\hat{x})$:

$$\hat{H}_+ = \frac{\hat{p}^2}{2m} + V_+(\hat{x}); \quad V_+(\hat{x}) \equiv \hat{W}^2(\hat{x}) + \frac{\hbar}{\sqrt{2m}}\hat{W}'(\hat{x}) \ . \tag{16.74}$$

We denote by E_n^+ its eigenenergies: $\hat{H}_+|n; +\rangle = E_n^+ |n; +\rangle$. The two hamiltonians $\hat{H}_-$ and $\hat{H}_+$ are supersymmetric partner hamiltonians. Their eigenvectors and eigenenergies are related:

$$E_n^+ = E_{n+1}^-; \quad E_0^- = 0 \ , \tag{16.75}$$

$$|n; +\rangle = \frac{1}{\sqrt{E_{n+1}^-}}\, \hat{A}^- \,|n + 1; -\rangle \ , \tag{16.76}$$

$$|n + 1; -\rangle = \frac{1}{\sqrt{E_n^+}}\, \hat{A}^+ \,|n; +\rangle \ . \tag{16.77}$$

Thus, with the exception of the ground state, the supersymmetric partner potentials $\hat{H}_-$ and $\hat{H}_+$ share the same energy spectrum (isospectrality), and have interconnected eigenvector sets. As also seen in Chapter 4, Eqs. (16.75)–(16.77) together with Eq. (16.71) represent a quick algebraic way of finding the complete spectrum and eigenvectors of $\hat{H}_+$ if we know the spectrum and eigenvectors of $\hat{H}_-$. In this way, if $\hat{H}_-$ corresponds to a very simple system, we have an elegant solution for solving more complicated cases associated to $\hat{H}_+$.

In the special case of shape invariance, the potential operators $\hat{V}_-(\hat{x})$ and $\hat{V}_+(\hat{x})$ obey the additional constraint

$$\hat{V}_+(\hat{x}; a_0) = \hat{V}_-(\hat{x}; a_1) + R(a_0) \ , \tag{16.78}$$

where the last term multiplies the identity operator, which was omitted in order to simplify writing. These operators differ only by a set of parameters (modeling the strength of the interaction) a_0, $a_1 = f(a_0)$, and a constant term $R(a_0)$, independent of $\hat{x}$.

---

[27] E. Gozzi, "Ground-state wave-function representation", *Phys. Lett. B* **129**, 432 (1983).

Let us rewrite the shape invariance condition as

$$H_+(\hat{x}, \hat{p}; a_0) + g(a_0) = H_-(\hat{x}, \hat{p}; a_1) + g(a_1) , \qquad (16.79)$$

where $R(a_0) = g(a_1) - g(a_0)$. For all values of n we have

$$E_n^+(a_0) = E_n^-(a_1) + g(a_1) - g(a_0) , \qquad (16.80)$$

$$|n, a_0; +\rangle = |n, a_1; -\rangle . \qquad (16.81)$$

For normalizable ground states (unbroken SUSY), the ground state energy of $H_-(\hat{x}, \hat{p}; a_i)$ is zero, $E_0^-(a_i) = 0$, for each iteration of the parameter $a_i = f(a_{i-1})$. By successively using shape invariance, Eq. (16.80), and the isospectrality (16.75) of the partner hamiltonians, in conjunction with the unbroken supersymmetry condition, we get

$$E_0^-(a_0) = 0,$$
$$E_1^-(a_0) = E_0^+(a_0) \quad \text{(isospectrality)}$$
$$= E_0^-(a_1) + g(a_1) - g(a_0) \quad \text{(shape invariance)}$$
$$= g(a_1) - g(a_0) \quad \text{(unbroken supersymmetry)}$$
$$\cdots$$
$$E_n^-(a_0) = g(a_n) - g(a_0) . \qquad (16.82)$$

Similarly,

$$|n, a_0; -\rangle = \frac{\hat{A}^+(a_0) \cdots \hat{A}^+(a_{n-1}) | 0, a_n; -\rangle}{\left[\prod_{j=0}^{n-1} E_{n-j}^-(a_j) \right]^{1/2}} . \qquad (16.83)$$

All eigenenergies and eigenvectors of the hamiltonian $\hat{H}_-$ can be determined iteratively by this algorithm. Thus, SUSYQM and shape invariance determine the entire spectrum of the system without any need to solve complicated differential equations.

16.3   SUSYQM and Shape Invariance in Phase Space

We are now ready to apply these ideas to the phase space formalism. SUSYQM on phase space is obtained by applying the Weyl map to the operator framework discussed above. Thus, the Weyl symbols (the complex functions on phase space) of the partner supersymmetric hamiltonians $\hat{H}_-$ and $\hat{H}_+$ are

$$H_-(x, p) = \frac{p^2}{2m} + V_-(x) \equiv A^+(x, p) * A^-(x, p) , \qquad (16.84)$$

$$H_+(x, p) = \frac{p^2}{2m} + V_+(x) \equiv A^-(x, p) * A^+(x, p) , \qquad (16.85)$$

where the annihilation and creation functions are given by

$$A^-(x,p) = \frac{ip}{2m} + W(x) \; ; \tag{16.86}$$

$$A^+(x,p) \equiv \overline{A^-}(x,p) = -\frac{ip}{2m} + W(x) \; . \tag{16.87}$$

The supersymmetric partner potentials are as always

$$V_{\mp}(x) = W^2(x) \mp \frac{\hbar}{\sqrt{2m}} W'(x) \; . \tag{16.88}$$

The last formula can be obtained either by applying the Weyl map on the quantum operators $V_{\mp}(\hat{x})$, or by direct calculations from (16.84) and (16.85) using the definitions (16.86), (16.87) and the property of Moyal's bracket $[W(x), p]_* = i\hbar \, \partial_x W(x)$.

Problem 16.4. *Prove that for any continuous and differentiable function* $f(x)$, *we have* $[f(x), p]_* = i\hbar \, \partial_x f(x)$.

The $*$-eigenvalue problem for the supersymmetric partner potentials $H_-(x,p)$ and $H_+(x,p)$ written in terms of their corresponding Wigner functions reads

$$H_-(x,p) * P_n^-(x,p) = E_n^- \, P_n^-(x,p) \; , \tag{16.89}$$

$$H_+(x,p) * P_n^+(x,p) = E_n^+ \, P_n^+(x,p) \; , \tag{16.90}$$

where

$$P_n^{\mp}(x,p) = \frac{1}{2\pi} \int dy \, e^{-ipy} \left\langle x + \frac{\hbar y}{2} | n; \mp \rangle \langle n; \mp | x - \frac{\hbar y}{2} \right\rangle \tag{16.91}$$

are the n-th excited state Wigner functions of the supersymmetric partner hamiltonians $H_{\mp}(x,p)$. The necessary condition for the ground state energy E_0^- of $H_-(x,p)$ to be zero, is

$$A(x,p) * P_0^-(x,p) = 0 \; . \tag{16.92}$$

This yields the $*$-product equation

$$W(x) * P_0^-(x,p) = -\frac{i}{\sqrt{2m}} p * P_0^-(x,p) \; , \tag{16.93}$$

which in terms of Bopp shifts becomes

$$W\left(x + \frac{i\hbar}{2} \overrightarrow{\partial_p}\right) P_0^-(x,p) = \frac{-i}{\sqrt{2m}} \left(p - \frac{i\hbar}{2} \overrightarrow{\partial_x}\right) P_0^-(x,p) \; . \tag{16.94}$$

By applying the Weyl map on Eqs. (16.76) and (16.77) we obtain[28] the connections between the *-eigenfunctions of the supersymmetric partner potentials $H_-(x,p)$ and $H_+(x,p)$

$$P_n^+(x,p) = \frac{1}{E_{n+1}^-} A(x,p) * P_{n+1}^-(x,p) * A^+(x,p) , \qquad (16.95)$$

$$P_{n+1}^-(x,p) = \frac{1}{E_n^+} A^+(x,p) * P_n^+(x,p) * A(x,p) . \qquad (16.96)$$

The only constraint we have imposed so far was that the hamiltonians $H_\mp(x,p)$ are supersymmetric partners. This and isospectrality lead to the *-product connection between the corresponding Wigner functions $P_n^\mp(x,p)$. It is almost the "carbon copy" of the SUSYQM case.

Next, we explore the implications of shape invariance. In phase space, Eq. (16.79) becomes

$$H_+(x,p;a_0) + g(a_0) = H_-(x,p;a_1) + g(a_1) , \qquad (16.97)$$

where a_0 and $a_1 = f(a_0)$. Because the hamiltonians in (16.97) differ by a constant, they have the same set of *-eigenfunctions. Hence, shape invariance implies

$$P_n^+(x,p;a_0) = P_n^-(x,p;a_1) , \qquad (16.98)$$

and the energy spectrum

$$E_n^-(a_0) = g(a_n) - g(a_0) , \quad \text{with } a_n = f^n(a_0) . \qquad (16.99)$$

In addition, shape invariance together with the supersymmetric condition, Eqs. (16.95) and (16.96), leads to a new[29] recursion formula between Wigner functions:

$$P_n^-(x,p;a_0) = A^+(x,p;a_0) * \frac{P_{n-1}^-(x,p;a_1)}{E_n^-(a_0)} * A(x,p;a_0) . \qquad (16.100)$$

The proof is immediate:

$$P_n^-(x,p;a_0) = A^+(x,p;a_0) * \frac{P_{n-1}^+(x,p;a_0)}{E_{n-1}^+(a_0)} * A(x,p;a_0) ,$$

$$= A^+(x,p;a_0) * \frac{P_{n-1}^-(x,p;a_1)}{E_n^-(a_0)} * A(x,p;a_0) ,$$

[28]T. Curtright, D. Fairlie, and C. Zachos,"Features of time-independent Wigner functions", *Phys. Rev. D* **58**, 025002 (1998).
[29]C. Rasinariu, "Shape invariance in phase space", *Fortschritte der Physik* **61**, 4–19 (2013).

where the first equality follows from SUSYQM, and the second one from shape invariance and isospectrality. We can now iterate Eq. (16.100) to obtain the expression for a general $P_n^-(x, p; a_0)$ starting from the ground state Wigner function $P_0^-(x, p; a_n)$. We have

$$P_n^-(x, p; a_0) = A^+(x, p; a_0) * \cdots * A^+(x, p; a_{n-1}) * \frac{P_0^-(x, p; a_n)}{\prod_{j=0}^{n-1} E_{n-j}^-(a_j)}$$
$$* A(x, p; a_{n-1}) * \cdots * A(x, p; a_0) . \tag{16.101}$$

This recursion formula together with Eq. (16.92) as its starting point, determines the entire set of Wigner functions for shape invariant systems. However, for concrete examples the multiple $*$-products can become prohibitively difficult to calculate. The harmonic oscillator is one of the few cases where we can determine analytically the entire sequence of Wigner functions.

16.3.1   *Example: The Morse Potential*

As another example, let us consider the one-dimensional Morse potential, which was listed as one of the shape invariant potentials in Chapter 5. The Morse oscillator is a good approximation of the oscillatory motion of diatomic molecules.[30] As usual, we take $\hbar = 1$ and $2m = 1$. Then, the hamiltonian reads

$$H_-(x, p; a) = p^2 + V_-(x; a) , \tag{16.102}$$

where the Morse potential

$$V_-(x; a) = a^2 + e^{-2x} - 2(a + 1/2)e^{-x} ; \quad a > 0 , \tag{16.103}$$

can be generated as $V_- = W^2 - W'$ from the superpotential[31]

$$W(x; a) = a - e^{-x} . \tag{16.104}$$

The corresponding annihilation and creation phase space functions are

$$A^-(x, p) = ip + a - e^{-x} ; \quad A^+(x, p) = -ip + a - e^{-x} . \tag{16.105}$$

The relevant parameters of the shape invariance are $a_0 = a$ and $a_1 \equiv f(a_0) = a - 1$ respectively. Thus

$$H_+(x, p; a) + g(a) = H_-(x, p; a - 1) + g(a - 1) ; \quad g(a) = a^2 , \tag{16.106}$$

[30]H.-W. Lee and M. O. Scully, "Wigner phase-space description of a Morse oscillator", *J. Chem. Phys.* **77**, 4604 (1982);

   B. Belchev and M. A. Walton, "Solving for the Wigner functions of the Morse potential in deformation quantization", *J. Phys. A* **43**, 225206 (2010).

[31]F. Cooper, A. Khare, and U. P. Sukhatme, "Supersymmetry and quantum mechanics", *Phys. Rep.* **251**, 268 (1995).

and consequently, the energy spectrum is

$$E_n^-(a) = a^2 - (a-n)^2 . \tag{16.107}$$

Let us find the ground state Wigner function $P_0^-(x,p)$ using

$$A^-(x,p) * P_0^-(x,p) = 0 . \tag{16.108}$$

Writing the $*$-products in (16.108) as Bopp shifts, we obtain

$$\left\{ i\left(p - \frac{i}{2}\overrightarrow{\partial_x}\right) + a - e^{-\left(x + \frac{i}{2}\overrightarrow{\partial_p}\right)} \right\} P_0^-(x,p) = 0 , \tag{16.109}$$

which becomes a mixed difference-differential equation

$$(a + ip) P_0^-(x,p) = e^{-x} P_0^-(x, p - i/2) - \frac{1}{2}\partial_x P_0^-(x,p) . \tag{16.110}$$

The above equation mixes the partial derivative with respect to one variable with a shift with respect to the other variable. This hints[32] at a difference-differential formula for the modified Bessel functions $K_\nu(x)$

$$2\,\partial_x K_\nu(x) + K_{\nu-1}(x) + K_{1+\nu}(x) = 0 . \tag{16.111}$$

The exponential term in the right hand side of Eq. (16.110) suggests that $K_\nu(x)$ depends on x via an exponential term. The shift in momentum variable suggests that p appears in the index ν of $K_\nu(x)$. Therefore we substitute in Eq. (16.110) the following

$$P_0^-(x,p) = C_1 e^{-C_2\,x} K_{\alpha(p)}\left(C_3\,e^{-x}\right) , \tag{16.112}$$

where the constants C_1, C_2, C_3, and the function $\alpha(p)$ are to be determined. Plugging Eq. (16.112) into Eq. (16.110), after tedious and courageous calculations with the help of Mathematica, we find that Eq. (16.110) is satisfied for

$$C_2 = 2a , \quad C_3 = 2 ; \quad \text{and} \quad \alpha(p) = 2ip . \tag{16.113}$$

The constant C_1 is obtained from the normalization of the Wigner function, Eq. (16.34): $C_1 = \frac{2}{\pi}(2/a)^{2a}$. Thus, the ground state Wigner function for the Morse oscillator is

$$P_0^-(x,p;a) = \frac{2}{\pi} \left(\frac{2}{a}\right)^{2a} e^{-2ax} K_{2ip}\left(2\,e^{-x}\right) . \tag{16.114}$$

---

[32] N. N. Lebedev, *Special Functions and Their Applications*, Dover, NY (1972).

Finding the excited state Wigner functions is now straightforward using shape invariance (16.100). For example, let us find $P_1^-(x, p; a)$

$$P_1^-(x, p; a) = A^+(x, p; a) * \frac{P_0^-(x, p; a - 1)}{E_1^-(a_0)} * A(x, p; a)$$

$$= [a - ip - e^{-x}] * \frac{P_0^-(x, p; a - 1)}{a^2 - (a - 1)^2} * [a + ip - e^{-x}]$$

$$= \left[a - i\left(p - \frac{i}{2}\overrightarrow{\partial_x}\right) - e^{-(x + \frac{i}{2}\overrightarrow{\partial_p})}\right] \frac{P_0^-(x, p; a - 1)}{2a - 1}$$

$$\left[a + i\left(p + \frac{i}{2}\overleftarrow{\partial_x}\right) - e^{-(x - \frac{i}{2}\overleftarrow{\partial_p})}\right].$$

After some algebra, we arrive to

$$P_1^-(x, p; a) = \alpha(x) K_\nu(y) - \beta(x) [K_{\nu-1}(y) + K_{\nu+1}(y)], \qquad (16.115)$$

where $\nu = 2ip; y = 2 e^{-x}$, and

$$\alpha(x) = \frac{2^{2a} e^{-2ax} \left(\frac{1}{a-1}\right)^{-2+2a}}{2\pi(2a - 1)} [4 + e^{2x}(2a - 1)^2],$$

$$\beta(x) = \frac{4^a e^{(-2a+1)x} \left(\frac{1}{a-1}\right)^{-2+2a}}{\pi}.$$

In Fig. 16.3 we illustrate the Wigner function $P_1(x, p)$ for the Morse oscillator. The graph is cut on the vertical axis to emphasize the details near the origin.

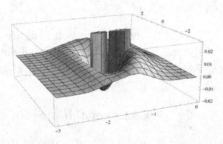

Fig. 16.3   The Wigner function $P_1(x, p)$ for the Morse oscillator, for $a = 5$.

Thus, we have shown that quantum mechanics can be expressed equivalently in the language of "normal" functions defined on phase space, endowed with the *-product. The Moyal bracket corresponds to the operator commutator, and one can use it as an heuristic tool for quantization. In the classical limit ($\hbar \to 0$) the Moyal bracket becomes simply the Poisson bracket, and the $*$-product becomes the normal commutative product of functions. In this picture, quantum mechanics appears as a deformation of classical mechanics, with the deformation parameter $\hbar$. Supersymmetric quantum mechanics induces additional relations between the Wigner functions, while shape invariance reveals a simple way to recursively obtain the Wigner functions of a system, starting from the ground state Wigner function, as exemplified by the Morse potential above.

Chapter 17

Solutions to Problems

Chapter 1

Problem 1.1

Let us consider the time dependent Schrödinger equation:

$$H(\mathbf{r})\Psi(\mathbf{r},t) \equiv -\frac{\hbar^2}{2m}\,\nabla^2\Psi(\mathbf{r},t) + V(\mathbf{r})\,\Psi(\mathbf{r},t) = i\hbar\,\frac{\partial}{\partial t}\Psi(\mathbf{r},t) \qquad (17.1)$$

for a time-independent hamiltonian[1] $H(\mathbf{r})$. In such cases, we can write $\Psi(\mathbf{r},t) = \psi(\mathbf{r})f(t)$. Substituting this trial solution, also known as an ansatz, into Eq. (1.1), we get

$$-\frac{\hbar^2}{2m}\,\frac{\nabla^2\psi(\mathbf{r})}{\psi(\mathbf{r})} + V(\mathbf{r}) = i\hbar\,\frac{df(t)/dt}{f(t)}\,. \qquad (17.2)$$

The above equation has to be valid for all values of $\mathbf{r}$ and t. Since the expression on the left hand side of Eq. (17.2) is a function of $\mathbf{r}$, while the right hand side is a function of t, the above equality can hold only if both sides are equal to a common constant, say E. Then

$$-\frac{\hbar^2}{2m}\,\frac{\nabla^2\psi(\mathbf{r})}{\psi(\mathbf{r})} + V(\mathbf{r}) = E = i\hbar\,\frac{df(t)/dt}{f(t)}\,.$$

Chapter 2

Problem 2.1

Let us consider the differential equation (2.21).

$$\frac{d^2\psi(x)}{dx^2} = \frac{1}{4}\,\omega^2\,x^2\psi(x)\,. \qquad (17.3)$$

---

[1]An isolated system that does not exchange energy with its surroundings is always time-independent.

Substituting the function $\psi(x) = e^{-\omega x^2/4}$, we get

$$\frac{d^2\psi(x)}{dx^2} = \left(\frac{\omega^2 x^2}{4} - \frac{\omega}{2}\right)\psi(x) .$$

For large x; i.e., $x \gg \sqrt{2/\omega}$, we get

$$\frac{d^2\psi(x)}{dx^2} \approx \frac{1}{4}\,\omega^2 x^2 \psi(x) ,$$

hence $\psi(x) = e^{-\omega x^2/4}$ is indeed an asymptotic solution of Eq. (17.5).

Problem 2.2

Let us substitute $\xi \equiv x\sqrt{\omega/2}$ and $K \equiv 2E/\omega$ into Eq. (17.4).

$$-\frac{d^2\psi(x)}{dx^2} + \frac{1}{4}\,\omega^2 x^2 \psi(x) = E\psi(x) . \qquad (17.4)$$

Since $\frac{d}{dx} = \sqrt{\frac{\omega}{2}}\frac{d}{d\xi}$, the above equation can be written as

$$-\frac{\omega}{2}\frac{d^2\psi(x)}{d\xi^2} + \frac{\omega}{2}\,\xi^2\psi(x) = \frac{\omega}{2}\,K\,\psi(x) . \qquad (17.5)$$

Now substituting $\psi(x) = C\,e^{-\xi^2/2}\,v(\xi)$, after some simplification we get the desired result

$$\frac{d^2v(\xi)}{d\xi^2} - 2\xi\,\frac{dv(\xi)}{d\xi} + (K-1)\,v(\xi) = 0 . \qquad (17.6)$$

Problem 2.3

To show that for $j \gg 1$, $v_j \approx C/(j/2)!$ obeys the recursion relation Eq. (2.27),

$$v_{j+2} \approx \frac{2}{j}v_j , \qquad (17.7)$$

we compute approximate forms for v_j and v_{j+2}. They are given by $\frac{C}{(\frac{j}{2})!}$ and $\frac{C}{(\frac{j+2}{2})!}$ respectively. Dividing v_{j+2} by v_j, we get

$$\frac{v_{j+2}}{v_j} = \frac{(\frac{j}{2})!}{(\frac{j+2}{2})!} = \frac{(\frac{j}{2})!}{(\frac{j}{2}+1)!} = \frac{1}{(\frac{j}{2}+1)} \approx \frac{2}{j} . \qquad (17.8)$$

Problem 2.4

From the Rodrigues formula

$$H_n(\xi) = (-1)^n\,e^{\xi^2}\,\frac{d^n}{d\xi^n}\left(e^{-\xi^2}\right) , \qquad (17.9)$$

we want to show that

$$H_0(\xi) = 1, \ H_1(\xi) = 2\xi, \ H_2(\xi) = 4\xi^2 - 2, \ H_3(\xi) = 8\xi^3 - 12\xi \ . \quad (17.10)$$

The Hermite polynomial $H_0(\xi)$ is given by $(-1)^0 \, e^{\xi^2} \left(e^{-\xi^2}\right)$, which is equal to 1.

The Hermite polynomial $H_1(\xi)$ is given by $(-1) \, e^{\xi^2} \frac{d}{d\xi} \left(e^{-\xi^2}\right)$. It is equal to $(-1) \, e^{\xi^2} \left(-2\xi \, e^{-\xi^2}\right) = 2\xi$.

The Hermite polynomial $H_2(\xi)$ is given by $(-1)^2 \, e^{\xi^2} \frac{d^2}{d\xi^2} \left(e^{-\xi^2}\right)$. It is equal to $e^{\xi^2} \left((4\xi^2 - 2)e^{-\xi^2}\right) = 4\xi^2 - 2$.

Similarly the other two Hermite polynomials are given by $H_3(\xi) = 8\xi^3 - 12\xi$ and $H_4 = 16\xi^4 - 48\xi^2 + 12$.

Problem 2.5

Let us start with the recursion relation:

$$H_{n+1}(\xi) - 2\xi H_n(\xi) + 2nH_{n-1}(\xi) = 0 \ .$$

Replacing ξ by $\sqrt{\frac{\omega}{2}} \, x$,

$$H_{n+1} - \sqrt{2\omega} x H_n + 2nH_{n-1} = 0 \ .$$

Using

$$\psi_n(x) = \left(\frac{\omega}{2}\right)^{1/4} \left[\sqrt{\pi} n! \, 2^n\right]^{-1/2} e^{-\omega x^2/4} H_n\left(x\sqrt{\frac{\omega}{2}}\right)$$

we get

$$\frac{\psi_{n+1}(x)}{\sqrt{(n+1)! \, 2^{(n+1)}}} - \frac{\sqrt{2\omega} \, x \, \psi_n(x)}{\sqrt{n! \, 2^n}} + \frac{2n \, \psi_{n-1}(x)}{\sqrt{(n-1)! \, 2^{(n-1)}}} = 0 \ . \quad (17.11)$$

Starting from the recursion relation

$$H'_n(\xi) = 2nH_{n-1}(\xi) \ ,$$

and following a similar procedure we get Eq. (2.37). Then, using Eqs. (17.11) and (2.37), we arrive at Eq. (2.38).

Problem 2.6

From Eq. (2.20), we have

$$-\frac{d^2\psi(x)}{dx^2} + \frac{1}{4}\omega^2 x^2 \, \psi(x) = E\psi(x) \ .$$

Redefining the potential,

$$V \equiv \frac{1}{4}\omega^2 x^2 - E_0$$

where E_0 is the ground state energy, we get

$$-\frac{d^2\psi(x)}{dx^2} + \underbrace{\left[\frac{1}{4}\omega^2 x^2 - E_0\right]}_{V_{\text{SUSY}}}\psi(x) = (E - E_0)\psi(x) \ .$$

Now, defining $E_{\text{SUSY}} \equiv E - E_0$,

$$-\frac{d^2\psi(x)}{dx^2} + V_{\text{SUSY}}\psi(x) = E_{\text{SUSY},0}\psi(x) \ .$$

From the definition of $E_{\text{SUSY},0}$, we have $E_{\text{SUSY},0} = E_0 - E_0 = 0$. Shifting a potential cannot change any physics, since the choice of zero potential is arbitrary. Hence, the wavefunctions do not change.

Problem 2.7

The potential for the half-harmonic oscillator (HHO) is

$$V(x) = \begin{cases} \infty & x < 0 \\ \frac{1}{4}\omega^2 x^2 & x \geq 0 \ . \end{cases} \qquad (17.12)$$

Let us denote the eigenstates of the full harmonic oscillator by ψ_n. The corresponding energies are given by

$$\left(n + \frac{1}{2}\right)\hbar\omega \qquad n = 0, 1, 2, \cdots \ .$$

Since the potential for the HHO for $x > 0$ is identical to that of the full harmonic oscillator (HO), the set of eigenfunctions of HO form a possible list of eigenfunctions for HHO. However, we should eliminate all the even states of HO that do not vanish at the origin. Hence, the states that are eigenfunctions of HHO are: ψ_1, ψ_3, ψ_5, $\cdots$; i.e., odd states of the full harmonic oscillator states. The ground state of HHO, ψ_1, is the first excited state of HO. Hence the ground state energy of HHO is $\frac{3}{2}\hbar\omega$. Energies of the higher states are

$$\frac{3}{2}\hbar\omega, \ \frac{7}{2}\hbar\omega, \ \frac{11}{2}\hbar\omega, \ldots, \ \frac{2n+3}{2}\hbar\omega \ .$$

Problem 2.8

From Eq. (2.43), we have

$$\frac{d}{dr}\left(r^2\frac{dR(r)}{dr}\right) + r^2\left(\frac{e^2}{r} + E\right)R(r) = l(l+1)R(r) \ . \qquad (17.13)$$

Let us substitute $R(r) = r^{-1} u(r)$. This gives $\frac{dR}{dr} = -r^{-2}u + r^{-1}\frac{du}{dr}$. Equivalently, $r^2 \frac{dR}{dr} = -u + r\frac{du}{dr}$. Hence we have

$$\frac{d}{dr}\left(r^2 \frac{dR}{dr}\right) = \frac{d}{dr}\left(-u + r\frac{du}{dr}\right) = r\frac{d^2u}{dr^2} .$$

Substituting the above in Eq. (17.13), the radial equation finally reduces to

$$-\frac{d^2u}{dr^2} + \left[\frac{-e^2}{r} + \frac{l(l+1)}{r^2}\right]u = Eu . \tag{17.14}$$

Problem 2.9

From Eq. (2.49), we have

$$\rho^2 \frac{d^2u}{d\rho^2} \approx l(l+1)u . \tag{17.15}$$

Using the ansatz $u = \rho^m$, we get

$$m^2 - m - l(l+1) = 0 .$$

The two solutions are $\frac{1}{2}(1 \pm (2l+1))$; i.e, $m = -l$ or $m = l+1$. Hence the general solution of Eq. (2.49) is

$$u \sim C\rho^{l+1} + D\rho^{-l} .$$

Problem 2.10

The general solution of Eq. (17.15) is given by $u \sim C\rho^{l+1} + D\rho^{-l}$. The second term blows up near the origin. The behavior of the complete solution of Eq. (2.47) near the origin is then given by ρ^{l+1}. The behavior at large values of ρ has the form e^ρ. Let us substitute $u \sim \rho^{l+1}\, e^{-\rho}\, v(\rho)$ into Eq. (2.47). This gives

$$-2l\, v - 2\, v + 2l\, v' + 2\, v' + r\, v - 2\rho v' + \rho v'' - \rho v + \rho_0\, v = 0 .$$

Simplifying it, we get

$$\rho\frac{d^2v}{d\rho^2} + 2(l+1-\rho)\frac{dv}{d\rho} + [\rho_0 - 2(l+1)]v = 0 ,$$

which is same as in Eq. (2.50).

Problem 2.11

We want to show that for large j, the ansatz $v_{j+1} \approx \frac{2^{j+m}}{(j+m)!} v_0$ solves the equation

$$v_{j+1} \approx \frac{2}{j} v_j .$$

From the above ansatz, we have $v_{j+1} = \frac{2^{j+m}}{(j+m)!} v_0$ and $v_j = \frac{2^{j+m-1}}{(j+m-1)!} v_0$. Dividing one by the other, we get

$$\frac{v_{j+1}}{v_j} = \frac{2(j+m-1)!}{(j+m)!} = \frac{2}{(j+m)} .$$

For $j \gg 1$, the above reduces to

$$\frac{v_{j+1}}{v_j} \approx \frac{2}{j} .$$

Problem 2.12

The exact recursion relation is given by Eq. (2.52)

$$v_{j+1} = \frac{2(j+l+1) - \rho_0}{(j+1)(j+2l+2)} v_j , \quad \text{where } \rho_0 > 0 .$$

Since

$$\frac{2(j+l+1) - \rho_0}{(j+1)(j+2l+2)} < \frac{2(j+l+1)}{(j+1)(j+2(l+1))} < \frac{2(j+l+1)}{(j+1)(j+l+1)}$$

$$= \frac{2}{j+1} < \frac{2}{j} ,$$

Eq. (2.53) indeed over estimates Eq. (2.52).

Problem 2.13

To get a solution that does not diverge, the series of Eq. (2.52) must truncate. This happens if there is a j_{max} such that

$$v_{j_{max}+1} = \frac{2(j_{max}+l+1) - \rho_0}{(j_{max}+1)(j_{max}+2(l+1))} = 0 .$$

This happens for $2(j_{max}+l+1) = \rho_0 \equiv 2n$. Since $\rho_0 = \frac{e^2}{\sqrt{-E}}$, we get

$$E \rightarrow E_n = -\frac{e^4}{4n^2} .$$

Problem 2.14

(a) For $2s$ and $2p$ states, the radial functions are given by

$$R_{2s} = \frac{1}{a\sqrt{a}} \left(1 - \frac{r}{2a}\right) e^{-r/2a} ; \quad \text{and} \quad R_{2p} = \frac{1}{2a^2\sqrt{6a}} r e^{-r/2a} .$$

Since the probability density is given by $P(r) = r^2 R^2(r)$, the point r_{max} where it would be maximum would be given by

$$\frac{dP}{dr} = 0 .$$

For the 2s state,

$$\frac{d}{dr} \left(r^2 R^2\right) = \frac{d}{dr} \left(r^2 \left(1 - \frac{r}{2a}\right)^2 e^{-r/a}\right) = 0 ,$$

implies $r(r - 2a) + 4a^2 - 4ra - 2r + r^2 = 0$. Its solution is $r = 2a$.

Similarly for the 2p state,

$$\frac{d}{dr} r^2 r^2 e^{-r/a} = \frac{d}{dr} r^4 e^{-r/a} = 0 ,$$

gives $4r^3 - \frac{r^4}{a} = 0$. Its solution is $r = 4a$.

(b) Bohr assumed circular orbits. Thus $L = mvr = n\hbar$. The energy of bound states for circular orbits is $E = T + V = \frac{mv^2}{2} - \frac{e^2}{r}$. Combining this with the force law

$$\frac{mv^2}{r} = \frac{e^2}{r^2}$$

we obtain

$$E = -\frac{L^2}{2mr^2}$$

classically. But for a given n, $E_n = -\frac{n^2\hbar}{2mr^2}$. This should be different for r_{2s} and r_{2p}, but it isn't. Thus the Bohr model with $E = E_n$ violates the classical model $E \sim \frac{1}{2mr^2}$.

Problem 2.15

The radial function for the 2s state is given by $R_{2s} = \frac{1}{a\sqrt{a}} \left(1 - \frac{r}{2a}\right) e^{-r/2a}$. Hence the function $u = rR$ is given by

$$u_{2s} = r R_{2s} = \frac{r}{a\sqrt{a}} \left(1 - \frac{r}{2a}\right) e^{-r/2a} .$$

Keeping terms to first order in r, we expand $e^{-r/2a} \approx 1 - \frac{r}{2a}$. Thus, we get

$$u_{2s} \sim r\left(1 - \frac{r}{2a}\right)^2 \approx r\left(1 - \frac{r}{a}\right) = r - \frac{r^2}{a} \sim \frac{\rho}{\kappa} - \frac{\rho^2}{\kappa^2 a} \ ,$$

where $\rho \equiv \kappa r$. But $a = \frac{1}{n\kappa} = \frac{1}{2\kappa}$ for $n = 2$, gives

$$u_{2s} \sim \frac{\rho}{\kappa} - \frac{\rho^2}{\kappa^2} \cdot 2\kappa = \frac{1}{\kappa}(\rho - 2\rho^2) \ .$$

This agrees with

$$u_{n_0} \sim \rho - n\rho^2 \ .$$

Chapter 3

Problem 3.1

Using matrix representation for states and linear operators, we want to show that $\langle m|An \rangle = \langle A^+m|\, n \rangle$.

$$\langle m|An \rangle = (m_1^* m_2^*) \, , \begin{pmatrix} a_{11} & a_{12} \\ a_{21} & a_{22} \end{pmatrix} \begin{pmatrix} n_1 \\ n_2 \end{pmatrix} = (m_1^* \ m_2^*) \begin{pmatrix} a_{11}n_1 & a_{12}n_2 \\ a_{21}n_1 & a_{22}n_2 \end{pmatrix}$$

$$= m_1^*(a_{11}n_1 + a_{12}n_2) + m_2^*(a_{21}n_1 + a_{22}n_2) \ . \tag{17.16}$$

$$\langle A^+m|\, n \rangle = (A^+m)^{*T} \cdot \begin{pmatrix} n_1 \\ n_2 \end{pmatrix} = \left(\begin{pmatrix} a_{11}^* & a_{21}^* \\ a_{12}^* & a_{22}^* \end{pmatrix} (m_1 \ m_2) \right)^{*T} \cdot \begin{pmatrix} n_1 \\ n_2 \end{pmatrix}$$

$$= \begin{pmatrix} a_{11}^* m_1 + a_{21}^* m_2 \\ a_{12}^* m_1 + a_{22}^* m_2 \end{pmatrix}^{*T} \cdot \begin{pmatrix} n_1 \\ n_2 \end{pmatrix} \tag{17.17}$$

$$= \begin{pmatrix} a_{11}m_1^* + a_{21}m_2^* \\ a_{12}m_1^* + a_{22}m_2^* \end{pmatrix}^{T} \cdot \begin{pmatrix} n_1 \\ n_2 \end{pmatrix}$$

$$= (a_{11}m_1^* + a_{21}m_2^* \quad a_{12}m_1^* + a_{22}m_2^*) \cdot \begin{pmatrix} n_1 \\ n_2 \end{pmatrix}$$

$$= a_{11}m_1^* n_1 + a_{21}m_2^* n_1 + a_{12}m_1^* n_2 + a_{22}m_2^* n_2$$

$$= m_1^*(a_{11}n_1 + a_{12}n_2) + m_2^*(a_{21}n_1 + a_{22}n_2) = \langle m|An \rangle \ . \tag{17.18}$$

Problem 3.2

Let us check the hermiticity of the operator $p = -i\frac{d}{dx}$. Recall that for a hermitian operator, $\int \psi^* A\phi \, dx = \int \left(A^\dagger \psi\right)^* \phi \, dx$.

Thus, we need to show that $\int \psi^* \left(-i\frac{d}{dx}\right) \phi \, dx = \int \left(-i\frac{d}{dx}\psi\right)^* \phi \, dx$.

Hence,

$$\int \psi^* p\phi \, dx = \int \psi^* \cdot \left(-i\frac{d\phi}{dx}\right) dx$$

$$= -i \int \left[\frac{d}{dx}(\psi^*\phi) - \left(\frac{d\psi^*}{dx}\right) \cdot \phi\right] dx$$

$$= -i \, (\psi^*\phi)|_{\text{Boundary}} + i \int \left(\frac{d\psi^*}{dx}\right) \cdot \phi \, dx$$

$$= -0 + \int \left(-i\frac{d\psi}{dx}\right)^* \cdot \phi \, dx$$

$$= \int (p\psi)^* \, \phi \, dx \, . \tag{17.19}$$

so p is hermitian. $\frac{d}{dx}$ is not hermitian. Integration by parts gives the wrong sign:

$$\int \psi^* \left(\frac{d\phi}{dx}\right) dx = \psi^*\phi|_{\text{Boundary}} - \int \left(\frac{d\psi}{dx}\right)^* \phi \, dx$$

$$= -\int \left(\frac{d\psi}{dx}\right)^* \phi \, dx \, .$$

Hence, the operator $\frac{d}{dx}$ is skew hermitian:

$$\left(\frac{d}{dx}\right)^\dagger = -\frac{d}{dx} \, .$$

Problem 3.3

(a) Assume operators A and B are hermitian $A^\dagger = A$ and $B^\dagger = B$. Then

$$(A+B)^\dagger = A^\dagger + B^\dagger = A^\dagger + B^\dagger = A + B \, .$$

So $A + B$, the sum of two hermitian operators A and B, is hermitian.

(b) Now let us see whether the same holds for products as well. Consider the product AB of two hermitian operators A and B.

$$\langle\psi|AB\,\phi\rangle = \langle A^\dagger\,\psi|B\,\phi\rangle$$
$$= \langle B^\dagger A^\dagger\,\psi|\phi\rangle$$
$$= \langle BA\,\psi|\phi\rangle \ .$$

Hence, $(AB)^\dagger = B^\dagger A^\dagger = BA$, which is not necessarily equal to AB unless they commute. Since an operator A commutes with itself, the $n-th$ power of a hermitian operator A; i.e., $(A)^n$, is a hermitian operator.

Problem 3.4

For an infinitely deep potential well of width π, the eigenfunctions are $|n\rangle \equiv \sqrt{\frac{2}{\pi}}\,\sin nx$. Then

$$\langle m|n\rangle = \frac{2}{\pi}\int_0^\pi \sin{(mx)}\sin{(nx)}dx = \delta_{nm}\ .$$

Thus, all eigenfunctions are orthonormal.

Problem 3.5

(a) Let us now check the normalization of the state $|\psi\rangle = \frac{1}{2}|1\rangle + \frac{\sqrt{3}}{2}|2\rangle$.

$$\langle\psi|\psi\rangle = \frac{1}{4}\,\langle 1|1\rangle + \frac{3}{4}\,\langle 2|2\rangle = \frac{1}{4} + \frac{3}{4} = 1\ .$$

The state $\langle\psi|\psi\rangle$ is normalized.

(b) The average value of energy in this state is given by

$$\langle E\rangle = \langle\psi|H\psi\rangle = \left(\frac{1}{2}\,\langle 1| + \frac{\sqrt{3}}{2}\,\langle 2|\right)\left(\frac{1}{2}H\,|1\rangle + \frac{\sqrt{3}}{2}H\,|2\rangle\right)$$

$$= \frac{1}{4}\,\langle 1|H|1\rangle + \frac{3}{4}\,\langle 2|H|2\rangle + 0 + 0$$

$$= \frac{1}{4}\times 1 + \frac{3}{4}\times 4 = 3\frac{1}{4}\ . \tag{17.20}$$

We have used $\langle 1|2\rangle = 0$. This does not correspond to any single measurement, since it is not an eigenvalue of any single state.

(c) The probability of finding the particle with $E = 4$ for state $|\psi\rangle = C_1 |1\rangle + C_2 |2\rangle$ is given by $|C_2|^2$. For this case, the probability of $E = 4$ is $|C_2|^2 = \frac{3}{4}$.

Problem 3.6

We want to show that x commutes with potential $V(x)$. We start with showing that x commutes an arbitrary power of x: x^n. $[x, x^n] = [x\, x^n - x^n\, x] = [x^{n+1} - x^{n+1}] = 0$.

The commutator of x with $V(x)$ is given by

$$[V(x), x] = \left[\sum_{n=0}^{\infty} \frac{x^n}{n!} \frac{d^n V}{dx^n} \bigg|_{x_0} , x \right]$$

$$= \sum_{n=0}^{\infty} \frac{1}{n!} \frac{d^n V}{dx^n} \bigg|_{x_0} [x^n, x] = 0 . \tag{17.21}$$

Problem 3.7

For any operator A, the evolution of its expectation value $\langle A \rangle$ is given by

$$\frac{d\langle A \rangle}{dt} = \langle [H, A] \rangle + \frac{\partial \langle A \rangle}{\partial t} .$$

For the momentum operator p, the rate of change of its expectation value with time is given by

$$\frac{d\langle p \rangle}{dt} = i \left\langle \left[\frac{p^2}{2m}, p \right] \right\rangle + i \langle [V, \rho] \rangle . \tag{17.22}$$

Since p commutes with any power of itself, the first term on the right hand side of the above equation is zero. Hence,

$$\frac{d\langle p \rangle}{dt} = i \langle [V, p] \rangle = i \langle Vp - pV \rangle$$

$$= \left\langle \psi \left| V \frac{d\psi}{dx} \right\rangle - \left\langle \psi \left| \frac{d}{dx} (V\psi) \right\rangle \right.$$

$$= \left\langle \psi | V \frac{d\psi}{dx} \right\rangle - \left\langle \psi \left| V \frac{d\psi}{dx} \right\rangle - \left\langle \psi \left| \frac{dV}{dx} \psi \right\rangle \right.$$

$$= -\left\langle \frac{dV}{dx} \right\rangle . \tag{17.23}$$

Problem 3.8

In Problem 3.3 we showed that for a product of two operators A and B, $(AB)^\dagger = B^\dagger A^\dagger$. For operators $A^\pm$, we know that $(A^\pm)^\dagger = A^\mp$. Hence,

$$\left(A^+A^-\right)^\dagger = \left(A^-\right)^\dagger \left(A^+\right)^\dagger = A^+A^-.$$

Thus, A^+A^- and A^-A^+ are self-adjoint operators. However, we will go beyond this proof and show it explicitly as well. The operators A^+, A^- are given by

$$A^- \equiv \left(\frac{d}{dx} + \frac{1}{2}\omega x\right); \quad A^+ \equiv \left(-\frac{d}{dx} + \frac{1}{2}\omega x\right).$$

The number operator and its conjugate are given by

$$N \equiv \frac{1}{\omega}A^+A^- = \frac{1}{\omega}\left(-\frac{d}{dx} + \frac{1}{2}\omega x\right)\left(\frac{d}{dx} + \frac{1}{2}\omega x\right)$$

$$N^\dagger = \frac{1}{\omega}\left(A^-\right)^\dagger \left(A^+\right)^\dagger = \frac{1}{\omega}\left(\frac{d}{dx} + \frac{1}{2}\omega x\right)^\dagger \left(-\frac{d}{dx} + \frac{1}{2}\omega x\right)^\dagger. \quad (17.24)$$

However, as per results of Problem 3.2, $\frac{d}{dx}$ is skew hermitian: $\left(\frac{d}{dx}\right)^\dagger = -\frac{d}{dx}$. Hence,

$$N^\dagger = \frac{1}{\omega}\left(-\frac{d}{dx} + \frac{1}{2}\omega x\right)\left(\frac{d}{dx} + \frac{1}{2}\omega x\right) = N.$$

N is hermitian.

Problem 3.9

$[A^-, A^+] = A^-A^+ - A^+A^- = \omega$; i.e., $\omega M - \omega N = \omega$. Thus, operators M and N are related by $M = N + 1$. Hence, they have a common set of eigenvectors and their eigenvalues differ by one.

Problem 3.10

From Eq. (3.26), we have

$$\left(\frac{d}{dx} + \frac{\omega x}{2}\right)\psi_0(x) = 0.$$

Thus,

$$\int \frac{d\psi_0}{\psi_0} = -\int \frac{\omega x}{2}\,dx + C.$$

Integrating the above,

$$\ln \psi_0(x) = -\frac{\omega x^2}{4} + C, \quad \text{so} \quad \psi_0(x) = N e^{-\frac{\omega x^2}{4}}.$$

Normalizing $\psi_0(x)$; i.e., demanding that $|N|^2 \int_{-\infty}^{\infty} e^{-\frac{\omega x^2}{4}} dx = 1$,

$$N = \left(\frac{\pi}{2\omega}\right)^{1/4}.$$

And hence

$$\psi_0 = \left(\frac{\pi}{2\omega}\right)^{1/4} e^{-\omega x^2/4}.$$

Chapter 4

Problem 4.1

To show that the ground state eigenfunction of $\psi_0^-(x, a)$ can be written as $\psi_0^-(x, a) \sim e^{-\int_{x_0}^{x} W(x,a)\, dx}$, we start with $\left(\frac{d}{dx} + W(x, a)\right) \psi^-(x, a) = 0$. We then have $\frac{1}{\psi^-} \frac{d\psi^-}{dx} = -W(x, a)$. Integrating this equation we get $\psi_0^-(x, a) \sim e^{-\int_{x_0}^{x} W(x,a)\, dx}$.

Problem 4.2

For the following superpotentials, we need to show that the ground state wave functions vanish at both end points of their domains.

(a) Finite domain: $W(x, \alpha) = \alpha \tan x$, where $\alpha > 0$. The domain for this system is $\left(-\frac{\pi}{2}, \frac{\pi}{2}\right)$. The ground state wave function is given by $\psi_0^-(x, a) \sim e^{-\int_0^x \alpha \tan x\, dx} \sim \cos^\alpha x$. Since $\cos x$ vanishes at $\pm \frac{\pi}{2}$, the wave function is well-behaved at both ends.

(b) Semi-infinite domain: $W(x, \alpha, \beta) = \alpha - \beta \coth x$ with $\alpha > \beta$. The domain for this system is $(0, \infty)$. The ground state wave function is given by $\psi_0^-(x, a) \sim e^{-\int_{x_0}^x (\alpha - \beta \coth x)\, dx} \sim e^{-\alpha x} \sinh^\beta x$. Since $\sinh x$ vanishes at the origin and the exponential function vanishes at infinity, the wave function is well-behaved at both ends.

(c) Infinite domain: $W(x, \gamma) = \gamma \tanh x$ with $\gamma > 0$. The domain for this system is $(-\infty, \infty)$. The ground state wave function is given by $\psi_0^-(x, a) \sim e^{-\int_{x_0}^x \gamma \tanh x\, dx} \sim \cosh^{-\gamma} x$. Since $\cosh x$ vanishes at $\mp \infty$, the wave function is well behaved at both ends.

Problem 4.3

For the three-dimensional harmonic oscillator, $W(r, \omega, \ell) = \omega\, r - \frac{\ell}{r}$.

(a) The plot of $W(r, 10, -1)$ as a function of r is given below.

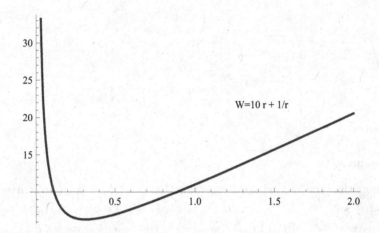

W=10 r + 1/r

Fig. 17.1   Plot of $W(r, \omega, \ell) = \omega\, r - \frac{\ell}{r}$ as a function of r for $\omega = 10$ and $\ell = -1$.

(b) Let us now check the behavior of the ground state wave function $\psi_0^-(r, \omega, \ell) \sim \exp\left(- \int_0^\infty W(r, 10, -1)\, dr\right)$. Integrating this expression, we get $\psi_0^-(r, \omega, \ell) = e^{-\left(\frac{1}{2}\omega r^2 + \ell \ln r\right)} = r^\ell e^{-\frac{1}{2}\omega r^2}$. For $\ell = -1$, the wave function blows up at the origin.

Problem 4.4

To determine the normalization constants of ψ_n^+, we start with $\psi_n^+ = N\, A^- \psi_{n+1}^-$, where N is the normalization constant. Taking a scalar product of this state with itself, we get

$$\langle \psi_n^+ | \psi_n^+ \rangle = N^2 \langle\, \psi_{n+1}^- | A^+ A^- | \psi_{n+1}^- \rangle .$$

Since $A^+ A^- |\psi_{n+1}^-\rangle \equiv H_- |\psi_{n+1}^-\rangle = E_{n+1}^- |\psi_{n+1}^-\rangle$, we get $N = \frac{1}{\sqrt{E_{n+1}}}$. We leave the second case as an exercise.

Problem 4.5

$\psi_3^-(x, 1)$ is simply $\sin 3x$. Since $\psi_{n-1}^+(x, 1) \sim \left(\frac{d}{dx} - \cot x\right) \sin nx$, $\psi_2^+(x, 1)$ is given by $\left(\frac{d}{dx} - \cot x\right) \sin 3x = 3 \cos 3x - \cot x \sin 3x = -8 \sin^2(x) \cos(x)$.

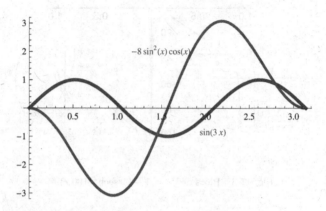

Fig. 17.2   Plots of $\psi_3^-(x, 1)$ and $\psi_2^+(x, 1)$.

Problem 4.6

The graphs show that A^- acting on function $\sin 3x$ generates a function $-8 \sin^2(x) \cos x$ that has one less node. Similarly, the operation of A^+ on function $\sin^2 x \cos x$ generates a function $-\sin 3x$ that has one additional node.

Problem 4.7

The superpotential is given by $W(x) = 2\Theta(x - x_0) - 1$, where $\Theta(x - x_0)$ is a step function; i.e., $\Theta(x - x_0) = 1$ if $x > x_0$, $1/2$ for $x = x_0$, and zero otherwise.

(a) The zero energy ground state wave function is given by

$$\psi_0^-(x, a) \sim e^{-\int_{x_0}^x (2\Theta(x-x_0)-1)\, dx} \sim \begin{cases} e^{-(x-x_0)} & \text{for } x > x_0 \\ e^{+(x-x_0)} & \text{for } x < x_0. \end{cases}$$

This ground state vanishes sufficiently rapidly as $x \to \infty$, and hence is a normalizable state. This implies that supersymmetry is unbroken.

(b) A good representation of $2\Theta(x-x_0)-1$ is given by $\lim_{\epsilon\to 0}\tanh(\frac{x-x_0}{\epsilon})$. We have plotted partner potentials for $\epsilon = 0.01$ and $x_0 = 0$. The potential $V_- = 1 - 99\,\mathrm{sech}^2(100\,x)$ is given in Fig. 17.3. The potential

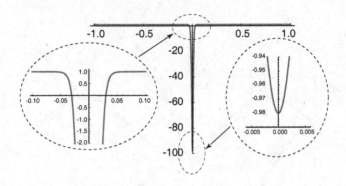

Fig. 17.3   Plots of $V_- = 1 - 99\,\mathrm{sech}^2(100\,x)$.

$V_+ = 1 + 99\,\mathrm{sech}^2(100\,x)$ is given in Fig. 17.4. As we see, the potential

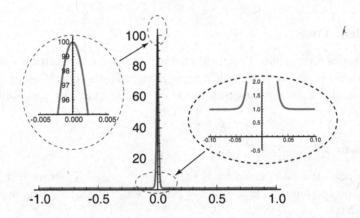

Fig. 17.4   Plots of $V_+ = 1 + 99\,\mathrm{sech}^2(100\,x)$.

V_+ does not hold any bound state. Thus, there are no states for the potential V_- above the ground state. Therefore, the attractive delta-function potential V_- holds only one bound state.

Chapter 5

Problem 5.1

From Chapter 5, the Coulomb superpotential is

$$W(r,\ell,e) = \frac{1}{2}\frac{e^2}{\ell+1} - \frac{\ell+1}{r} .$$

The resulting partner potentials are shape invariant since we have

$$V_+(r,\ell,e) = V_-(r,\ell+1,e) + \underbrace{\underbrace{\frac{1}{4}\left(\frac{e^2}{\ell+1}\right)^2}_{-g(\ell+1)} - \underbrace{\frac{1}{4}\left(\frac{e^2}{\ell+2}\right)^2}_{-g(\ell+2)}}_{R(a_0)} . \qquad (17.25)$$

The above equation implies

$$\psi_n^{(+)}(r,\ell,e) = \psi_n^{(-)}(r,\ell+1,e) .$$

The ground state of the hydrogen atom is

$$\psi_0^{(-)}(r,\ell,e) = r^{\ell+1}\exp\left(-\frac{1}{2}\frac{e^2}{\ell+1}r\right) . \qquad (17.26)$$

Now let us determine $\psi_1^{(-)}(r,\ell,e)$. Using SUSYQM, $\psi_1^{(-)}(r,\ell,e) = A^+\psi_0^{(+)}(r,\ell,e)$. From Eq. (17.25) we have $\psi_0^{(-)}(r,\ell+1,e) = \psi_0^{(+)}(r,\ell,e)$. Hence,

$$\begin{aligned}
\psi_1^{(-)}(r,\ell,e) &= A^+(r,\ell+1,e)\,\psi_0^{(-)}(r,\ell+1,e) \\
&= \left(-\frac{d}{dr} + W(r,\ell,e)\right)\psi_0^{(-)}(r,\ell+1,e) \\
&= \left(-\frac{d}{dr} + \frac{1}{2}\frac{e^2}{\ell+1} - \frac{\ell+1}{r}\right)r^{\ell+2}\exp\left(-\frac{1}{2}\frac{e^2}{\ell+2}r\right) .
\end{aligned}$$
$$(17.27)$$

Similarly,

$$\psi_2^{(-)}(r,\ell,e) = A^+(r,\ell+2,e)A^+(r,\ell+1,e)\,r^{\ell+3}\exp\left(-\frac{1}{2}\frac{e^2}{\ell+3}r\right) .$$
$$(17.28)$$

Now let us check the conditions for the breaking of supersymmetry. From Eq. (17.26), we have $\left|\psi_0^{(-)}(r,\ell,e)\right|^2 = r^{2\ell+2}\exp\left(-\frac{e^2}{\ell+1}r\right)$. The integral of this function diverges near the origin if $2\ell + 3 < -1$, or if $\ell < -2$. This is the condition for the breaking of supersymmetry.

Problem 5.2

The superpotential for the Hulthen potential is

$$W(r,\ell) = \frac{a}{\ell} - \frac{\ell}{a}\left[\frac{1+e^{-\frac{r}{a}}}{1-e^{-\frac{r}{a}}}\right] = \frac{a}{\ell} - \frac{\ell}{2a}\coth\left(r/2a\right), \qquad (17.29)$$

and corresponding coefficient $g(a_0)$ is

$$g(a_0) \equiv g(\ell) = -\left(\frac{\ell^2}{4a^2} + \frac{a^2}{\ell^2}\right).$$

The ground state $\sim e^{-\int W\, dx}$ is

$$\psi_0^{(-)} = e^{-\frac{a}{\ell}r}\left(\sinh\frac{r}{2a}\right)^{\frac{\ell}{2a}}.$$

The first excited state is

$$\psi_1^{(-)} = \left[\frac{a}{\ell} - \frac{\ell}{2a}\coth\left(r/2a\right)\right]e^{-\frac{a}{\ell+1}r}\left(\sinh\frac{r}{2a}\right)^{\frac{\ell+1}{2a}}.$$

The energies are

$$\begin{aligned}
E_n &= g(a_n) - g(a_0) \\
&= \left(\frac{(\ell+n)^2}{4a^2} + \frac{a^2}{(\ell+n)^2}\right) - \left(\frac{\ell^2}{4a^2} + \frac{a^2}{\ell^2}\right).
\end{aligned} \qquad (17.30)$$

Hence, $E_0 = 0$ and $E_1 = \left(\frac{(\ell+1)^2}{4a^2} + \frac{a^2}{(\ell+1)^2}\right) - \left(\frac{\ell^2}{4a^2} + \frac{a^2}{\ell^2}\right)$.

Problem 5.3

(a) Since $W(x,a) = a\tanh x$, the partner potentials are given by

$$\begin{aligned}
V_\pm(x,a) &= a^2\tanh^2 x \pm a\,\mathrm{sech}^2 x = a^2\tanh^2 x \pm a\left(1-\tanh^2 x\right) \\
&= a(a \mp 1)\,\mathrm{sech}^2 x + a^2.
\end{aligned} \qquad (17.31)$$

Thus, we have

$$V_+(x,a) = V_-(x,a-1) + \underbrace{a^2 - (a-1)^2}_{R(a_0)\equiv R(a)}.$$

(b) *Ground state:* The energy for the ground state is zero and the corresponding eigenfunction is

$$\psi_0^{(-)}(x,a) = N e^{-\int a \tanh x \, dx} = N \left(\cosh x \right)^{-a} .$$

First excited state: The energy for the first excited state is $R(a_0) = a^2 - (a-1)^2$, and the corresponding eigenfunction is given by

$$\psi_1^{(-)}(x,a) \sim A^+(x,a)\psi_0^{(-)}(x,a-1)$$

$$= N \left(-\frac{d}{dx} + a \tanh x \right) \left(\cosh x \right)^{-(a-1)} .$$

(c) We can proceed this way to determine eigenvalues and eigenfunctions of this system. The eigenvalues are given by $E_n = a^2 - (a-n)^2$ and the eigenfunctions are given by

$$\psi_n^{(-)}(x,a) \sim A^+(x,a)A^+(x,a-1)A^+(x,a-2)$$

$$\cdots A^+(x,a-n+1)\psi_0^{(-)}(x,a-n) .$$

These eigenfunctions are normalizable as long as $a > n$. Thus, the number of bound states for this system is limited by the value of parameter a.

Problem 5.4

$$W(x,a,b) = a \tan x - b \cot x ,$$

with $0 < x < \frac{\pi}{2}$ and $a > 0$, $b > 0$.

$$W(x,a,b) = a \tan x - b \cot x = a \frac{\sin x}{\cos x} + b \frac{\cos x}{\sin x}$$

$$= \frac{a \sin^2 x + b \cos^2 x}{\sin x \cos x}$$

$$= \frac{(b-a)\cos^2 x + a}{\sin x \cos x}$$

$$= (a+b) \operatorname{cosec} 2x + (b-a) \cot 2x$$

$$= c \operatorname{cosec} z + d \cot z$$

where $c = a+b$, $d = b-a$, and $z = 2x$.

Problem 5.5

The superpotential and the eigenenergies for the Eckart potential are given by $W = -A \coth r + \frac{B}{A}$ and $E_n^{(-)} = A^2 - (A+n)^2 + B^2 \left[\frac{1}{A^2} - \frac{1}{(A+n)^2} \right]$ respectively. In order for the supersymmetry to remain unbroken, we must have $B > A^2$. To have two excited states, we need $B > (A+2)^2$.

To have n bound states, we must have $B > (A+n)^2$. With this relationship between parameters A and B, we must have following conditions satisfied: $E_n^{(-)} > 0$; $E_n^{(-)} > E_{n-1}^{(-)}$, etc.

To show that $E_n^{(-)} > 0$, we first simplify $E_n^{(-)}$ to get

$$E_n^{(-)} = (2A+n)n \left[\frac{B^2}{A^2(A+n)^2} - 1 \right].$$

Since $B > (A+n)^2$, the expression within the above square bracket is positive, and hence $E_n^{(-)} > 0$.

Similarly, for the energy levels to progressively increase, we must have $E_n^{(-)} > E_{n-1}^{(-)}$. To show this, we start with the expression

$$E_n^{(-)} - E_{n-1}^{(-)} = (2A + 2n - 1) \left[\frac{B^2}{A^2(A+n)^2} - 1 \right].$$

Now using $\frac{B^2}{A^2(A+n)^2} > 1$, we get the desired result.

Problem 5.6

From Eq. (5.15) we have

$$\left(\frac{d}{dx} + W(x) \right) \psi_E^-(x) = N_1 \, \psi_E^+(x) \text{ and } \left(-\frac{d}{dx} + W(x) \right) \psi_E^+(x) = N_2 \, \psi_E^-(x) ,$$

where N_1 and N_2 are constants. Applying these conditions to the asymptotic forms given in Eq. (5.14), we obtain

$$\left(\frac{d}{dx} + W_{-\infty} \right) \left(e^{ikx} + r^-(k) e^{-ikx} \right) = N_1 \left(e^{ikx} + r^+(k) e^{-ikx} \right) ,$$

and

$$\left(\frac{d}{dx} + W_\infty \right) t^- e^{ik'x} = N_1 t^+(k) e^{ik'x} .$$

These lead to

$$(ik + W_{-\infty}) e^{ikx} + r^- (-ik + W_{-\infty}) e^{-ikx} = N_1 \left(e^{ikx} + r^+ e^{-ikx} \right) ,$$

$$(17.32)$$

and

$$(ik' + W_\infty) t^-(k') = N_1 t^+(k') . \qquad (17.33)$$

Comparing the coefficients of e^{ikx} in Eq. (17.32), and using Eq. (17.33), we obtain the following relationships:

$$(ik + W_{-\infty}) = N_1 \quad \text{and} \quad t^+(k') = \frac{(ik' + W_\infty)}{N_1} t^-(k') . \qquad (17.34)$$

Eliminating the coefficient N_1,

$$t^+(k') = \left(\frac{ik' + W_\infty}{ik + W_{-\infty}} \right) t^-(k') . \qquad (17.35)$$

Problem 5.7

From Eq. (5.17), we have

$$(-ik + W_{-\infty}) e^{ikx} + r^+ (ik + W_{-\infty}) e^{-ikx} = N_2 \left(e^{ikx} + r^- e^{-ikx} \right) . \qquad (17.36)$$

Comparing the coefficients of $e^{\pm ikx}$ in Eq. (5.16), we get the following relationships:

$$(-ik + W_{-\infty}) = N_2 \quad \text{and} \quad r^+ (ik + W_{-\infty}) = N_2 r^- . \qquad (17.37)$$

Elimination of N_2 between these two equations leads to

$$r^+ = \left(\frac{W_{-\infty} - ik}{W_{-\infty} + ik} \right) r^- ; \qquad r^- = \left(\frac{W_{-\infty} + ik}{W_{-\infty} - ik} \right) r^+ . \qquad (17.38)$$

These are same as those in Eq. (5.19).

Problem 5.8

(a) The partner potentials $V_\pm(x, a_0)$ generated by the superpotential $W(x, a_0) = a_0 \tanh x$ are:

$$\begin{aligned} V_\pm(x, a_0) &= a_0^2 \tanh^2 x \pm a_0 \text{sech}^2 x \\ &= a_0^2 (1 - \text{sech}^2 x) \pm a_0 \text{sech}^2 x \\ &= a_0^2 - a_0(a_0 \mp 1) \text{sech}^2 x) . \end{aligned}$$

(b) Using Mathematica, we have plotted $W(x, 1)$ and $V_\pm(x, 1)$ for $-5 < x < 5$. The potential $V_+(x, 1)$ is equal to unity.

(c) For $a_0 = 1$, since the potential $V_+(x, a_0)$ is a constant, the coefficients $r^+(k, a_0)$ is zero. This implies that $r^-(k, a_0)$ is zero as well, and hence $t^-(k, a_0) = 1$. Thus, the reflection amplitude for $V_-(x, a_0)$ is zero for all values of energy, provided $a_0 = 1$.

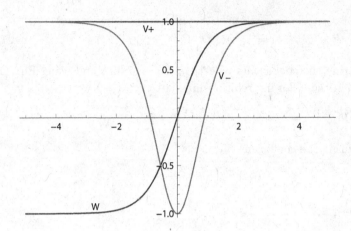

Chapter 6

Problem 6.1

(a) From Chapter 6, we have $(Q^-) = \begin{pmatrix} 0 & 0 \\ A^- & 0 \end{pmatrix}$. From this definition,

we find $(Q^-)^2 = \begin{pmatrix} 0 & 0 \\ A^- & 0 \end{pmatrix}\begin{pmatrix} 0 & 0 \\ A^- & 0 \end{pmatrix} = \begin{pmatrix} 0 & 0 \\ 0 & 0 \end{pmatrix}$. Similarly, $(Q^+)^2 =$

$\begin{pmatrix} 0 & A^+ \\ 0 & 0 \end{pmatrix}\begin{pmatrix} 0 & A^+ \\ 0 & 0 \end{pmatrix} = \begin{pmatrix} 0 & 0 \\ 0 & 0 \end{pmatrix}$.

(b) Since the hamiltonian H is given by $\{Q^-,\, Q^+\} \equiv Q^-Q^+ + Q^+Q^-$, commutator of H with $Q^\mp$ can be computed as follows:

$$[Q^-, H] = [Q^-, Q^-Q^+ + Q^+Q^-]$$
$$= [Q^- \cdot (Q^-Q^+ + Q^+Q^-) - (Q^-Q^+ + Q^+Q^-) \cdot Q^-]$$
$$= [Q^-Q^+Q^- - Q^-Q^+Q^-] = 0.$$

Similarly, $[Q^+, H] = 0$.

Problem 6.2

Since H is given by $\begin{pmatrix} H_- & 0 \\ 0 & H_+ \end{pmatrix} = \begin{pmatrix} A^+A^- & 0 \\ 0 & A^-A^+ \end{pmatrix}$, the action of H

on $\begin{pmatrix} \psi_n^-(x) \\ 0 \end{pmatrix}$ leads to

$$\begin{pmatrix} A^+A^- & 0 \\ 0 & A^-A^+ \end{pmatrix}\begin{pmatrix} \psi_n^-(x) \\ 0 \end{pmatrix} = \begin{pmatrix} A^+A^-\psi_n^-(x) \\ 0 \end{pmatrix} = E_n^{(-)}\begin{pmatrix} \psi_n^-(x) \\ 0 \end{pmatrix}.$$

Similarly,

$$\begin{pmatrix} A^+A^- & 0 \\ 0 & A^-A^+ \end{pmatrix} \begin{pmatrix} 0 \\ \psi^+_{n-1}(x) \end{pmatrix} = \begin{pmatrix} 0 \\ A^-A^+\psi^+_{n-1}(x) \end{pmatrix} = E^{(+)}_{n-1} \begin{pmatrix} 0 \\ \psi^+_{n-1}(x) \end{pmatrix}.$$

From SUSYQM, we have $E^{(+)}_{n-1} = E^{(-)}_n$, hence $\begin{pmatrix} \psi^-_n(x) \\ 0 \end{pmatrix}$ and $\begin{pmatrix} 0 \\ \psi^+_{n-1}(x) \end{pmatrix}$ are degenerate eigenstates of the hamiltonian H.

Problem 6.3

Since $H = \{Q^-, Q^+\}$, the non-vanishing of $\langle 0|H|0\rangle$ implies $\langle 0|Q^-Q^+ + Q^+Q^-|0\rangle \neq 0$. However, both $\langle 0|Q^-Q^+|0\rangle$ and $\langle 0|Q^+Q^-|0\rangle$ are positive semi-definite, since Q^- is the adjoint of Q^+; at least one of them has to be positive. Let us assume $\langle 0|Q^+Q^-|0\rangle \neq 0$. Since $\langle 0|Q^+Q^-|0\rangle = |Q^-|0\rangle|^2$, we get $|Q^-|0\rangle| \neq 0$. Thus, the vacuum is not invariant under supersymmetry.

Problem 6.4

$A^+A^-\phi^-(x) = 0$ implies $\int dx \, (\phi^-(x))^* A^+A^-\phi^-(x) = |A^-|\phi^-\rangle|^2 = 0$. This implies $|A^-|\phi^-\rangle| = 0$; hence, $A^-|\phi^-\rangle = 0$, or in coordinate space $A^-\phi^-(x) = 0$.

Problem 6.5

$N_B - N_F = 0$ implies that we have broken supersymmetry. In such a case, the eigenstates of H_- and H_+ are in one-to-one correspondence. For every state of H_-, there is an eigenstate of H_+ with the same eigenvalue. Hence,

$$\begin{aligned}
Tr\left((-)^F e^{-\beta H}\right) &= (-1)^0 \left\langle \psi^{(-)}_0|e^{-\beta H}|\psi^{(-)}_0\right\rangle + (-1)^1 \left\langle \psi^{(+)}_0|e^{-\beta H}|\psi^{(+)}_0\right\rangle \\
&\quad + (-1)^0 \left\langle \psi^{(-)}_1|e^{-\beta H}|\psi^{(-)}_1\right\rangle \\
&\quad + (-1)^1 \left\langle \psi^{(+)}_1|e^{-\beta H}|\psi^{(+)}_1\right\rangle + \cdots \\
&= \left\langle \psi^{(-)}_0|e^{-\beta E^{(-)}_0}|\psi^{(-)}_0\right\rangle - \left\langle \psi^{(+)}_0|e^{-\beta E^{(+)}_0}|\psi^{(+)}_0\right\rangle \\
&\quad + \left\langle \psi^{(-)}_1|e^{-\beta E^{(-)}_1}|\psi^{(-)}_1\right\rangle - \left\langle \psi^{(+)}_1|e^{-\beta E^{(+)}_1}|\psi^{(+)}_1\right\rangle + \cdots.
\end{aligned}$$

Since for broken supersymmetric cases, $E^{(-)}_n = E^{(+)}_n$, the sum in the above equation vanishes.

Problem 6.6

$N_B - N_F = 1$ implies that we have unbroken supersymmetry. In this case, the excited states of H_- and the eigenstates H_+ are in one-to-one correspondence. The ground state of H_-, corresponding to eigenvalue zero, has no partner among the states of H_+. Hence,

$$Tr\left((-)^F e^{-\beta H}\right) = (-1)^0 \left\langle \psi_0^{(-)} | e^{-\beta H} | \psi_0^{(-)} \right\rangle = e^{-\beta E_0^{(-)}} = 1 \ .$$

Chapter 7

Problem 7.1

The components of the angular momentum operator $\vec{L} = \vec{r} \times \vec{p}$ are:

$$L_x = yp_z - zp_y$$
$$L_y = zp_x - xp_z$$
$$L_z = xp_y - yp_x \ .$$

We write them in a compact notation by $L_i = x_j p_k - x_k p_j$, where i, j, k are in cyclic order.

In this problem, we will make extensive use of the identity

$$[AB,C] = A\,[B,C] + [A,C]\,B \ .$$

(a) Let us start with $[L_i, x_i]$. We assume i, j, k are a cyclic permutation of $1, 2, 3$.

$$[L_i, x_i] = [x_j p_k, x_i] - [x_k p_j, x_i]$$
$$= x_j\,[p_k, x_i] + [x_j, x_i]\,p_k - x_k\,[p_j, x_i] - [x_k, x_i]\,p_j \ .$$

Using $[x_i, p_j] = i\delta_{i,j} = 0$, and $[x_i, x_i] = 0$, we get $[L_i, x_i] = 0$. Similarly,

$$[L_i, x_j] = [x_j p_k - x_k p_j, x_j]$$
$$= [x_j p_k, x_j] - [x_k p_j, x_j]$$
$$= -[x_k p_j, x_j]$$
$$= -x_k\,[p_j, x_j] - [x_k, x_j]\,p_j = -x_k(-i) = ix_k \ .$$

Commutators $[L_i, p_i]$, $[L_i, p_j]$ are determined by very similar steps.

(b) Now let us determine commutators of the type: $[L_i, L_j]$.

$$[L_i, L_j] = [x_j p_k - x_k p_j, L_j]$$
$$= [x_j p_k, L_j] - [x_k p_j, L_j]$$
$$= x_j [p_k, L_j] + [x_j, L_j] p_k - x_k [p_j, L_j] - [x_k, L_j] p_j .$$

From the results of part (a):

1st Term: $x_j [p_k, L_j] = -x_j [L_j, p_k] = -\imath x_j p_i$

2nd Term: $[x_j, L_j] p_k = -[L_j, x_j] p_k = 0$

3rd Term: $-x_k [p_j, L_j] = 0$

4th Term: $-[x_k, L_j] p_j = +[L_j, x_k] p_j = +\imath x_i p_j$

Putting all of these together, we get $[L_i, L_j] = \imath (x_i p_j - x_j p_i) = \imath L_k$. This can be written as

$$\left[\vec{L} \times \vec{L} \right] = \imath \vec{L} .$$

(c) Now we show that the square of the angular momentum operator L^2 commutes with each component L_i; i.e., $[L^2, L_i] = 0$.

$$[L^2, L_i] = [L_i^2 + L_j^2 + L_k^2, L_i]$$
$$= 0 + [L_j^2, L_i] + [L_k^2, L_i]$$
$$= L_j [L_j, L_i] + [L_j, L_i] L_j + L_k [L_k, L_i] + [L_k, L_i] L_k$$
$$= -\imath L_j L_k - \imath L_k L_j + \imath L_k L_j + \imath L_j L_k$$
$$= 0 .$$

Problem 7.2

(a) Up to a normalization constant, the action of the lowering operator $L_- = -e^{-\imath\phi} \left(\frac{\partial}{\partial\theta} - \imath \cot\theta \frac{\partial}{\partial\phi} \right)$ on $Y_2^2 \sim \sin^2\theta e^{2\imath\phi}$ gives:

$$L_- Y_2^2 \sim -e^{-\imath\phi} \left(\frac{\partial}{\partial\theta} - \imath \cot\theta \frac{\partial}{\partial\phi} \right) \sin^2\theta e^{2\imath\phi}$$
$$\sim -e^{-\imath\phi} \left[e^{2\imath\phi} 2\sin\theta\cos\theta - \imath\cot\theta\sin^2\theta \cdot 2\imath e^{2\imath\phi} \right]$$
$$\sim e^{-\imath\phi} [2\sin\theta\cos\theta + 2\sin\theta\cos\theta]$$
$$\sim e^{\imath\theta} \cdot 4\sin\theta\cos\theta$$
$$\sim e^{\imath\theta} \cdot \sin\theta\cos\theta$$
$$\sim Y_2^1 .$$

(b) We compute the actions of $L_\mp$ and L_3 on Y_2^2.

$$L_- Y_2^2 = 4Ne^{i\phi} \sin\theta \cos\theta; \quad \text{where } N \text{ is the normalization constant.}$$

$$
\begin{aligned}
L_+ Y_2^2 &\sim -e^{-i\phi}\left(\frac{\partial}{\partial\theta} + i\cot\theta\frac{\partial}{\partial\phi}\right)\sin^2\theta e^{2i\phi} \\
&\sim -e^{-i\phi}\left[e^{2i\phi}2\sin\theta\cos\theta + i\cot\theta\sin^2\theta \cdot 2ie^{2i\phi}\right] \\
&\sim e^{-i\phi}\left[2\sin\theta\cos\theta - 2\sin\theta\cos\theta\right] \\
&= 0 \, .
\end{aligned}
$$

$$L_3 Y_2^2 = -i\frac{\partial}{\partial\phi}N\sin^2\theta e^{2i\phi} = 2\,Ne^{2i\phi}\sin^2\theta = 2\,Y_2^2 \, .$$

(c) Now we compute the action of L^2 on Y_2^2. For this, we use $L^2 = L_+L_- + L_3^2 - L_3$.

$$
\begin{aligned}
L_+(L_- Y_2^2) &= -e^{i\phi}\left(\frac{\partial}{\partial\theta} + i\cot\theta\frac{\partial}{\partial\phi}\right)4Ne^{i\phi}\sin\theta\cos\theta \\
&= -4Ne^{i\phi}\left[e^{i\phi}\frac{\partial}{\partial\theta}\sin\theta\cos\theta + i\cot\theta\sin\theta\cos\theta\frac{\partial}{\partial\phi}e^{i\phi}\right] \\
&= -4Ne^{i\phi}\left[e^{i\phi}(\cos^2\theta - \sin^2\theta) + i\cos^2\theta i\phi e^{i\phi}\right] \\
&= -4Ne^{2i\phi}\left[\cos^2\theta - \sin^2\theta - \cos^2\theta\right] \\
&= 4Ne^{2i\phi}\sin^2\theta \, .
\end{aligned}
$$

$$\left(L_3\right)^2 Y_2^2 = i\frac{\partial}{\partial\phi}\left(2Ne^{2i\phi}\sin^2\theta\right) = 4Ne^{2i\phi}\sin^2\theta \, .$$

Combining these operations, we get

$$
\begin{aligned}
\left(L_+L_- + L_3^2 - L_3\right)Y_2^2 &= 4Ne^{2i\phi}(-\sin^2\theta) + 4Ne^{2i\phi}\sin^2\theta - 2Ne^{2i\phi}\sin^2\theta \\
&= Ne^{2i\phi}\left[4\sin^2\theta + 4\sin^2\theta - 2\sin^2\theta\right] \\
&= 6Ne^{2i\phi}\sin^2\theta \\
&= 2\times 3Y_2^2 \\
&= 2(2+1)Y_2^2 \quad \text{as expected.}
\end{aligned}
$$

Chapter 8

Problem 8.1

Apart from normalization constants, the eigenfunctions are

$$\psi_n(\xi) = N_n \, H_n(\xi) \, e^{-\frac{\xi^2}{2}}, \tag{17.39}$$

where $N_n = \left(\frac{1}{\pi}\right)^{\frac{1}{4}} \frac{1}{\sqrt{2^n n!}}$. We will now apply the SUSY operators $A^+(\xi, a_n) \equiv -\frac{d}{d\xi} + \xi$, and $A^-(\xi, a_n) \equiv \frac{d}{d\xi} + \xi$ to these eigenstates, and use the following relationship we had derived from SUSYQM and shape invariance:

$$\psi_n^{(-)}(\xi, a_j) = \frac{1}{\sqrt{E_{n-1}^{(+)}}} \, A^+(\xi, a_j) \, \psi_{n-1}^{(-)}(\xi, a_{j+1}) \, .$$

Substituting for A^+ we obtain

$$N_n e^{-\frac{\xi^2}{2}} H_n(\xi) = \frac{1}{\sqrt{2n}} \left[-\frac{d}{d\xi} + \xi \right] N_{n-1} e^{-\frac{\xi^2}{2}} H_{n-1}(\xi) \, .$$

After some simplification we obtain the following recursion relation:

$$H_n(\xi) = \left[2\xi - \frac{d}{d\xi} \right] H_{n-1}(\xi) \, .$$

Similarly, using the identity that relates eigenfunctions of partner potentials via A^-, the following recursion relation can be generated:

$$2n H_{n-1}(\xi) = \left[\frac{d}{d\xi} \right] H_n(\xi) \, .$$

Problem 8.2

We substitute $\theta = f(z)$ into

$$\frac{d^2 P_{\ell,m}}{d\theta^2} + \cot\theta \frac{dP_{\ell,m}}{d\theta} + \left[\ell(\ell+1) - \frac{m^2}{\sin^2\theta} \right] P_{\ell,m} = 0 \, . \tag{17.40}$$

We use

$$\frac{dP_{\ell,m}}{d\theta} = \frac{dP_{\ell,m}}{dz} \frac{dz}{d\theta} = \frac{1}{f'} \frac{dP_{\ell,m}}{dz} \, ,$$

where $f' = \frac{d\theta}{dz}$. The second derivative is

$$\frac{d^2 P_{\ell,m}}{d\theta^2} = \frac{1}{f'} \frac{d}{dz} \left(\frac{1}{f'} \frac{dP_{\ell,m}}{dz} \right) = -\frac{f''}{f'^3} \frac{dP_{\ell,m}}{dz} + \frac{1}{f'^2} \frac{d^2 P_{\ell,m}}{dz^2} \, .$$

Substituting in Eq. (17.40), we obtain

$$\frac{1}{f'^2}\frac{d^2 P_{\ell,m}}{dz^2} + \left(\frac{\cot f}{f'} - \frac{f''}{f'^3}\right)\frac{dP_{\ell,m}}{dz} + \left[\ell(\ell+1) - \frac{m^2}{\sin^2 f}\right]P_{\ell,m} = 0 \ .$$

Multiplying the above equation by f'^2,

$$\frac{d^2 P_{\ell,m}}{dz^2} + \left[-\frac{f''}{f'} + f'\cot f\right]\frac{dP_{\ell,m}}{dz} + f'^2\left[\ell(\ell+1) - \frac{m^2}{\sin^2 f}\right]P_{\ell,m} = 0 \ .$$

Problem 8.3

Denoting differentiation with respect to x by prime, and suppressing sub- and superscripts,

$$xL'' + (\alpha + 1 - x)L' + nL = 0 \ .$$

We choose the ansatz $L \equiv Uf$, and demand that the coefficient of U' in the resulting differential equation vanish. $L' = Uf' + fU'$ and $L'' = Uf'' + 2f'U' + fU''$.

Then the differential equation becomes

$$x[fU'' + 2f'U' + Uf''] + (\alpha + 1 - x)[fU' + Uf'] + nUf = 0 \ .$$

Collecting terms in orders of U, the coefficient of U' is $2xf' + (\alpha + 1 - x)f$. $2xf' + (\alpha + 1 - x)f = 0$ has the solution

$$f = cx^{-\frac{(\alpha+1)}{2}} e^{\frac{x}{2}} \ ,$$

suppressing α and n.

Chapter 9

Problem 9.1

Because $\hat{W}(x) = W(x) + f(x)$, where

$$f(x) = \frac{\left[\psi_0^{(-)}(x)\right]^2}{\int_{-\infty}^{x}\left[\psi_0^{(-)}(t)\right]^2 dt + \lambda}$$

in order to show that the two superpotentials have the same asymptotic limits we need to prove that $\lim_{x\to\pm\infty} f(x) = 0$. At $x \to \pm\infty$ the denominator of $f(x)$ is finite, while $\psi_0^{(-)}(x)$ vanishes.

Problem 9.2

Given that

$$\hat{W}(x) = W(x) + \frac{d}{dx}\ln[I(x) + \lambda] \ .$$

we want to show

$$\hat{V}_-(\lambda, x) = \hat{W}(x)^2 - \hat{W}(x)' = V_-(x) - 2\frac{d^2}{dx^2}\ln[I(x) + \lambda] .$$

$$\hat{W}(x)^2 = \left(W(x) + \frac{d}{dx}\ln[I(x) + \lambda]\right)^2$$

$$= W(x)^2 + \underbrace{\left(\frac{d}{dx}\ln[I(x) + \lambda]\right)^2}_{f^2} + \underbrace{2W(x)\frac{d}{dx}\ln[I(x) + \lambda]}_{2Wf} .$$

$$= W(x)^2 - \frac{d^2}{dx^2}\ln[I(x) + \lambda]$$

where we used the results of Eq. (9.7). Thus, for $\hat{W}(x)^2 - \hat{W}(x)'$ we obtain

$$\hat{W}(x)^2 - \hat{W}(x)' = W(x)^2 - W(x)' - 2\frac{d^2}{dx^2}\ln[I(x) + \lambda]$$

$$\hat{V}_-(\lambda, x) = = V_-(x) - 2\frac{d^2}{dx^2}\ln[I(x) + \lambda] .$$

Problem 9.3

We want to show that these deformations do not change the reflection and transmission coefficients. From Eq. (9.3), we have

$$r_-(k) = \frac{W_- + ik}{W_- - ik}\, r_+(k) , \qquad t_-(k) = \frac{W_+ - ik'}{W_- - ik}\, t_+(k) , \qquad (17.41)$$

where $W_\pm$ are the asymptotic values of the superpotential. Let us compute the asymptotic values of the deformed superpotential $\hat{W}(x)$. Since the difference between the two superpotentials, $\hat{W}(x)$ and $W(x)$, is $\frac{d}{dx}\ln[I(x) + \lambda]$, we will focus on the asymptotic values of this function. From Eq. (9.11), this difference is given by

$$\frac{d}{dx}\ln[I(x) + \lambda] = \frac{\left[\psi_0^{(-)}(x)\right]^2}{\int_{-\infty}^{x}\left[\psi_0^{(-)}(t)\right]^2 dt + \lambda} .$$

where $\psi_0^{(-)}(x)$ is the ground state wave function of the undeformed system. Since this must vanish for large distances, asymptotic values for this difference function are zero; i.e., $\hat{W}_\pm = W_\pm$. Hence, deformation does not alter the reflection and transmission amplitudes.

Problem 9.4

From Eq. (9.15), we have

$$\hat{\psi}_0^{(-)}(x) = \frac{\sqrt{\lambda(\lambda + 1)}}{I(x) + \lambda}\, \psi_0^{(-)}(x) . \qquad (17.42)$$

We assume that $\psi_0^{(-)}(x)$ is a normalized ground state wave function. We want to show that $\hat{\psi}_0^{(-)}(x)$ as given by Eq. (17.42) is normalized.

$$\int_{-\infty}^{\infty} \left| \hat{\psi}_0^{(-)}(x) \right|^2 dx = \int_{-\infty}^{\infty} \left| \frac{\sqrt{\lambda(\lambda+1)}}{I(x)+\lambda} \psi_0^{(-)}(x) \right|^2 dx$$

$$= \lambda(\lambda+1) \int_{-\infty}^{\infty} \frac{\left(\psi_0^{(-)}(x) \right)^2}{(I(x)+\lambda)^2} dx$$

$$= \lambda(\lambda+1) \int_{\lambda}^{\lambda+1} \frac{1}{\xi^2} d\xi$$

$$= \lambda(\lambda+1) \left(\frac{1}{\lambda} - \frac{1}{\lambda+1} \right) = 1 .$$

In the above derivation, $\xi \equiv I(x)+\lambda$. The limits of integration are obtained by noticing that $I(-\infty) = 0$, and $I(\infty) = 1$.

Problem 9.5

Without loss of generality we set $B = A$. We also choose the scale factor $\alpha = 1$. Then, the ground state of Morse potential has the form

$$\psi_0^{(-)}(x) = (2A)^A e^{-Ae^{-x}} e^{-Ax} . \tag{17.43}$$

This then leads to

$$I(x) \equiv \int_{-\infty}^{x} \left[\psi_0^{(-)}(t) \right]^2 dt = (2A)^{2A} \int_{-\infty}^{x} e^{-2Ae^{-t}} \left(e^{-2At} \right) dt . \tag{17.44}$$

The deformed ground state wave function is given by

$$\hat{\psi}_0^{(-)}(x) = \frac{\sqrt{\lambda(\lambda+1)}}{I(x)+\lambda} (2A)^A e^{-Ae^{-x}} e^{-Ax} .$$

For $\lambda = \infty$, the deformed wave function reduces to the undeformed $\psi_0^{(-)}(x)$. To plot the deformed wave function for other values of λ, let us set $A = 1/2$. Thus,

$$\psi_0^{(-)}(x) = e^{-\frac{1}{2}\left(x+e^{-x} \right)}, \text{ and } I(x) = \int_{-\infty}^{x} e^{-(t+e^{-t})} dt = e^{-e^{-x}} .$$

We see that the form of $I(x)$ obeys $I(-\infty) = 0$, and $I(\infty) = 1$. Wave function $\hat{\psi}_0^{(-)}(x)$ is now given by

$$\hat{\psi}_0^{(-)}(x) = \frac{\sqrt{\lambda(\lambda+1)}}{e^{-e^{-x}}+\lambda} e^{-\frac{1}{2}\left(x+e^{-x} \right)} .$$

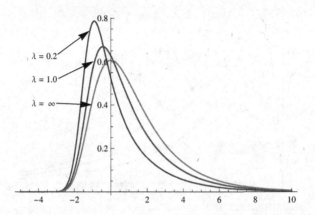

Fig. 17.5   Deformed ground state wave functions for $\lambda = 0.2, 1.0$ and ∞.

For $\lambda = 0.2, 1.0$ and ∞ the deformed ground state wave functions are shown in Fig. 17.5.

Problem 9.6

This problem is similar to the previous one for the Morse potential. For Coulomb, the ground state wave function is

$$\psi_0^{(-)}(r, q) = \frac{q^3}{\sqrt{2}} \, r \exp\left(-\frac{q^2}{2} r\right) .$$

This then gives

$$I(r) \equiv \int_0^r \left[\psi_0^{(-)}(t)\right]^2 dt = \frac{1}{2}\left(2 - e^{-q^2 r}\left(q^2 r \left(q^2 r + 2\right) + 2\right)\right) .$$

We see that, as expected, $I(0) = 0$ and $I(\infty) = 1$. The deformed wave function $\hat{\psi}_0^{(-)}(r, q)$ is then given by

$$\hat{\psi}_0^{(-)}(r, q) = \frac{\sqrt{\lambda(\lambda + 1)}}{I(r) + \lambda} \, \psi_0^{(-)}(r, q)$$

$$= \frac{\sqrt{\lambda(\lambda + 1)}}{\frac{1}{2}\left(2 - e^{-q^2 r}\left(q^2 r \left(q^2 r + 2\right) + 2\right)\right) + \lambda} \, \frac{q^3}{\sqrt{2}} \, r \exp\left(-\frac{q^2}{2} r\right) .$$

In Fig. 17.6 we have drawn deformed ground state wave functions for $\lambda = 0.2, 1.0$ and ∞.

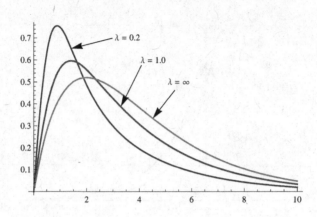

Fig. 17.6   Deformed ground state wave functions for $\lambda = 0.2, 1.0$ and ∞.

Chapter 10

Problem 10.1

A state $\psi_n(x, a)$ is derived by applying a product of operators $\mathcal{A}^+(x, a)\, \mathcal{A}^+(x, a + 1) \cdots \mathcal{A}^+(x, a + n)$ on the ground state $\psi_0(x, a + n)$. In our case, the value of a is -3 and since we want to derive ψ_3, our n is 3. So, our starting ground state is $\psi_0(x, a + n) = \psi_0(x, 0)$ which, from Eq. (10.31), is given by

$$\psi_0(x, 0) = N \exp \left[\frac{\left(1 - 2\hbar^2 P\right) \tan^{-1}\left(\frac{e^x}{\hbar \sqrt{Q}} \right)}{\hbar \sqrt{Q}} \right]. \qquad (17.45)$$

The asymptotic value of this state for $x \to \infty$ is $N \exp \left[\frac{(1 - 2\hbar^2 P)}{\hbar \sqrt{Q}} \frac{\pi}{2} \right]$. Since the wave function does not vanish as $x \to \infty$, it is not normalizable. Hence, $\psi_3(x, -3)$ built upon this state would not be normalizable either.

Chapter 11

Problem 11.1

Near the origin, Rosen-Morse, Eckart and Pöschl-Teller potentials are described by $-A/x$, $-A/r$ and $-(A + B)/r$ respectively. The behavior of the corresponding potentials (V_-) are

$$\text{Rosen-Morse:} \quad V_-(x, A, B) \approx \frac{A(A+1)}{x^2}$$

$$\text{Eckart:} \quad V_-(x, A, B) \approx \frac{A(A+1)}{r^2}$$

$$\text{Pöschl-Teller:} \quad V_-(x, A, B) \approx \frac{(A+B)(A+B+1)}{r^2} .$$

Hence, the corresponding α's are $A(A+1)$, $A(A+1)$ and $(A+B)(A+B+1)$ respectively.

Problem 11.2

The normalization condition for wave function ψ is given by

$$\int_0^{r_0} |\psi|^2 \, dr = 1 .$$

Let us assume that the behavior of ψ near the origin is of the form r^m. Then the above integral yields

$$\frac{r^{2m+1}}{2m+1} \Big|_0^{r_0} .$$

For $m > -\frac{1}{2}$, $2m + 1$ is a positive number, hence the integral does not blow up near the origin. $m = -\frac{1}{2}$ leads to a logarithmic divergence near the origin.

Problem 11.3

We want to solve the following approximate differential equation:

$$-\frac{d^2\psi}{dr^2} + \frac{\alpha}{r^2} \, \psi \approx 0 . \tag{17.46}$$

Multiplying Eq. (17.46) by r^2, we obtain

$$-r^2 \frac{d^2\psi}{dr^2} + \alpha \, \psi \approx 0 .$$

Let us substitute the ansatz: $\psi = r^m$. This leads to $m^2 - m - \alpha = 0$. Its solutions are

$$m = \frac{1}{2} \left(1 \pm \sqrt{1 + 4\alpha} \right) .$$

Thus, the approximate solution for ψ near the origin is a linear combination of two possible solutions:

$$\psi \approx c_1 \, r^{\left(\frac{1}{2}-\sqrt{\alpha+\frac{1}{4}}\right)} + c_2 \, r^{\left(\frac{1}{2}+\sqrt{\alpha+\frac{1}{4}}\right)} . \tag{17.47}$$

Problem 11.4

For $-\frac{1}{4} \leq \alpha < \frac{3}{4}$, we want to check the behavior of both terms of Eq. (17.47) near the origin. Let us consider the exponent $\left(\frac{1}{2} - \sqrt{\alpha + \frac{1}{4}}\right)$ at the two extreme values of α. For $-\frac{1}{4} \leq \alpha < \frac{3}{4}$, the exponent $-\frac{1}{2} < \frac{1}{2} - \sqrt{\alpha + \frac{1}{4}} \leq \frac{1}{2}$ and the exponent $\frac{1}{2} \leq \frac{1}{2} + \sqrt{\alpha + \frac{1}{4}} < \frac{3}{2}$. Thus, neither of the terms of Eq. (17.47) diverges near the origin.

Problem 11.5

From Eq. (11.5), we have

$$-\frac{d^2\psi}{dr^2} + \frac{\alpha}{r_0{}^2}\psi = 0 \, . \tag{17.48}$$

It is a linear homogeneous differential equation with constant coefficients. For such equations, the trial solution used is $\psi(r) = e^{mr}$. Substituting this ansatz in the above equation we obtain

$$-m^2 e^{mr} + \frac{\alpha}{r_0{}^2} e^{mr} = 0.$$

The constant m is then given by $\pm\frac{\sqrt{\alpha}}{r_0}$. Hence, a general solution of the differential equation is given by

$$\begin{aligned}
\psi(r) &= C e^{\frac{\sqrt{\alpha}}{r_0} r} + D e^{-\frac{\sqrt{\alpha}}{r_0} r} \\
&= \frac{1}{2} C \left(e^{\frac{\sqrt{\alpha}}{r_0} r} + e^{-\frac{\sqrt{\alpha}}{r_0} r} + e^{\frac{\sqrt{\alpha}}{r_0} r} - e^{-\frac{\sqrt{\alpha}}{r_0} r} \right) \\
&\quad + \frac{1}{2} D \left(e^{\frac{\sqrt{\alpha}}{r_0} r} + e^{-\frac{\sqrt{\alpha}}{r_0} r} - e^{\frac{\sqrt{\alpha}}{r_0} r} + e^{-\frac{\sqrt{\alpha}}{r_0} r} \right) \\
&= C \left(\cosh \frac{\sqrt{\alpha}}{r_0} r + \sinh \frac{\sqrt{\alpha}}{r_0} r \right) + D \left(\cosh \frac{\sqrt{\alpha}}{r_0} r - \sinh \frac{\sqrt{\alpha}}{r_0} r \right) \\
&= (C + D) \cosh \frac{\sqrt{\alpha}}{r_0} r + (C - D) \sinh \frac{\sqrt{\alpha}}{r_0} r \\
&= A \cosh \frac{\sqrt{\alpha}}{r_0} r + B \sinh \frac{\sqrt{\alpha}}{r_0} r \, . \tag{17.49}
\end{aligned}$$

Problem 11.6

Since we will be taking a limit $r_0 \to 0$, Eq. (11.3) still governs the wave function just outside the well of radius r_0. We have derived its solution in

Problem 11.3. It is given by the linear combination

$$\psi \approx c_1 \, r^{\left(\frac{1}{2} - \sqrt{\alpha + \frac{1}{4}}\right)} + c_2 \, r^{\left(\frac{1}{2} + \sqrt{\alpha + \frac{1}{4}}\right)}. \tag{17.50}$$

Problem 11.7

Since $W(r, \alpha) = -\frac{(1 + \sqrt{1 + 4\alpha})}{2r}$, the partner potentials are

$$V_-(r, \alpha) = \left(-\frac{(1 + \sqrt{1 + 4\alpha})}{2r}\right)^2 + \frac{d}{dr}\frac{(1 + \sqrt{1 + 4\alpha})}{2r},$$

$$= \frac{(1 + 1 + 4\alpha + 2\sqrt{1 + 4\alpha})}{4r^2} - \frac{(1 + \sqrt{1 + 4\alpha})}{2r^2},$$

$$= \frac{\alpha}{r^2}.$$

Similarly,

$$V_+(r, \alpha) = \left(-\frac{(1 + \sqrt{1 + 4\alpha})}{2r}\right)^2 - \frac{d}{dr}\frac{(1 + \sqrt{1 + 4\alpha})}{2r},$$

$$= \frac{(1 + 1 + 4\alpha + 2\sqrt{1 + 4\alpha})}{4r^2} + \frac{(1 + \sqrt{1 + 4\alpha})}{2r^2},$$

$$= \frac{(1 + \sqrt{1 + 4\alpha})(3 + \sqrt{1 + 4\alpha})}{4r^2}.$$

Problem 11.8

From Eqs. (11.24) and (11.25), we have

$$W^2(x, a_0) + W'(x, a_0) = W^2(x, a_1) - W'(x, a_1) + \omega_0, \tag{17.51}$$
$$W^2(x, a_1) + W'(x, a_1) = W^2(x, a_0) - W'(x, a_0) + \omega_1.$$

Adding these equations we obtain

$$W'(x, a_0) + W'(x, a_1) = \frac{1}{2}(\omega_0 + \omega_1) \equiv \omega$$

I.e.,

$$W(x, a_0) + W(x, a_1) = \omega x \; ; \quad \text{or} \quad W(x, a_1) = \omega x - W(x, a_0) .$$

Substituting $W(x, a_1)$ in Eq. (17.51), we obtain

$$W^2(x, a_0) + W'(x, a_0) = (\omega x - W(x, a_0))^2 - (\omega - W'(x, a_0)) + \omega_0 .$$

This yields

$$2\omega x \, W(x, a_0) = \omega^2 x^2 - \omega + \omega_0 .$$

Dividing this equation by $2\omega x$, we obtain

$$W(x, a_0) = \frac{1}{2}\omega x + \frac{1}{2}\frac{\Omega}{\omega}\frac{1}{x} , \qquad (17.52)$$

where $\omega = \frac{1}{2}(\omega_0 + \omega_1)$ and $\Omega = \frac{1}{2}(\omega_0 - \omega_1)$.

Problem 11.9

We want to show that for all positive values of ω_0 and ω_1, the quantity $\frac{1}{4}\frac{\Omega(\Omega+2\omega)}{\omega^2}$ obeys the constraint

$$-\frac{1}{4} \leq \frac{1}{4}\frac{\Omega(\Omega+2\omega)}{\omega^2} < \frac{3}{4} ,$$

where $\omega = \frac{1}{2}(\omega_0 + \omega_1)$ and $\Omega = \frac{1}{2}(\omega_0 - \omega_1)$. This implies $-\infty < \Omega < \infty$ and $0 < \omega < \infty$. The fraction $\frac{\Omega(\Omega+2\omega)}{\omega^2}$ can be written as

$$\frac{\Omega^2}{\omega^2} + 2\frac{\Omega}{\omega} = \left(\frac{\Omega}{\omega} + 1\right)^2 - 1 . \qquad (17.53)$$

Since $-1 \leq \frac{\Omega}{\omega} \leq 1$, the quantity $\frac{\Omega^2}{\omega^2} + 2\frac{\Omega}{\omega}$ obeys

$$-1 \leq \frac{\Omega^2}{\omega^2} + 2\frac{\Omega}{\omega} \leq 3 .$$

Thus,

$$-\frac{1}{4} \leq \frac{1}{4}\frac{\Omega(\Omega+2\omega)}{\omega^2} < \frac{3}{4} .$$

Problem 11.10

For a homogeneous linear differential equation of the type

$$\frac{d^2\psi}{dr^2} + p(r)\frac{d\psi}{dr} + q(r)\psi = 0 ,$$

if one solution is $f(r)$, then the other solution can be obtained[2] by setting $\psi(r) = f(r)v(r)$. The substitution of $\psi(r) = f(r)v(r)$ reduces the equation to a first order in $\frac{dv}{dr}$. This is then integrated to yield

$$v(r) = \int \frac{e^{-\int p(r)dr}}{[f(r)]^2} dr . \qquad (17.54)$$

In our case the differential equation is

$$-\frac{d^2\psi}{dr^2} + \frac{\alpha}{r^2}\psi = 0 .$$

---

[2]R. K. Nagle, E. B. Saff and A. D. Snider, *Fundamentals of Differential Equations*, 5th Ed., Reading, PA: Addison Wesley (2001).

Two linearly independent solutions for ψ are

$$r^{\left(\frac{1}{2}-\sqrt{\alpha+\frac{1}{4}}\right)} \quad \text{and} \quad r^{\left(\frac{1}{2}+\sqrt{\alpha+\frac{1}{4}}\right)} .$$

For $\alpha = -\frac{1}{4}$, both solutions are identical: $\psi = \sqrt{r}$. Hence, from Eq. (17.54), since $p(r) = 0$ and $f(r) = \sqrt{r}$, the two linearly independent solutions are $\sqrt{r}$ and $\sqrt{r} \log r$.

Chapter 12

Problem 12.1 The WKB condition is

$$\int_0^a \sqrt{E_n - V(x)}\, dx = n\pi .$$

In this case we break the domain into three parts, and the integral becomes

$$\int_0^a \sqrt{E_n - V(x)}\, dx = \int_0^{\frac{a}{3}} \sqrt{E_n - V(x)}\, dx + \int_{\frac{a}{3}}^{\frac{2a}{3}} \sqrt{E_n - V(x)}\, dx$$

$$+ \int_{\frac{2a}{3}}^a \sqrt{E_n - V(x)}\, dx$$

$$n\pi = \frac{2a}{3} \sqrt{E_n} + \frac{a}{3} \sqrt{E_n - V_0} .$$

In the absence of the perturbation; i.e., for the infinite well, the energy $E_n^0 = \frac{n^2\pi^2}{a^2}$. Hence $a = \frac{n\pi}{\sqrt{E_n^0}}$. Substituting this value of a into the above equation, we get

$$\frac{2}{3}\frac{n\pi}{\sqrt{E_n^0}}\sqrt{E_n} + \frac{n\pi}{\sqrt{3E_n^0}}\sqrt{E_n - V_0} = n\pi .$$

Simplifying,

$$E_n = 5E_n^0 \pm \frac{4}{3}\sqrt{9(E_n^0)^2 - 3E_n^0 V_0} - \frac{V_0}{3} .$$

As $V_0 \to 0$, this reduces to $E_n = E_n^0$ if we choose the minus sign.

Problem 12.2

The half-harmonic oscillator potential is

$$V(x) = \begin{cases} \infty & x < 0 \\ \frac{1}{4}\omega^2 x^2 & x > 0 . \end{cases}$$

For this case the WKB condition is

$$\int \sqrt{E_n - V(x)}\, dx = \int_0^{x_{cl}} \sqrt{E_n - \frac{1}{4}\omega^2 x^2}\, dx$$

$$= \left(n - \frac{1}{4}\right)\pi \quad n = 1, 2, 3, \ldots .$$

x_{cl} is the classical turning point: $E_n = \frac{1}{4}\omega^2 x_{cl}^2$. From the above equation we obtain

$$\frac{\omega}{2}\int_0^{x_{cl}} \sqrt{x_{cl}^2 - x^2}\, dx = \frac{\omega\pi}{8} x_{cl} = \frac{\omega\pi}{8} \cdot \frac{4E_n}{\omega^2} = \frac{E_n \pi}{2\omega} = \left(n - \frac{1}{4}\right)\pi ,$$

which gives $E_n = \left(2n - \frac{1}{2}\right)\omega$. This is the exact result.

Problem 12.3

The potential is given by

$$V = \frac{\omega^2 x^2}{4} , \qquad -\infty < x < \infty .$$

In this case, $E_n = \frac{\omega^2}{4}x_{cl1}^2 = \frac{\omega^2}{4}x_{cl2}^2$, or equivalently $x_{cl1}^2 = x_{cl2}^2 = \frac{4E_n}{\omega^2}$. WKB condition is

$$\int_{x_{cl1}}^{x_{cl2}} \sqrt{E_n - \frac{1}{4}\omega^2 x^2}\, dx = 2\int_0^{x_{cl2}} \sqrt{E_n - \frac{1}{4}\omega^2 x^2}\, dx = \left(n - \frac{1}{2}\right)\pi .$$

This integral was evaluated in Problem (12.2). Hence, $\frac{E_n \pi}{\omega} = \left(n - \frac{1}{2}\right)\pi$, or

$$E_n = \left(n - \frac{1}{2}\right)\omega \quad n = 1, 2, 3, \ldots .$$

the exact solution for the harmonic oscillator.

Problem 12.4

Our objective is to derive Eq. (12.19):

$$\frac{\partial I(\alpha)}{\partial \alpha} = -\frac{\sin^{-1}\left(\sqrt{\frac{\alpha}{E}}\tan y\right)}{2\sqrt{\alpha}} + \frac{\sin^{-1}\left(\sqrt{\frac{E+\alpha}{E}}\tan y\right)}{2\sqrt{E+\alpha}} .$$

We start from

$$\frac{\partial I(\alpha)}{\partial \alpha} = -\frac{1}{2} \int_{-\tan^{-1}\sqrt{E}}^{\tan^{-1}\sqrt{E}} \left(\frac{\tan^2 y}{\sqrt{E - \alpha \cdot \tan^2 y}} \right) dy$$

$$= -\frac{1}{2} \int_{-\tan^{-1}\sqrt{E}}^{\tan^{-1}\sqrt{E}} \left(\frac{\sec^2 y}{\sqrt{E - \alpha \cdot \tan^2 y}} \right) dy$$

$$+ \frac{1}{2} \int_{-\tan^{-1}\sqrt{E}}^{\tan^{-1}\sqrt{E}} \left(\frac{dy}{\sqrt{E - \alpha \cdot \tan^2 y}} \right)$$

$$= -\frac{1}{2\sqrt{\alpha}} \int \left(\frac{d(\tan y)}{\sqrt{\frac{E}{\alpha} - \tan^2 y}} \right) + \frac{1}{2} \int \left(\frac{\cos y \, dy}{\sqrt{E - (E + \alpha) \cdot \sin^2 y}} \right)$$

$$= -\frac{1}{2\sqrt{\alpha}} \sin^{-1} \left(\sqrt{\frac{\alpha}{E}} \tan y \right) + \frac{1}{2\sqrt{E + \alpha}} \sin^{-1} \left(\sqrt{\frac{E + \alpha}{E}} \tan y \right).$$

Problem 12.5

The SWKB condition is $\int_{x_1}^{x_2} \sqrt{E_s - W^2} = n\pi$, $n = 1, 2, 3 \cdots$. E_s is the SWKB energy eigenvalue, and x_1, x_2 are the turning points, where $E_s - W^2 = 0$.

The superpotential for the harmonic oscillator is $W = \omega/2$; thus, the potential is $V_s = W^2 - W' = \omega^2 x^2/4 - \omega/2$. Note that this is shifted down from the standard V used in the WKB method: $V_s = V - \omega/2$.

Now the SWKB integral, $\int_{x_1}^{x_2} \sqrt{E_s - W^2}$ looks like the WKB integral, $\int_{x_1}^{x_2} \sqrt{E - V}$. This is the same integral as in Problem (12.4), hence each SWKB energy is given by $E_{s,n}/\omega$. Setting this equal to $n\pi$, $E_{s,n} = n\omega$, $n = 0, 1, 2, 3 \ldots$. Note that this yields a zero ground state energy, as is required for (unbroken) SUSY.

Comparing this to the WKB result of Problem (12.4), $E_n = (n - 1/2)\omega$, $n = 1, 2, 3 \ldots$, the SWKB result is, as expected, shifted down by $\omega/2$ from the WKB result.

Problem 12.6

The WKB condition for the Morse potential is given by

$$\int \left[E - A^2 \left(1 - e^{-z} \right)^2 + A e^{-z} \right]^{\frac{1}{2}} dz = n\pi . \tag{17.55}$$

With a change of variables $e^{-z} = y$, the left hand side of Eq. (17.55) becomes

$$- \int \left[E - A^2 \left(1 - y \right)^2 + Ay \right]^{\frac{1}{2}} dy/y$$

$$= - \int \left[E - A^2 \left(1 - y \right)^2 - A(1 - y) + A \right]^{\frac{1}{2}} dy/y$$

$$= - \int \left[E + A + \frac{1}{4} - \left\{ A \left(1 - y \right) + \frac{1}{2} \right\}^2 \right]^{\frac{1}{2}} dy/y$$

$$= - A \int \frac{dy}{y} \left[\underbrace{\left(\frac{E + A + \frac{1}{4}}{A^2} \right)}_{\beta^2} - \left\{ (1 - y) + \frac{1}{2A} \right\}^2 \right]^{\frac{1}{2}} .$$

With two more changes of variables: $(1 - y) + \frac{1}{2A} = u$ and $y = (1 - u) + \frac{1}{2A}$, the integral becomes

$$A \int \frac{\sqrt{\beta^2 - u^2}}{(1 - y) + \frac{1}{2A}} du .$$

Now setting $u = \beta \sin \theta$, the above integral becomes

$$A \beta^2 \int_{-\pi/2}^{\pi/2} \left(\frac{\cos^2 \theta}{1 - \beta \sin \theta + \frac{1}{2A}} \right) d\theta .$$

An integration then leads to

$$\frac{1}{2A\beta^2} \left[2\sqrt{-4A^2 \left(\beta^2 - 1 \right) + 4A + 1} \tan^{-1} \left(\frac{2A\beta - (2A + 1) \tan \left(\frac{\theta}{2} \right)}{\sqrt{-4A^2 \left(\beta^2 - 1 \right) + 4A + 1}} \right) \right]$$

$$+ \left[\frac{-2A\beta \cos(\theta) + 2A\theta + \theta}{2A\beta^2} \right] . \tag{17.56}$$

Setting $-4A^2 \left(\beta^2 - 1\right) + 4A + 1 = 4A^2(1 - \alpha^2)$, where $\alpha = \frac{\sqrt{E}}{A}$, we get

$$\frac{4A\sqrt{1-\alpha^2}\tan^{-1}\left(\frac{2A\beta-(2A+1)\tan\left(\frac{\theta}{2}\right)}{2A\sqrt{1-\alpha^2}}\right) - 2A\beta\cos(\theta) + 2A\theta + \theta}{2A\beta^2}. \quad (17.57)$$

Substituting the limits for θ; i.e., $\pm\pi/2$, the term containing $\tan^{-1}$ in Eq. (17.57) gives

$$\underbrace{\tan^{-1}\left(\frac{2A\beta - (2A+1)}{2A\sqrt{1-\alpha^2}}\right)}_{\theta_1} - \underbrace{\tan^{-1}\left(\frac{2A\beta + (2A+1)}{2A\sqrt{1-\alpha^2}}\right)}_{\theta_2}.$$

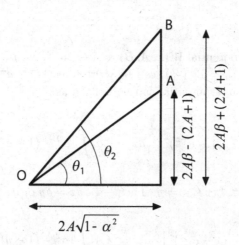

From the diagram above, the angle $\theta_1 - \theta_2$ is the angle between vectors $\overrightarrow{OA}$ and $\overrightarrow{OB}$. Since

$$\overrightarrow{OA} = [2A\beta - (2A + 1)]\hat{i} + [2A\sqrt{1 - \alpha^2}]\hat{j}$$

and

$$\overrightarrow{OB} = [2A\beta + (2A + 1)]\hat{i} + [2A\sqrt{1 - \alpha^2}]\hat{j},$$

we find that their scalar product is zero, hence the angle between them is $\theta_1 - \theta_2 = -\pi/2$. Thus, Eq. (17.57), with limits put in, becomes

$$\frac{4A\sqrt{1 - \alpha^2}(-\pi/2) + (2A + 1)\pi}{2A\beta^2}. \quad (17.58)$$

Substituting the above into our original integral of Eq. (17.56) and setting it equal to $\left(n + \frac{1}{2}\right)\pi$, we obtain

$$\frac{1}{2}\left(4A\sqrt{1-\alpha^2}(-\pi/2) + (2A+1)\pi\right) = \left(n + \frac{1}{2}\right)\pi .$$

Solving the above equation for energy, we get

$$E_n = A^2 - (A - n)^2 , \tag{17.59}$$

the correct eigenvalues for the Morse potential.

Problem 12.7

The Morse superpotential $W(x, A, B) = A - Be^{-x}$, with a shift of variable $z = x - x_0$ with $x_0 = \log\left(\frac{B}{A}\right)$, can be written as $A\left(1 - e^{-z}\right)$. The SWKB integral for this superpotential can be written as

$$A\int\left[\frac{E}{A^2} - \left(1 - e^{-z}\right)^2\right]^{\frac{1}{2}} dz = A\int\left[\alpha^2 - \left(1 - e^{-z}\right)^2\right]^{\frac{1}{2}} dz$$

where $\alpha^2 \equiv E/A^2$. Let $e^{-z} = y$. Then $-e^{-z}dz = dy$ or $dz = -dy/y$.

$$A\int\left[\alpha^2 - \left(1 - e^{-z}\right)^2\right]^{\frac{1}{2}} dz = -A\int\left[\alpha^2 - (1-y)^2\right]^{\frac{1}{2}} dy/y$$

$$= -A\int\frac{\sqrt{\alpha^2 - u^2}}{1 - u}(-du) , \tag{17.60}$$

where we have used $1 - y = u$. Now setting $u = \alpha\sin\theta$, we get

$$I(\alpha) \equiv \int\frac{\sqrt{\alpha^2 - u^2}}{1 - u} du = \alpha^2\int\frac{\cos^2\theta}{1 - \alpha\sin\theta} d\theta$$

$$= \frac{\sqrt{1-\alpha^2}(x - \alpha\cos(x)) - 2\left(\alpha^2 - 1\right)\tan^{-1}\left(\frac{\alpha - \tan\left(\frac{x}{2}\right)}{\sqrt{1-\alpha^2}}\right)}{\sqrt{1 - \alpha^2}} .$$

Since the limits are $\pm\pi/2$,

$$I(\alpha) = \frac{\sqrt{1-\alpha^2}(\pi/2 - \alpha\cos(\pi/2)) - 2(\alpha^2-1)\tan^{-1}\left(\frac{\alpha-\tan\left(\frac{\pi}{4}\right)}{\sqrt{1-\alpha^2}}\right)}{\sqrt{1-\alpha^2}}$$

$$- \frac{\sqrt{1-\alpha^2}(-\pi/2 - \alpha\cos(-\pi/2)) - 2(\alpha^2-1)\tan^{-1}\left(\frac{\alpha-\tan\left(\frac{-\pi}{4}\right)}{\sqrt{1-\alpha^2}}\right)}{\sqrt{1-\alpha^2}}$$

$$= \frac{\sqrt{1-\alpha^2}(\pi/2) - 2(\alpha^2-1)\tan^{-1}\left(\frac{\alpha-1}{\sqrt{1-\alpha^2}}\right)}{\sqrt{1-\alpha^2}}$$

$$- \frac{\sqrt{1-\alpha^2}(-\pi/2) - 2(\alpha^2-1)\tan^{-1}\left(\frac{\alpha+1}{\sqrt{1-\alpha^2}}\right)}{\sqrt{1-\alpha^2}}$$

$$= \pi + \frac{-2(\alpha^2-1)\tan^{-1}\left(\frac{\alpha-1}{\sqrt{1-\alpha^2}}\right)}{\sqrt{1-\alpha^2}} - \frac{-2(\alpha^2-1)\tan^{-1}\left(\frac{\alpha+1}{\sqrt{1-\alpha^2}}\right)}{\sqrt{1-\alpha^2}}$$

$$= \pi - \frac{2(\alpha^2-1)}{\sqrt{1-\alpha^2}}\left(\tan^{-1}\left(\frac{\alpha-1}{\sqrt{1-\alpha^2}}\right) - \tan^{-1}\left(\frac{\alpha+1}{\sqrt{1-\alpha^2}}\right)\right)$$

$$= \pi + \frac{2(\alpha^2-1)}{\sqrt{1-\alpha^2}}\left(\tan^{-1}\left(\frac{-\alpha+1}{\sqrt{1-\alpha^2}}\right) + \tan^{-1}\left(\frac{\alpha+1}{\sqrt{1-\alpha^2}}\right)\right)$$

$$= \pi + \frac{2(\alpha^2-1)}{\sqrt{1-\alpha^2}}\underbrace{\left(\tan^{-1}\left(\frac{\sqrt{1-\alpha}}{\sqrt{1+\alpha}}\right) + \tan^{-1}\left(\frac{\sqrt{1+\alpha}}{\sqrt{1-\alpha}}\right)\right)}_{\pi/2}$$

$$= \pi + \frac{(\alpha^2-1)}{\sqrt{1-\alpha^2}}\pi = \pi\left(1 - \sqrt{1-\alpha^2}\right).$$

The SWKB condition then yields

$$A\pi\left(1 - \sqrt{1-\alpha^2}\right) = n\pi. \tag{17.61}$$

Replacing α^2 by E_n/A^2, we obtain the energy for the Morse potential:

$$E_n = A^2 - (A-n)^2.$$

Chapter 13

Problem 13.1

Since $\sigma_x = \frac{S_x}{2}$, we want to show that $\chi_x^+ = \frac{1}{\sqrt{2}}\begin{pmatrix}1\\1\end{pmatrix}$ and $\chi_x^- = \frac{1}{\sqrt{2}}\begin{pmatrix}1\\-1\end{pmatrix}$
are eigenvectors of $\sigma_x = \begin{pmatrix}0 & 1\\1 & 0\end{pmatrix}$ with eigenvalues 1 and -1 respectively.

$$\sigma_x \chi_x^\pm = \begin{pmatrix}0 & 1\\1 & 0\end{pmatrix}\frac{1}{\sqrt{2}}\begin{pmatrix}1\\\pm 1\end{pmatrix} = \pm\frac{1}{\sqrt{2}}\begin{pmatrix}1\\\pm 1\end{pmatrix} = \pm\chi_x^\pm.$$

Similarly, $\chi_y^+ = \frac{i}{\sqrt{2}}\begin{pmatrix}1\\i\end{pmatrix}$ and $\chi_y^+ = \frac{i}{\sqrt{2}}\begin{pmatrix}i\\1\end{pmatrix}$ are eigenvectors of $\sigma_y = \begin{pmatrix}0 & -i\\i & 0\end{pmatrix}$ with eigenvalues ± 1.

Problem 13.2

We want to derive the anti-commutation relations:

$$\alpha_i\alpha_j + \alpha_j\alpha_i = 0, \quad j \neq i \quad \text{and} \quad \alpha_i\beta + \beta\alpha_i = 0, \quad i = 1,\, 2,\, 3\,.$$

In Eq. (15.4) we obtained

$$p^2 + m^2 = \sum_{i=1}^{3}\alpha_i p_i \sum_{j=1}^{3}\alpha_j p_j + m^2\beta^2 + m\left[\left(\sum_{i=1}^{3}\alpha_i p_i\right)\beta + \beta\left(\sum_{j=1}^{3}\alpha_j p_j\right)\right]$$

$$= \sum_{i=1}^{3}(\alpha_i p_i)^2 + \sum_{j\neq i}\alpha_i\alpha_j p_i p_j + m^2\beta^2 + m\sum_{i=1}^{3}p_i(\alpha_i\beta + \beta\alpha_i)\,.$$

$$(17.62)$$

Since $p^2 + m^2$ only has diagonal terms of the form $\sum_i (p_i)^2$, the coefficients of the cross terms must vanish. Hence, $\sum_{j\neq i}\alpha_i\alpha_j p_i p_j = 0$. We need to be careful about this constraint to insure that we do not double count.

$$\sum_{j\neq i}\alpha_i\alpha_j p_i p_j = \left(\sum_{j<i}\alpha_i\alpha_j + \sum_{j>i}\alpha_i\alpha_j\right)p_i p_j\,.$$

$$= \left(\sum_{j<i}\alpha_i\alpha_j + \sum_{j<i}\alpha_j\alpha_i\right)p_i p_j$$

$$= \sum_{j<i}(\alpha_i\alpha_j + \alpha_j\alpha_i)p_i p_j\,, \qquad (17.63)$$

where we have used dumminess of the indices i and j, and the symmetry of the factor $p_i p_j$ under the interchange of indices. For the last sum of Eq. (17.63) to be zero, we must have

$$\alpha_i \alpha_j + \alpha_j \alpha_i = 0 \ .$$

Similarly, since $p^2 + m^2$ does not have a term linear in momentum, the last term of Eq. (17.62) must vanish as well. This implies

$$\alpha_i \beta + \beta \alpha_i = 0 \ .$$

Problem 13.3

We want to show that $[\vec{L}, H] = i(\vec{\alpha} \times \vec{p})$.

$$[L_x, H] = [L_x, \alpha_x p_x + \alpha_y p_y + \alpha_z p_z] + [L_x, \beta m] + \left[L_x, -\frac{e^2}{r} \right] \ .$$

The operator L_x contains coordinates and their derivatives, and hence commutes with β; i.e., $[L_x, \beta m] = 0$. The angular momentum operator L_x commutes with any spherically symmetric function of coordinates; i.e., $[L_x, V(r)] = 0$. So

$$[L_x, H] = \alpha_x \underbrace{[L_x, p_x]}_{0} + \alpha_y \underbrace{[L_x, p_y]}_{i\,p_z} + \alpha_z \underbrace{[L_x, p_z]}_{-i\,p_y}$$

$$= i(\alpha_y p_z + \alpha_z p_y) \ . \tag{17.64}$$

Generalizing to the other components of $\vec{L}$

$$\left[\vec{L}, H \right] = i(\vec{\alpha} \times \vec{p}) \ .$$

Problem 13.4

We want to show that

$$[\vec{\sigma}, H] = [\vec{\sigma}, \vec{\alpha} \cdot \vec{p} + \beta m + V(\vec{r})] = -2i(\vec{\alpha} \times \vec{p}) \ .$$

Since the operator $\sigma_i \equiv \begin{pmatrix} \sigma_i & 0 \\ 0 & \sigma_i \end{pmatrix}$ commutes with the operator $\beta = \begin{pmatrix} 1 & 0 \\ 0 & -1 \end{pmatrix}$, and with the scalar potential $V(\vec{r})$, we have

$$[\vec{\sigma}, H] = [\vec{\sigma}, \vec{\alpha} \cdot \vec{p}] = [\vec{\sigma}, \alpha_x p_x + \alpha_y p_y + \alpha_z p_z] \ .$$

Now

$$[\sigma_x, \alpha_x] = \begin{pmatrix} \sigma_x & 0 \\ 0 & \sigma_x \end{pmatrix} \begin{pmatrix} 0 & \sigma_x \\ \sigma_x & 0 \end{pmatrix} - \begin{pmatrix} 0 & \sigma_x \\ \sigma_x & 0 \end{pmatrix} \begin{pmatrix} \sigma_x & 0 \\ 0 & \sigma_x \end{pmatrix}$$

$$= \begin{pmatrix} 0 & \sigma_x^2 \\ \sigma_x^2 & 0 \end{pmatrix} - \begin{pmatrix} 0 & \sigma_x^2 \\ \sigma_x^2 & 0 \end{pmatrix} = 0 \,. \tag{17.65}$$

Similarly,

$$[\sigma_x, \alpha_y] = \begin{pmatrix} \sigma_x & 0 \\ 0 & \sigma_x \end{pmatrix} \begin{pmatrix} 0 & \sigma_y \\ \sigma_y & 0 \end{pmatrix} - \begin{pmatrix} 0 & \sigma_y \\ \sigma_y & 0 \end{pmatrix} \begin{pmatrix} \sigma_x & 0 \\ 0 & \sigma_x \end{pmatrix}$$

$$= \begin{pmatrix} 0 & \sigma_x\sigma_y \\ \sigma_x\sigma_y & 0 \end{pmatrix} - \begin{pmatrix} 0 & \sigma_y\sigma_x \\ \sigma_y\sigma_x & 0 \end{pmatrix}$$

$$= \begin{pmatrix} 0 & [\sigma_x,\sigma_y] \\ [\sigma_x,\sigma_y] & 0 \end{pmatrix} = i \begin{pmatrix} 0 & 2\sigma_z \\ 2\sigma_z & 0 \end{pmatrix} = 2i\,\alpha_z \,, \tag{17.66}$$

and

$$[\sigma_x, \alpha_z] = \begin{pmatrix} 0 & [\sigma_x,\sigma_z] \\ [\sigma_x,\sigma_z] & 0 \end{pmatrix} = -i \begin{pmatrix} 0 & 2\sigma_y \\ 2\sigma_y & 0 \end{pmatrix} = -2i\,\alpha_y \,.$$

Similarly, we obtain

$$[\sigma_y, \alpha_y] = 0, \quad [\sigma_y, \alpha_x] = -2i\,\sigma_z, \quad \text{and} \quad [\sigma_y, \alpha_z] = 2i\,\alpha_x \,,$$

and

$$[\sigma_z, \alpha_z] = 0, [\sigma_z, \alpha_x] = 2i\,\alpha_y, \quad \text{and} \quad [\sigma_z, \alpha_y] = -i\,\alpha_x \,.$$

Putting all of these together

$$[\sigma_x, H] = -2i\,(\vec{\alpha} \times \vec{p})_x \,.$$

Generalizing this result, we obtain

$$[\vec{\sigma}, H] = -2i\,(\vec{\alpha} \times \vec{p}) \,.$$

Problem 13.5

The exact fine structure is

$$E_{nj} = m \left\{ \left[1 + \left(\frac{\alpha}{n - (j + \frac{1}{2}) + \sqrt{(j + \frac{1}{2})^2 - \alpha^2}} \right)^2 \right]^{-\frac{1}{2}} \right\}. \tag{17.67}$$

Since $\alpha \ll 1$, $\sqrt{(j + 1/2)^2 - \alpha^2} = (j + 1/2)\sqrt{1 - \frac{\alpha^2}{(j+1/2)^2}} \approx (j + 1/2) - \frac{\alpha^2}{2(j+1/2)}$.

Canceling the $j + 1/2$s in the denominator, and pulling n out, the term in large round brackets of Eq. 17.67 becomes $\frac{\alpha}{n}\left[1 + \frac{\alpha^2}{2n(j+1/2)}\right]$. Invoking again the smallness of α, the entire term in the curly brackets becomes, to order α^4, $1 - \frac{\alpha^2}{2n^2}\left[1 + \frac{\alpha^2}{n(j+1/2)}\right] + \frac{3\alpha^4}{8n^4}$. Collecting terms in powers of α, this is $1 - \frac{\alpha^2}{2n^2} + \frac{\alpha^4}{2n^4}\left(\frac{-n}{j+1/2} + \frac{3}{4}\right)$. This is now reduced to the Bohr fine structure:

$$\frac{-13.6eV}{n^2}\left[1 + \frac{\alpha^2}{n^2}\left(\frac{n}{j+1/2} - \frac{3}{4}\right)\right].$$

Subtracting the rest energy, the Bohr-model fine structure is

$$\frac{-13.6eV}{n^2}\left[\frac{\alpha^2}{n^2}\left(\frac{n}{j+1/2} - \frac{3}{4}\right)\right].$$

Problem 13.6

$$
\begin{aligned}
A^-A^+\mathcal{F} &= \left(-\frac{d}{d\mu} + \frac{s}{\mu} - \frac{\gamma}{s}\right)\left(\frac{d}{d\mu} + \frac{s}{\mu} - \frac{\gamma}{s}\right)\mathcal{F} \\
&= -\frac{d^2\mathcal{F}}{d\mu^2} + \left(\frac{s}{\mu} - \frac{\gamma}{s}\right)^2 + \left[\left(\frac{s}{\mu} - \frac{\gamma}{s}\right) - \left(\frac{s}{\mu} - \frac{\gamma}{s}\right)\right]\frac{d\mathcal{F}}{d\mu} - \frac{s}{\mu^2}\mathcal{F} \\
&= -\frac{d^2}{d\mu^2}\mathcal{F} + \left(\frac{s(s+1)}{\mu^2} - \frac{2\gamma^2}{\mu} + \frac{2s}{\mu^2}\right)\mathcal{F}.
\end{aligned}
$$

Since $A^-A^+\mathcal{F} = \left(\frac{\gamma^2}{s^2} + 1 - \frac{m^2}{E^2}\right)\mathcal{F}$

$$A^-A^+\mathcal{F} - \left(\frac{\gamma^2}{s^2} + 1 - \frac{m^2}{E^2}\right)\mathcal{F} = -\frac{d^2}{d\mu^2}\mathcal{F} + \left(\frac{s(s+1)}{\mu^2} - \frac{2\gamma^2}{\mu} - 1 + \frac{m^2}{E^2}\right)\mathcal{F} = 0.$$

A similar calculation yields

$$-\frac{d^2\mathcal{G}}{d\mu^2} + \left(\frac{s(s+1)}{\mu^2} - \frac{2\gamma^2}{\mu} - 1 + \frac{m^2}{E^2}\right)\mathcal{G} = 0.$$

Chapter 14

Problem 14.1

The constraint on values of α and β comes from Eq. (14.5).

$$z^{1+\beta}(1 - z)^{-\alpha-\beta} = \frac{2z(1 - z)}{\sqrt{R(z)}},$$

where $R(z) = az^2 + bz + c$. Solving for $R(z)$, we obtain

$$R(z) = 4z^2(1-z)^2 z^{-2(1+\beta)}(1-z)^{2(\alpha+\beta)} = 4z^{-2\beta}(1-z)^{2(1+\alpha+\beta)}.$$

We must have

$$az^2 + bz + c = 4z^{-2\beta}(1-z)^{2(1+\alpha+\beta)}.$$

First of all, since $R(z)$ is a quadratic function of z, β cannot be positive. The only allowed values for β are 0, $-\frac{1}{2}$ and -1. We will consider these two cases separately.

(1) $\beta = 0$. In this case, there are three possibilities for the combination $2(1 + \alpha + \beta) = 2(1 + \alpha)$. It can be 0, 1 or 2. Corresponding values for α are -1, $-\frac{1}{2}$, 0.

(2) $\beta = -\frac{1}{2}$. In this case, there are two possibilities for the combination $2(1 + \alpha + \beta) = 2(\frac{1}{2} + \alpha)$. It can be 0 or 1. Corresponding values for α are $-\frac{1}{2}$, 0.

(3) $\beta = -1$. In this case, there is only one allowed value for the combination $2(1 + \alpha + \beta) = 2(\alpha)$. It has to be equal to zero. Hence, $\alpha = 0$.

Summary of possibilities:
$\alpha = 0$, $\beta = 0$, $-\frac{1}{2}$, -1 ; $\alpha = -\frac{1}{2}$, $\beta = 0$, $-\frac{1}{2}$; $\alpha = -1$, $\beta = 0$.

Problem 14.2

Comparing this case of $\alpha = 0$, $\beta = -1$ with the case $\alpha = \beta = 0$ described in the text, we see that z and $1 - z$ simply exchange roles. Define $q \equiv 1 - z$ and the identical calculation occurs, producing the Eckart potential.

Problem 14.3

For $\alpha = -1$, $\beta = 0$. Eq. (14.5) becomes,

$$z(1-z) = \frac{2z(1-z)}{\sqrt{R(z)}} ,$$

whence $R(z) = 4$. The Natanzon potential is then

$$U[z(r)] = \frac{-fz(1-z) + h_0(1-z) + h_1 z}{4} + \frac{1}{4} . \tag{17.68}$$

Defining $A \equiv \frac{-f}{4}, B \equiv \frac{h_0}{4}, C \equiv \frac{h_1}{4}$, with $z = [1 + \tanh\left(\frac{r}{2}\right)]/2$,

$$U = A \left(\frac{1 + \tanh\left(\frac{r}{2}\right)}{2} \right) \left(\frac{1 - \tanh\left(\frac{r}{2}\right)}{2} \right) + B \left(\frac{1 - \tanh\left(\frac{r}{2}\right)}{2} \right)$$
$$+ C \left(\frac{1 + \tanh\left(\frac{r}{2}\right)}{2} \right) + \frac{1}{4} .$$

Expanding the first term and collecting terms,

$$U = \frac{A}{4} - \frac{A}{4} \tanh^2 \left(\frac{r}{2} \right) + \frac{B + C}{2} + \frac{C - B}{2} \tanh \left(\frac{r}{2} \right) + \frac{1}{4} ,$$

$$= \frac{A + 1}{4} + \frac{B + C}{2} + \frac{1}{4} - \frac{A}{4} \tanh^2 \left(\frac{r}{2} \right) + \frac{C - B}{2} \tanh \left(\frac{r}{2} \right) .$$

Since this must have the form $W^2 - \frac{dW}{dr}$, we examine various possibilities, and eventually choose

$$W = \tilde{m}_1 \tanh \left(\frac{r}{2} \right) + \tilde{m}_2 .$$

This gives U, with the identifications

$$\tilde{m}_2^2 - \tilde{m}_2 = \frac{A + 2B + 2C + 1}{4}, \quad \tilde{m}_1^2 + 2\tilde{m}_2 = -\frac{A}{4}, \quad 2\tilde{m}_1 \tilde{m}_2 + \tilde{m}_2^2 = \frac{C - B}{2} .$$

This is the Rosen-Morse potential.

Chapter 15

Problem 15.1

From Eq. (15.39), we have

$$W^2(r, \alpha_0) + \frac{dW(r, \alpha_0)}{dr} = W^2(r, \alpha_1) - \frac{dW(r, \alpha_1)}{dr} + g(\alpha_1) - g(\alpha_0) .$$

To see the singularity structure at infinity, we change variables to $u = \frac{1}{r}$.

$$\frac{dW(r, \alpha_0)}{dr} = \frac{dW(r(u), \alpha_0)}{du} \frac{du}{dr} = -u^2 \frac{dW(1/u, \alpha_0)}{du} .$$

This leads to the following shape invariance condition:

$$W^2 \left(\frac{1}{u}, \alpha_0 \right) - u^2 \frac{dW\left(\frac{1}{u}, \alpha_0\right)}{du} = W^2 \left(\frac{1}{u}, \alpha_1 \right) + u^2 \frac{dW\left(\frac{1}{u}, \alpha_1\right)}{du} + R(\alpha_0) ,$$

$$(17.69)$$

where $R(\alpha_0) = g(\alpha_1) - g(\alpha_0)$.

Problem 15.2

From Chapter 15, we obtain

$$W^2\left(1/u, \alpha_0\right) = \frac{(c_{-1}(\alpha_0))^2}{u^2} + \frac{(2\,c_{-1}(\alpha_0)\,c_0(\alpha_0))}{u}$$
$$+ \left((c_0(\alpha_0))^2 + 2\,c_{-1}(\alpha_0)\,c_1(\alpha_0)\right) + \cdots$$

and

$$-u^2\,\frac{d\,W(1/u, \alpha_0)}{du} = c_{-1}(\alpha_0) - c_1(\alpha_0)\,u^2 - 2\,c_2(\alpha_0)\,u^3 + \cdots .$$

Substituting these in Eq. (17.69) we obtain

$$
\begin{bmatrix}
c_{-1}(\alpha_0)^2 u^{-2} + \\
2\,c_{-1}(\alpha_0)\,c_0(\alpha_0)u^{-1} + \\
c_0(\alpha_0)^2 + 2\,c_{-1}(\alpha_0)\,c_1(\alpha_0) \\
+ c_{-1}(\alpha_0) - c_1(\alpha_0)\,u^2 + \cdots
\end{bmatrix}
-
\begin{bmatrix}
c_{-1}(\alpha_1)^2 u^{-2} + \\
2\,c_{-1}(\alpha_1)\,c_0(\alpha_1)u^{-1} + \\
c_0(\alpha_1)^2 + 2\,c_{-1}(\alpha_1)\,c_1(\alpha_1) \\
- c_{-1}(\alpha_1) + c_1(\alpha_1)\,u^2 + \cdots \\
+ g(\alpha_1) - g(\alpha_0)
\end{bmatrix}
= 0 .
$$

Setting the coefficient of u^{-2} to zero, we obtain $c_{-1}(\alpha_0)^2 = c_{-1}(\alpha_1)^2$. The solution of this constraint is $c_{-1}(\alpha_0) = c_{-1}(\alpha_1)$. Then, the coefficient of u^{-1} gives $c_0(\alpha_0) = c_0(\alpha_1)$. Now from the u-independent terms we get

$$2\,c_{-1}(\alpha_0)\,c_1(\alpha_0) + c_{-1}(\alpha_0) + g(\alpha_0) = 2\,c_{-1}(\alpha_1)\,c_1(\alpha_1) - c_{-1}(\alpha_1) + g(\alpha_1) .$$

Problem 15.3

In this problem we are asked to derive the energy assuming that the origin was not a singular point. We assume the domain to be $(-\infty, \infty)$. This allows us to translate the origin by a finite amount without affecting the system. The expansion of W in powers of the coordinate r would then have the form

$$W(r, \alpha_0) = \sum_{k=1}^{\infty} b_k(\alpha_0)\,r^k = b_1(\alpha_0)\,r + b_2(\alpha_0)\,r^2 + \cdots$$

The derivative of the superpotential is then given by

$$\frac{d\,W(r, \alpha_0)}{dr} = b_1(\alpha_0) + 2\,b_2(\alpha_0)\,r + \cdots ;$$

and the square of the superpotential is

$$W^2(r, \alpha_0) = (b_1(\alpha_0))^2 r^2 + 2 b_1(\alpha_0))b_2(\alpha_0))r^3 \cdots$$

Substituting these expansions into the shape invariance condition (15.39), we arrive at the following constraints on coefficients of the superpotential $W(x, \alpha_0)$:

$$b_1(\alpha_0) + g(\alpha_0) = -b_1(\alpha_1) + g(\alpha_1) .$$

$$b_2(\alpha_0) = -b_2(\alpha_1) . \tag{17.70}$$

These constraints are solved by

$$b_1(\alpha_0) = \omega , \text{ (a constant)} . \tag{17.71}$$

Then, $g(\alpha_0)$ is given by

$$g(\alpha_0) = \omega \, \alpha_0 , \tag{17.72}$$

and hence, the eigenvalues of the system are given by

$$E_n = g(\alpha_n) - g(\alpha_0) = g(\alpha_0 + n) - g(\alpha_0) = \omega \, (\alpha_0 + n) - \omega \, \alpha_0 = n\omega .$$

Problem 15.4

We want to show that the superpotential with singularity at the origin and no singularity at infinity is equivalent to the superpotential that is regular at the origin and has singularity at infinity. Here we consider the case that $b_{-1} = 0$ and $c_{-1} \neq 0$.

Near $r = 0$, the superpotential is given by

$$W(r, \alpha_0) = b_0(\alpha_0) + b_1(\alpha_0)r + \cdots . \tag{17.73}$$

From W, we obtain

$$\frac{dW}{dx} = \frac{dW}{dr}\frac{dr}{dx} = r\frac{dW}{dr} = b_1(\alpha_0)\, r + 2b_2(\alpha_0)\, r^2 + \cdots , \tag{17.74}$$

and

$$W^2 = b_0(\alpha_0)^2 + 2b_0(\alpha_0)\, b_1(\alpha_0)\, r + \cdots . \tag{17.75}$$

The shape invariance condition is

$$W^2(\alpha_0) + r\frac{dW(\alpha_0)}{dr} = W^2(\alpha_1) - r\frac{dW(\alpha_1)}{dr} + R(\alpha_0) . \tag{17.76}$$

Substituting the expansion of W into this equation and equating powers of r, we obtain

$$b_0^2(\alpha_0) + g(\alpha_0) = b_0^2(\alpha_1) + g(\alpha_1) . \tag{17.77}$$

From Eq. (17.77), we get $b_0(\alpha_0) = \pm\sqrt{\gamma - g(\alpha_0)}$. We carry out a similar analysis in the region $r \to \infty$, where the superpotential has the form

$$\widetilde{W}(u, \alpha_0) = \frac{c_{-1}(\alpha_0)}{u} + c_0(\alpha_0) + c_1(\alpha_0)\, u + c_1(\alpha_0)\, u^2 + \cdots , \quad (17.78)$$

and we obtain

$$c_{-1}(\alpha_0)^2 = c_{-1}(\alpha_1)^2 \Rightarrow c_{-1}(\alpha_0) \text{ is independent of } \alpha_0 , \quad (17.79)$$

$$2\, c_{-1}(\alpha_0)\, c_0(\alpha_0) + c_{-1}(\alpha_0) = 2\, c_{-1}(\alpha_1)\, c_0(\alpha_1) - c_{-1}(\alpha_1) ,$$

$$\Rightarrow \quad c_0(\alpha_0) + 1 = c_0(\alpha_0 + 1)$$

$$\Rightarrow \quad c_0(\alpha_0) = \alpha_0 . \quad (17.80)$$

Thus, near $r = 0$

$$W(r, \alpha_0) = \pm\sqrt{\gamma - g(\alpha_0)} + b_1(\alpha_0)r + \cdots \quad (17.81)$$

and near $r = \infty$,

$$\widetilde{W}(u, \alpha_0) = \frac{c_{-1}}{u} + \alpha_0 + \cdots . \quad (17.82)$$

Near $r = 0$, the QHJ equation is given by

$$p^2 - i\, r\frac{dp}{dr} = E - W^2(r, \alpha_0) + r\frac{dW(r, \alpha_0)}{dr} , \quad (17.83)$$

where the superpotential $W(r, \alpha_0)$ is given by Eq. (15.83). Expanding the quantum momentum function as

$$p(r) = \frac{p_{-1}}{r} + p_0 + p_1 r + \cdots , \quad (17.84)$$

and substituting it in the above QHJ equation, we obtain

$$p_{-1}^2 = 0; \quad 2p_{-1}p_0 + i\, p_{-1} = 0; \quad p_0^2 = E - b_0(\alpha_0)^2 \quad \text{etc.} \quad (17.85)$$

Thus, we obtain $p_{-1} = 0$ and $p_0 = \pm\sqrt{E - b_0(\alpha_0)^2} = \pm\sqrt{E - (\gamma - g(\alpha_0))}$.

Near $r \to \infty$, we expand the QMF as $\widetilde{p}(u) = \widetilde{p}_{-1}u^{-1} + \widetilde{p}_0 + \widetilde{p}_1 u + \cdots$. We substitute this expansion for $\widetilde{p}(u)$ and the superpotential given by Eq. (15.84) into the QHJ equation

$$\widetilde{p}^2 + i\, u\frac{d\widetilde{p}}{du} = E - \left(\widetilde{W}^2(u, \alpha_0) + u\frac{d\widetilde{W}(u, \alpha_0)}{du} \right) . \quad (17.86)$$

Now comparing various powers of u, we obtain $\widetilde{p}_{-1} = ic_{-1}$, $\widetilde{p}_0 = i(\alpha_0 + 1)$, etc.

The quantization condition for this system then yields

$$\oint_{x=-\infty} p\, dx + \oint_{x=\infty} p\, dx = -2\pi n \ . \qquad (17.87)$$

Since near $x = -\infty$, the variable $r = e^x \to 0$ and as $x \to \infty$, the variable $u \equiv 1/r = e^{-x} \to 0$, we write the above condition as

$$\oint_{r=0} \left(\frac{p_0}{r}\right) dr - \oint_{u=0} \left(\frac{\tilde{p}_0}{u}\right) du = -2\pi n \ .$$

This then gives

$$\pm 2\pi\, i \sqrt{E_n - (\gamma - g(\alpha_0))} - 2\pi\, i \cdot i(\alpha_0 + 1) = -2\pi n \ .$$

Solving for E_n we obtain

$$E_n = (\gamma - g(\alpha_0)) - (1 + \alpha_0 + n\ )^2 \ .$$

Since we assume that supersymmetry is unbroken, we must have $E_0 = 0$. This implies $g(\alpha_0) = \gamma - (1 + \alpha_0)^2$, hence $g(\alpha_n) = \gamma - (1 + \alpha_0 + n)^2$. Thus, we have

$$E_n = g(\alpha_n) - g(\alpha_0) = (1 + \alpha_0)^2 - (1 + \alpha_0 + n)^2 \ . \qquad (17.88)$$

If we identify $1 + \alpha_0$ with the parameter $-A$, we obtain the energy $E_n = A^2 - (A - n)^2$.

Chapter 16

Problem 16.1

Using the definition of the Poisson bracket, $\{A, B\}_P = \frac{\partial A}{\partial x}\frac{\partial B}{\partial p} - \frac{\partial A}{\partial p}\frac{\partial B}{\partial x}$, the first term of Eq. (16.3) becomes

$$\{x^3, p^3\}_P = 3x^2\, 3p^2 = 9\, x^2\, p^2$$

and the second

$$3\,\{xp^2, x^2 p\} = 3\,(p^2\, x^2 - 4x^2\, p^2) = -9x^2\, p^2 \ .$$

For Eq. (16.4) we use $\hat{x} \cdot \hat{p} = i\hbar + \hat{p} \cdot \hat{x}$ to express all products $\hat{x} \cdot \hat{p}$ in terms of $\hat{p} \cdot \hat{x}$. Then, from the commutator $[\hat{x}^3, \hat{p}^3] = \hat{x}^3 \cdot \hat{p}^3 - \hat{p}^3 \cdot \hat{x}^3$ in Eq. (16.4) we expand the term $\hat{x}^3 \cdot \hat{p}^3$ as follows:

$$\begin{aligned}
\hat{x}^3 \cdot \hat{p}^3 &= \hat{x}^2 \cdot \hat{x} \cdot \hat{p} \cdot \hat{p}^2 = \hat{x}^2\,(i\hbar + \hat{p} \cdot \hat{x}) \cdot \hat{p}^2 = i\hbar \hat{x}^2 \cdot \hat{p}^2 + \hat{x}^2 \cdot \hat{p} \cdot \hat{x} \cdot \hat{p}^2 \\
&= i\hbar \hat{x} \cdot (\hat{x} \cdot \hat{p}) \cdot \hat{p} + \hat{x}^2 \cdot \hat{p} \cdot (\hat{x} \cdot \hat{p}) \cdot \hat{p} \\
&= 3i\hbar \hat{p}^2 \cdot \hat{x}^2 + 3i\hbar\,(\hat{p}^2 \cdot \hat{x}^2 + 2i\hbar \hat{p} \cdot \hat{x}) \\
&\quad + 3i\hbar\,(2i\hbar \hat{p} \cdot \hat{x} + 2i\hbar(\hat{p} \cdot \hat{x} + i\hbar) + \hat{p}^2 \cdot \hat{x}^2) + \hat{p}^3 \cdot \hat{x}^3 \\
&= -6i\hbar^3 - 18\hbar^2 \hat{p} \cdot \hat{x} + 9i\hbar \hat{p}^2 \cdot \hat{x}^2 + \hat{p}^3 \cdot \hat{x}^3 \ .
\end{aligned}$$

Consequently, the first term of Eq. (16.4) becomes

$$\frac{1}{i\hbar}[\hat{x}^3, \hat{p}^3] = \frac{1}{i\hbar}\left(-6i\hbar^3 - 18\hbar^2\hat{p}\cdot\hat{x} + 9i\hbar\hat{p}^2\cdot\hat{x}^2 + \hat{p}^3\cdot\hat{x}^3 - \hat{p}^3\cdot\hat{x}^3\right)$$
$$= -6\hbar^2 + 18i\hbar\hat{p}\cdot\hat{x} + 9\hat{p}^2\cdot\hat{x}^2.$$

Similarly, for the second term of Eq. (16.4) we obtain

$$\frac{3}{i\hbar}\left[\frac{\hat{x}\cdot\hat{p}^2 + \hat{p}^2\cdot\hat{x}}{2}, \frac{\hat{x}^2\cdot\hat{p} + \hat{p}\cdot\hat{x}^2}{2}\right] = 3\hbar^2 - 18i\hbar\hat{p}\cdot\hat{x} - 9\hat{p}^2\cdot\hat{x}^2.$$

Putting everything together, we get

$$\frac{1}{i\hbar}[\hat{x}^3, \hat{p}^3] + \frac{3}{i\hbar}\left[\frac{\hat{x}\cdot\hat{p}^2 + \hat{p}^2\cdot\hat{x}}{2}, \frac{\hat{x}^2\cdot\hat{p} + \hat{p}\cdot\hat{x}^2}{2}\right] = -3\hbar^2,$$

yielding the deficiency.

Problem 16.2

Making the substitution $\hbar y = t$ in the third line below, we obtain for the Weyl symbol of $\hat{x}$:

$$\mathscr{W}(\hat{x}) = \hbar\int dy\, e^{-ipy}\langle x + \hbar y/2\,|\,\hat{x}\,|\,x - \hbar y/2\rangle$$
$$= \hbar\int dy\, e^{-ipy}\,(x - \hbar y/2)\,\langle x + \hbar y/2\,|\,x - \hbar y/2\rangle$$
$$= \hbar\int dy\, e^{-ipy}\,(x - \hbar y/2)\,\delta(\hbar y)$$
$$= \hbar\int \frac{1}{\hbar}dt\, e^{-ipt/\hbar}\,(x - t/2)\,\delta(t)$$
$$= x.$$

To calculate the Weyl symbol for $\hat{p}$, in the first line below we use the identity $\langle x''|\hat{p}|x'\rangle = i\hbar\frac{\partial}{\partial x'}\delta(x'' - x')$ and then we make the substitution $\hbar y = t$ in the second line. We obtain

$$\mathscr{W}(\hat{p}) = \hbar\int dy\, e^{-ipy}\langle x + \hbar y/2\,|\,\hat{p}\,|\,x - \hbar y/2\rangle$$
$$= \hbar\int dy\, e^{-ipy}(i\hbar)\frac{\partial}{\partial(\hbar y)}\delta(\hbar y)$$
$$= \hbar\int \frac{1}{\hbar}dt\, e^{-ipt/\hbar}(i\hbar)\frac{\partial}{\partial t}\delta(t) = (i\hbar)\frac{-ip}{\hbar}e^0$$
$$= p.$$

Problem 16.3

Let us consider the first term of Eq. (16.48), and suppress the explicit (x, p) dependence. On the last line we use the fact that ∂_x and p commute.

$$\left(p - \frac{i\hbar}{2}\overrightarrow{\partial_x}\right)^2 P = \left(p - \frac{i\hbar}{2}\partial_x\right)\left(pP - \frac{i\hbar}{2}\partial_x P\right)$$

$$= p^2 P - \frac{i\hbar}{2}p\,\partial_x P - \frac{i\hbar}{2}\partial_x\, p\, P - \frac{\hbar^2}{4}\partial_x{}^2 P$$

$$= \left(p^2 - i\hbar\, p\, \partial_x - \frac{\hbar^2}{4}\partial_x{}^2\right) P .$$

Similarly,

$$\left(x + \frac{i\hbar}{2}\overrightarrow{\partial_p}\right)^2 P = \left(x^2 + i\hbar\, x\, \partial_p - \frac{\hbar^2}{4}\partial_p{}^2\right) P .$$

Then, Eq. (16.49) follows.

Problem 16.4

The Moyal bracket is $[f(x), p]_* = f(x) * p - p * f(x)$. Using the Bopp shifts representation, Eq. (16.13) for the $*$-product, we obtain

$$f(x) * p = f(x)\left(p + \frac{i\hbar}{2}\overleftarrow{\partial_x}\right) = pf(x) + \frac{i\hbar}{2}\partial_x f(x) ,$$

$$p * f(x) = \left(p - \frac{i\hbar}{2}\overrightarrow{\partial_x}\right) f(x) = pf(x) - \frac{i\hbar}{2}\partial_x f(x) .$$

Thus

$$f(x) * p - p * f(x) = i\hbar\, \partial_x f(x) .$$

Bibliography

[1] R. Adhikari, R. Dutt, A. Khare, and U. Sukhatme, "Higher order WKB approximations in supersymmetric quantum mechanics", Phys. Rev. **A 38**, 1679 (1986).

[2] R. Adhikari, R. Dutt, Y. P. Varshni, "On the averaging of energy eigenvalues in the supersymmetric WKB method", Phys. Lett. **A 131**, 217–221 (1988).

[3] Z. Ahmed, "Quasi-bound state in supersymmetric quantum mechanics", Phys. Lett. **A 281**, 213–217 (2001).

[4] E. A. Akhundova, V. V. Dodonov, V. I. Man'ko, "Wigner Functions of Quadratic Systems", *Physica*, **115 A**, 215, (1982).

[5] A. N. F Aleixo and A. B. Balantekin, "A bosonized formulation of the supersymmetric quantum mechanics for shape-invariant potential systems", Jour. Phys. A: Math. and Gen. **47**, 335305 (2014). doi:10.1088/1751-8113/47/33/335305

[6] V. De Alfaro, S. Fubini, G. Furlan, M. Roncadelli, "Operator ordering and supersymmetry", Nucl. Phys. **B 296**, 402–430 (1988).

[7] V. de Alfaro, Giuliano M. Gavazzi, "Point canonical transformations in the path integral formulation of supersymmetric quantum mechanics", Nucl. Phys. **B 335**, 655–676 (1990).

[8] Y. Alhassid, F. Gürsey and F. Iachello, "Group Theory Approach to Scattering", Ann. Phys. **148** (1983) 346.

[9] V. Andreev, S. Kharchev, M. Shmakova, "Closed chains of supersymmetrical Hamiltonians", Phys. Lett. **A 181**, 142–148 (1993).

[10] A. A. Andrianov, N. V. Borisov, M. V. Ioffe, "The factorization method and quantum systems with equivalent energy spectra", Phys. Lett. **A 105**, 19–22 (1984).

[11] A. A. Andrianov, N. V. Borisov, M. I. Eides, M.V. Ioffe, "Supersymmetric origin of equivalent quantum systems", Phys. Lett. **A 109**, 143–148 (1985).

[12] A. A. Andrianov, M. V. Ioffe, V. P. Spiridonov, "Higher-derivative supersymmetry and the Witten index", Phys. Lett. **A 174**, 273–279 (1993).

[13] A. A. Andrianov, M. V. Ioffe, D. N. Nishnianidze, "Polynomial SUSY in quantum mechanics and second derivative Darboux transformations", Phys. Lett. **A 201**, 103–110 (1995).

[14] A. A. Andrianov, M. V. Ioffe and D. N. Nishnianidze, "Classical integrable two-dimensional models inspired by SUSY quantum mechanics", Jour. Phys. A: Math. and Gen. **32** 4641–4654 (1999). doi:10.1088/0305-4470/32/25/307.

[15] A. Andrianov, F. Cannata, M. Ioffe, D. Nishnianidze, "Systems with higher-order shape invariance: spectral and algebraic properties", Phys. Lett. **A 266**, 341–349 (2000).

[16] A. A. Andrianov, A. V. Sokolov, "Nonlinear supersymmetry in quantum mechanics: algebraic properties and differential representation", Nucl. Phys. **B 660**, 25–50 (2003).

[17] A. A .Andrianov and M. V. Ioffe, "Nonlinear supersymmetric quantum mechanics: concepts and realizations", J. Phys. A: Math. Theor. 45 503001 (2012). doi:10.1088/1751-8113/45/50/503001

[18] H. Aoyama, M. Sato, T. Tanaka, "N-fold supersymmetry in quantum mechanics: general formalism", Nucl. Phys. **B 619**, 105–127 (2001).

[19] G. B. Arfken, H. J. Weber, F. Harris, *Mathematical Methods for Physicists*, 5th edition, San Diego: Harcourt-Academic Press (2001).

[20] W. Arveson, "Quantization and the uniqueness of invariant structures", Comm. Math. Phys. **89**, 77 (1983).

[21] J. A. de Azcárraga, J. M. Izquierdo, A. J. Macfarlane, "Hidden supersymmetries in supersymmetric quantum mechanics", Nucl. Phys. **B 604**, 75–91 (2001).

[22] B. Bagchi, K. Samanta, "Deformation in supersymmetric quantum mechanics", Phys. Lett. **A 179**, 59–62 (1993).

[23] B. Bagchi, "Generalized quantum condition and supersymmetric quantum mechanics", Phys. Lett. **A 189**, 439–441 (1994).

[24] B. Bagchi, A. Ganguly, A. Sinha, "Supersymmetry across nanoscale heterojunction", Phys. Lett. **A 374**, 2397–2400 (2010).

[25] G. Baker, "Formulation of Quantum Mechanics Based on the Quasi-Probability Distribution Induced on Phase Space", Phys. Rev. **109**, 2198, (1958).

[26] A. B. Balantekin, "Accidental degeneracies and supersymmetric quantum mechanics", Ann. of Phys. **164**, 277 (1985).

[27] A. B. Balantekin, "Algebraic approach to shape invariance", Phys. Rev. **A 57**, 4188 (1998).

[28] A. B. Balantekin, M. A. Cândido Ribeiro and A. N. F. Aleixo, Algebraic nature of shape-invariant and self-similar potentials, Jour. Math. Phys. **32**, 2785 (1999).

[29] D. Barclay and C. J. Maxwell, "Shape invariance and the SWKB series", Phys. Lett. **A 157**, 351 (1991).

[30] D. Barclay, R. Dutt, A. Gangopadhyaya, A. Khare, A. Pagnamenta and U. Sukhatme, "New exactly solvable Hamiltonians: Shape invariance and self-similarity", Phys. Rev. **A 48**, 2786 (1993).

[31] D. Barclay, A. Khare and U. Sukhatme, "Is the Lowest Order Supersymmetric WKB Approximation Exact for All Shape Invariant Potentials?", Phys. Lett. **A 183**, 263–266 (1993).

[32] M. S. Bartlett and J. E. Moyal, "The Exact Transition Probabilities of Quantum-Mechanical Oscillators Calculated by the Phase-Space Method", Proc. Camb. Phil. Soc., **45**, 545 (1949).

[33] A. Barut, A. Inomata and R. Wilson, "Algebraic treatment of second Poschl-Teller, Morse-Rosen and Eckart equations", Jour. Phys. **A 20**, 4075 and 4083 (1987).

[34] D. Baye, G. Lévai, J.-M. Sparenberg, "Phase-equivalent complex potentials", Nucl. Phys. **A 599**, 435–456 (1996).

[35] F. Bayen, M. Flato, C. Fronsdal, A. Lichnerowicz, and D. Sternheimer, "Deformation Theory and Quantization. I. Deformations of Symplectic Structures", Ann. of Phys. **111**, 61 (1978).

[36] J. Beckers, N. Debergh, "Parastatistics and supersymmetry in quantum mechanics", Nucl. Phys. **B 340**, 767–776 (1990).

[37] J. Beckers, Y. Brihaye, and N. Debergh, "On realizations of nonlinear Lie algebras by differential operators", Jour. Math. Phys. **32**, 2791 (1999).

[38] B. Belchev and M. A. Walton, "Solving for the Wigner functions of the Morse potential in deformation quantization", J. Phys. **A 43**, 225206 (2010).

[39] A. V. Belitsky, "Supersymmetric quantum mechanics of the flux tube", Nucl. Phys. **B 913**, 551–592 (2016).

[40] M. Bentaïba, L. Chetouani, T. F. Hammann, "Feynman-Kleinert treatment of the supersymmetric generalization of the Morse potential", Phys. Lett. **A 189**, 433–438 (1994).

[41] S. Bera, B. Chakrabarti, T. K. Das, "Application of conditional shape invariance symmetry to obtain the eigen-spectrum of the mixed potential", Phys. Lett. **A 381**, 1356–1361 (2017).

[42] F. A. Berezin, "Feynman path integrals in a phase space", Sov. Phys. Usp. **23**, 763 (1980).

[43] V. P. Berezovoj, G. I. Ivashkevych, M. I. Konchatnij, "Exactly solvable diffusion models in the framework of the extended supersymmetric quantum mechanics", Phys. Lett. **A 374**, 1197–1200 (2010).

[44] D. Bermudez, D. J. Fernández C., N. Fernández-García, "Wronskian differential formula for confluent supersymmetric quantum mechanics", Phys. Lett. **A 376**, 692–696 (2012).

[45] H. Bethe and E. E. Salpeter, *Quantum Mechanics of One- and Two- Electron Atoms.* NY: Springer (1977).

[46] R. K. Bhaduri, J. Sakhr, D. W. L. Sprung, R. Dutt, and A. Suzuki, "Shape invariant potentials in SUSY quantum mechanics and periodic orbit theory", arXiv:quant-ph/0410041.

[47] D. Bhaumik, B. Dutta-Roy, B. K. Bagchi, A. Khare, "Berry phase for supersymmetric shape-invariant potentials", Phys. Lett. **A 193**, 11–14 (1994).

[48] F. Bopp, *Werner Heisenberg und die Physik unserer Zeit*, Vieweg, Braunschwieg, (1961).

[49] J. Bougie, A. Gangopadhyaya, J. V. Mallow, "Generation of a complete set of additive shape-invariant potentials from an Euler equation", Phys. Rev. Lett. 210402:1–210402:4 (2010).

[50] J. Bougie, A. Gangopadhyaya, J. V. Mallow, C. Rasinariu, "Supersymmetric Quantum Mechanics and Solvable Models", Symmetry 4(3), 452–473 (2012).

[51] A. S. Bruev, "Modification of the JWKB approximation for finding bound state energies", Phys. Lett. **A 161**, 407–410 (1992).

[52] F. Cannata, M. V. Ioffe, D. N. Nishnianidze, "Double shape invariance of the two-dimensional singular Morse model", Phys. Lett. **A 340**, 31–36 (2005).

[53] J. F. Carinena, A. Ramos, "Riccati equation, factorization method and shape invariance", Reviews in Mathematical Physics **12**, 1279–1304 (2000).

[54] J. F. Cariñena, A. Ramos, "Shape-invariant potentials depending on n parameters transformed by translation", Jour. Phys. A: Math. and Gen. **33**, 3467 (2000).

[55] J. F. Carinena, A Ramos, "Group theoretical approach to the intertwined Hamiltonians", Ann. of Phys. **292**, 42–66 (2001).

[56] J. Casahorran, "A family of supersymmetric quantum mechanics models with singular superpotentials", Phys. Lett. **A 156**, 425–428 (1991).

[57] E. Cattaruzza, E. Gozzi, C. Pagani, "Entanglement, superselection rules and supersymmetric quantum mechanics", Phys. Lett. **A 378**, 2501–2504 (2014).

[58] J. M. Cerveró, "Supersymmetric quantum mechanics: another nontrivial quantum superpotential", Phys. Lett. **A 153**, 1–4 (1991).

[59] B. Chakrabarti, "Spectral properties of supersymmetric shape invariant potentials", Pramana **70**, 41–50 (2008).

[60] S. Chaturvedi, R. Dutt, A. Gangopadhyaya, P. Panigrahi, U. Sukhatme, "Algebraic shape invariant models", Phys. Lett. **A 248**, 109–113 (1998).

[61] C.-Y. Chen, C.-L. Liu, F.-L. Lu, "Exact solutions of Schrödinger equation for the Makarov potential", Phys. Lett. **A 374**, 1346–1349 (2010).

[62] H. Y. Cheung, Nuovo Cimento B **101**, 193 (1988).

[63] T. Chow, *Mathematical Methods for Physicists - A Concise Introduction*, Cambridge Univ. Press, (2000).

[64] M. Claudson, M. B. Halpern, "Supersymmetric ground state wave functions", Nucl. Phys. **B 250**, 689–715 (1985).

[65] A. Comtet, A. D. Bandrauk and D. K. Campbell, "Exactness of Semiclassical Bound State Energies f or Supersymmetric Quantum Mechanics", Phys. Lett. **150 B**, 159–162 (1985).

[66] F. Cooper and B. Freedman, "Aspects of supersymmetric quantum mechanics", Ann. Phys. **146**, 262–288 (1983).

[67] F. Cooper, J. Ginocchio, A. Khare, "Relationship between supersymmetry and solvable potentials", Phys. Rev. D **36**, 2458 (1987).

[68] F. Cooper, A. Khare, R. Musto, and A. Wipf, "Supersymmetry and the Dirac equation", Ann. of Phys. **187**, 1–28 (1988).

[69] F. Cooper, J. N. Ginnocchio, and A. Wipf, "Supersymmetry, operator

transformations and exactly solvable potentials", J. Phys. A: Math. Gen. **22**, 3707 (1989).

[70] F. Cooper, P. Roy, "δ-expansion for the superpotential", Phys. Lett. **A 143**, 202–206 (1990).

[71] F. Cooper, J. Dawson, H. Shepard, "SUSY-based variational method for the anharmonic oscillator", Phys. Lett. **A 187**, 140–144 (1994).

[72] F. Cooper, A. Khare, and U. P. Sukhatme, "Supersymmetry and quantum mechanics", Phys. Rep. **251**, 268 (1995).

[73] F. Cooper, A. Khare and U. P. Sukhatme *Supersymmetry in Quantum Mechanics*; World Scientific: Singapore, 2001.

[74] I. L. Cooper, "Supersymmetric quantum-mechanical approach to atomic closed shells in a bare Coulomb field", Phys. Rev. **A 50**, 1040 (1994).

[75] P. Cordero and S. Salamó, Foundations of Physics **23**, 675; (1993) Jour. Math. Phys. **35**, 3301 (1994).

[76] P. Cordero and S. Salamó, "Algebraic solution for the Natanzon hypergeometric potentials", J. Math. Phys. **35**, 3301 (1994).

[77] T. Curtright, T. Uematsu, and C. Zachos,"Generating all Wigner functions", J. Math. Phys. **42**, 2396 (2001).

[78] T. Curtright, D. Fairlie, and C. Zachos,"Features of time-independent Wigner functions", Phys. Rev. **D 58**, 025002 (1998).

[79] M. Daoud, M. Kibler,"Fractional supersymmetric quantum mechanics as a set of replicas of ordinary supersymmetric quantum mechanics", Phys. Lett. **A 321**, 147–151 (2004).

[80] A. Das, S. A. Pernice, "Supersymmetry and singular potentials", Nucl. Phys. **B 561**, 357–384 (1999).

[81] A. J. Davies, "Supersymmetric quaternionic quantum mechanics", Phys. Rev. **A 49**, 714 (1994).

[82] E. J. de Vries, B. J. Schroers, "Supersymmetric quantum mechanics of magnetic monopoles: A case study", Nucl. Phys. **B 815**, 368–403 (2009).

[83] R. De, R. Dutt, and U. Sukhatme, "Path-integral solutions for shape-invariant potentials using point canonical transformations", Phys. Rev. **A 46**, 6869 (1992).

[84] R. De, R. Dutt, U.P. Sukhatme, "Role of caustics in the supersymmetric semiclassical approach to the path integral", Phys. Lett. **A 191**, 352–356 (1994).

[85] N. C. Dias and J. N. Prata, "Wigner functions with boudaries" J. Math. Phys. **43**, 4602 (2002).

[86] N. C. Dias and J. N. Prata, "Admissible states in quantum phase space", Ann. of Phys. **313**, 110 (2004).

[87] N. C. Dias and J. N. Prata, "Formal solutions of stargenvalue equations", Ann. of Phys. **311**, 120 (2004).

[88] P. A. M. Dirac, Communications of the Dublin Institute of Advanced Study, **A1**, 5–7 (1943).

[89] P. A. M. Dirac, *The Principles of Quantum Mechanics*, 4th edition, Oxford University Press (1958).

[90] P. A. M. Dirac, "The relation of Classical to Quantum Mechanics", 2nd Can. Math. Congress, Vancover 1949, U. Toronto Press, (1951).

[91] G. Dunne and J. Feinberg, "Self-isospectral periodic potentials and super-symmetric quantum mechanics", Phys. Rev. **D 57**, 1271 (1998).

[92] R. Dutt, A. Khare, U. Sukhatme, "Exactness of Supersymmetry WKB Spectra for Shape-Invariant Potentials; Phys. Lett. **181 B**, 295–298 (1986).

[93] R. Dutt, A. Khare, and U. P. Sukhatme: "Supersymmetry, shape invariance and exactly eolvable potentials", Am. Jour. Phys. **56**, 163 (1988).

[94] R. Dutt, R. De, R. Adhikari and A. Comtet, "Supersymmetric WKB approach to scattering problems", Phys. Lett. **A 152**, 381–387 (1991).

[95] R. Dutt, A. Gangopadhyaya, and U. Sukhatme, "Noncentral Potentials And Spherical Harmonics Using Supersymmetry And Shape Invariance", Am. J. Phys. **65**, 400–403 (1997).

[96] R. Dutt, A. Gangopadhyaya, C. Rasinariu and U. P. Sukhatme; "New solvable singular potentials", Jour. Phys. A-Math. & Gen. **34**, 4129 (2001).

[97] C. J. Efthimiou and D. Spector, "Shape invariance in the Calogero and Calogero-Sutherland models", Phys. Rev. **A 56**, 208 (1997).

[98] M. J. Englefield, "Algebra representations on eigenfunctions of the Rosen-Morse potential", J. Math. Phys. **28**, 827 (1987).

[99] M. J. Englefield and C. Quesne, "Dynamical potential algebras for Gendenshtein and Morse potentials", Jour. Phys. A: Math. Gen. **24**, 3557 (1987).

[100] M. J. Englefield and C. Quesne, "Dynamical potential algebras for Gendenshtein and Morse potentials," J. Phys. **A 24**, 3557–3574 (1991).

[101] F. Bagarello, "Extended SUSY quantum mechanics, intertwining operators and coherent states", Phys. Lett. **A 372**, 6226–6231 (2008).

[102] D. Fairlie, "The formulation of quantum mechanics in terms of phase space functions", Proc. Camb. Phil. Soc. **60**, 581 (1964).

[103] M. A. Jafarizadeh, H. Fakhri, "Supersymmetry and shape invariance in differential equations of mathematical physics", Phys. Lett. **A 230**, 164–170 (1997).

[104] H. Fakhri, "Relations between 1D shape invariant potentials and the commutation relations of the Lie algebra sl(2,c)", Phys. Lett. **A 308**, 120–130 (2003).

[105] H. Fakhri, A. Chenaghlou, "Shape invariant Natanzon potential hierarchy and its twin: A novel approach", Phys. Lett. **A 337**, 374–383 (2005).

[106] M. Faux and D. Spector, "Duality and central charges in supersymmetric quantum mechanics", Phys. Rev. **D 70**, 085014 (2004).

[107] D. J. Fernández C, M. L. Glasser, L. M. Nieto, "New isospectral oscillator potentials", Phys. Lett. **A 240**, 15–20 (1998).

[108] D. J. Fernández C, J. Negro, L. M. Nieto, "Second-order supersymmetric periodic potentials", Phys. Lett. **A 275**, 338–349 (2000).

[109] D. J. Fernández C., A. Ganguly, "New supersymmetric partners for the associated Lamé potentials", Phys. Lett. **A 338**, 203–208 (2005).

[110] D. J. Fernández C., E. Salinas-Hernández, "Wronskian formula for confluent second-order supersymmetric quantum mechanics", Phys. Lett. **A 338**, 13–18 (2005).

[111] D. J. Fernández and V. S. Morales-Salgado, "Supersymmetric partners of the harmonic oscillator with an infinite potential barrier", Jour. Phys A: Math. and Gen. **47**, 035304 (2014). doi:10.1088/1751-8113/47/3/035304

[112] E. D. Filho, R. M. Ricotta, "Morse potential energy spectra through the variational method and supersymmetry", Phys. Lett. **A 269**, 269–276 (2000)

[113] E. D. Filho and R. M. Ricotta, "The Hierarchy of Hamiltonians for a Restricted Class of Natanzon Potentials", Braz. J. Phys. 31:334–339,2001, arXiv:hep-th/9904038

[114] E. D. Filho, R. M. Ricotta, "Supersymmetric variational energies of 3d confined potentials", Phys. Lett. **A 320**, 95–102 (2003).

[115] M. Flato, A. Lichnerowicz, and A. Sternheimer, "Deformations of Poisson brackets, Dirac brackets and applications", J. Math. Phys. **17**, 1754, (1976).

[116] V. A. Fock, Zeits. f. Phys. **49**, 339 (1928).

[117] Dirac references Fock, Zeits. f. Phys. **49** 339 (1928).

[118] W. M. Frank, D. J. Land and R. M. Spector, "Singular potentials", Rev. Mod. Phys. **43**, 36 (1971).

[119] D. Z. Freedman, P. F. Mende, "An exactly solvable N-particle system in supersymmetric quantum mechanics", Nucl. Phys **B 344**, 317–343 (1990).

[120] S. H. Fricke, A. B. Balantekin, P. J. Hatchell, and T. Uzer, "Uniform semiclassical approximation to supersymmetric quantum mechanics", Phys. Rev. **A 37**, 2797 (1988).

[121] C. Fronsdal, "Some Ideas About Quantization", Rep. Math. Phys. **15**, 111 (1978).

[122] T. Fukui, "Shape-invariant potentials for systems with multi-component wave functions", Phys. Lett. **A 178**, 1–6 (1993).

[123] T. Fukui, N. Aizawa, "Shape-invariant potentials and an associated coherent state", Phys. Lett. **A 180**, 308–313 (1993).

[124] K. Funahashi, T. Kashiwa, S. Sakoda, K. Fujii, "Coherent states, Path integral, and Semiclassical approximation", J. Math. Phys. **36**, 3232–3253, (1995).

[125] K. Funahashi, T. Kashiwa, S. Sakoda, K. Fujii, "Exactness in the Wentzel-Kramers-Brillouin approximation for some homogeneous spaces", J. Math. Phys. **36**, 4590–4611 (1995).

[126] K. Fujii and K. Funahashi, "Multi-periodic coherent states and the WKB exactness", J. Math. Phys. **37**, 5987 (1996).

[127] A. Gangopadhyaya, P. K. Panigrahi, and U. P. Sukhatme, "Supersymmetry and tunneling in an asymmetric double well", Phys. Rev. **A 47**, 2720 (1993).

[128] R. Dutt, A. Gangopadhyaya, A. Khare, A. Pagnamenta, U. Sukhatme, "Solvable quantum mechanical examples of broken supersymmetry", Phys. Lett. **A 174**, 363–367 (1993).

[129] A. Gangopadhyaya, P. K. Panigrahi and U. P. Sukhatme, "Analysis of inverse-square potentials using supersymmetric quantum mechanics", Jour. Phys. A-Math. & Gen. **27**, 4295 (1994).

[130] A. Gangopadhyaya, P. Panigrahi and U. Sukhatme, "Inter-Relations of Solvable Potentials", Helv. Phys. Acta **67**, 363 (1994).

[131] A. Gangopadhyaya, A. Khare, U. P. Sukhatme, "Methods for generating quasi-exactly solvable potentials", Phys. Lett. **A 208**, 261–268 (1995).

[132] A. Gangopadhyaya and U. P. Sukhatme, "Potentials with two shifted sets of equally spaced eigenvalues and their Calogero spectrum", Phys. Lett. **A 224**, 5 (1996).

[133] A. Gangopadhyaya, J. Mallow and U. P. Sukhatme, "Translational shape invariance and the inherent potential algebra", Phys. Rev. **A 58**, 4287 (1998).

[134] A. Gangopadhyaya, J. V. Mallow and U. P. Sukhatme, "Broken supersymmetric shape invariant systems and their potential algebras", Phys. Lett. **A 283**, 279 (2001).

[135] A. Gangopadhyaya and J. V. Mallow, "Generating shape invariant potentials", Int. Jour. of Mod. Phys. **A 23**, 4959 (2008).

[136] L. E. Gendenshtein, "Derivation of exact spectra of the Schrödinger equation by means of supersymmetry", Pismah. Eksp. Teor. Fiz. **38**, (1983) 299–302. [English: JETP Lett. **38**, 356–359 (1983)].

[137] L. E. Gendenshtein and I. V. Krive, "Supersymmetry in quantum mechanics", Usp. Fiz. Nauk **146**, 553 (1985); [English: Sov. Phys. Usp. **28**, 645–666 (1985).]

[138] P. K. Ghosh, "Supersymmetric many-particle quantum systems with inverse-square interactions", Jour. Phys. A: Math. and Theor. **45**, 183001 (2012). doi:10.1088/1751-8113/45/18/183001

[139] B. Gönül, O. Özer, Y. Cançelik, M. Koçak, "Hamiltonian hierarchy and the Hulthén potential", Phys. Lett. **A 275**, 238–243.

[140] M. de Gosson, *Symplectic Geometry and Quantum Mechanics*, Birkhauser, Basel, (2006).

[141] M. de Gosson and F. Luef, "A New Approach to the *-Genvalue Equation", Lett. Math. Phys. **85**, 173 (2008).

[142] E. Gozzi, "Ground-state wave-function 'representation'", Phys. Lett. **B 129**, 432 (1983).

[143] E. Gozzi, "Classical and quantum adiabatic invariants", Phys. Lett. **B 165**, 351 (1985).

[144] E. Gozzi, M. Reuter, W. D. Thacker, "Variational methods via supersymmetric techniques", Phys. Lett. **A 183**, 29–32 (1993).

[145] E. Gozzi and M. Reuter, "A proposal for a differential calculus in quantum mechanics", Int. J. Mod. Phys. **A 9**, 2191 (1994).

[146] Ya. I. Granovskii, "Sommerfeld formula and Dirac's theory," UFN, **174**, 577 (2004).

[147] D. J. Griffiths, Quantum Mechanics, 2nd ed., CA: Benjamin-Cummings (2004).

[148] H. J. Groenewold, "On the Principles of Elementary Quantum Mechanics", Physica **12**, 405 (1946).

[149] T. Hakobyan and A. Nersessian, "Runge-Lenz vector in the Calogero-Coulomb problem", Phys. Rev. **A 92**, 022111 (2015).

[150] T. Hakobyan, A. Nersessian, and H. Shmavonyan, "Symmetries in superintegrable deformations of oscillator and Coulomb systems: Holomorphic factorization", Phys. Rev. **D 95**, 025014 (2017).

[151] J. Hancock, M. A. Walton, and B. Wynder, "Quantum mechanics another way", Eur. J. Phys. **25**, 525–534 (2004).

[152] Y. He, Z. Cao, Q. Shen, "Bound-state spectra for supersymmetric quantum mechanics", Phys. Lett. **A 326**, 315–321 (2004).

[153] P. Henselder, "Deformed geometric algebra and supersymmetric quantum mechanics", Phys. Lett. **A 363**, 378–380 (2007).

[154] M. Hillery, R. F. O'Connell, M. O. Scully, and E. P. Wigner, "Distribution Functions in Physics: Fundamentals", Phys. Rep. **106**, 121 (1983).

[155] A. C. Hirshfeld and P. Henselder, "Deformation quantization in the teaching of quantum mechanics", Am. J. Phys. **70**, 537 (2002).

[156] J. W. van Holten, "Propagators and path integrals", Nucl. Phys. **B 457**, 375–407 (1995).

[157] S.-T. Hong, J. Lee, T. H. Lee, and P. Oh, "Supersymmetric monopole quantum mechanics on a sphere", Phys. Rev. **D 72**, 015002 (2005).

[158] L. van Hove, "Sur certaines représentations unitaires", Proc. R. Acad. Sci. Belgium, **26**, 1 (1951).

[159] M. Hruska, W.-Y. Keung, and U. Sukhatme, "Accuracy of semiclassical methods for shape-invariant potentials, Phys. Rev. **A 55**, 3345 (1997).

[160] R. L. Hudson, "When is the Wigner quasi-probability density non-negative?" Rep. Math. Phys. **6**, 249 (1974).

[161] R. J. Hughes, V. A. Kostelecky, and M. M. Nieto, "Supersymmetric quantum mechanics in a first-order Dirac equation", Phys. Rev. **D 34**, 1100 (1986).

[162] R. J. Hughes, V. A. Kostelecký, and M. M. Nieto, "Supersymmetric quantum mechanics in a first-order Dirac equation", Phys. Rev. **D 34**, 1100 (1986).

[163] S. Hyun, Y. Kiem, H. Shin, "Supersymmetric completion of supersymmetric quantum mechanics", Nucl. Phys. **B 558**, 349–370 (1999).

[164] F. Iachello, "Algebraic models of hadronic structure", Nucl. Phys. **A 497**, 23–42 (1989).

[165] G. J. Iafrate, H. L. Grubin, and D. K. Ferry, "The Wigner Distribution Function", Phys. Lett. **A 87**, 145 (1982).

[166] L. Infeld and T. E. Hull, "The factorization method", Rev. Mod. Phys., **23**, 21–68 (1951).

[167] A. Inomata and G. Junker, "Quasiclassical path-integral approach to supersymmetric quantum mechanics", Phys. Rev. **A 50**, 3638 (1994).

[168] M. V. Ioffe, J. Negro, L. M. Nieto and D. N. Nishnianidze, "New two-dimensional integrable quantum models from SUSY intertwining", Jour. Phys A: Math. and Gen. **39**, 9297 (2006). doi:10.1088/0305-4470/39/29/020

[169] M. V. Ioffe and D. N. Nishnianidze, "Exact solvability of a two-dimensional real singular Morse potential", Phys. Rev. **A 76**, 052114 (2007).

[170] M. V. Ioffe, J. Mateos Guilarte, P. A. Valinevich, "A class of partially solvable two-dimensional quantum models with periodic potentials", Nucl. Phys. **B 790**, 414–431 (2008).

[171] M. V. Ioffe, E. V. Kolevatova, D. N. Nishnianidze, "SUSY method for the

three-dimensional Schrödinger equation with effective mass", Phys. Lett. **A 380**, 3349–3354 (2016).

[172] I. A. Ivanov, "WKB quantization of the Morse hamiltonian and periodic meromorphic functions", Jour. Phys. A-Math. & Gen. **30**, 3997 (1997).

[173] A. Alonso Izquierdo, M. A. Gonzalez Leon, M. de la Torre Mayado and J. Mateos Guilarte, "On two-dimensional superpotentials: from classical Hamilton-Jacobi theory to 2D supersymmetric quantum mechanics", Jour. Phys. A: Math. and Gen. **37** 10323 (2004). doi:10.1088/0305-4470/37/43/020

[174] T. Jana, P. Roy, "Shape invariance approach to exact solutions of the Klein-Gordon equation", Phys. Lett. **A 361**, 55–58 (2007).

[175] A. Jevicki and J. Rodrigues, "Singular potentials and supersymmetry breaking", Phys. Lett. **B 146**, 55 (1984).

[176] G. Junker, *Supersymmetric Methods in Quantum and Statistical Physics*, Springer Verlag, Berlin, (1996).

[177] G. Junker, P. Roy, "Conditionally exactly solvable problems and non-linear algebras", Phys. Lett. **A 232**, 155–161 (1997).

[178] D. Kabat, G. Lifschytz, "Approximations for strongly-coupled supersymmetric quantum mechanics", Nucl. Phys. **B 571**, 419–456 (2000).

[179] N. Kan, K. Kobayashi and K. Shiraishi, "Simple models in supersymmetric quantum mechanics on a graph", Jour. Phys A: Math. and Gen. **46**, 365401 (2013). doi:10.1088/1751-8113/46/36/365401

[180] A. Khare, J. Maharana, "Supersymmetric quantum mechanics in one, two and three dimensions", Nucl. Phys. **B 244**, 409–420 (1984).

[181] A. Khare, "How good is the supersymmetry-inspired WKB quantization condition?", Phys. Lett. B **161**, 131–135 (1985).

[182] A. Khare and U. P. Sukhatme, "Scattering amplitudes for supersymmetric shape-invariant potentials by operator methods", Jour. Phys. A-Math. & Gen. **21**, L501 (1988).

[183] A. Khare and U. Sukhatme, "Phase equivalent potentials obtained from supersymmetry", Jour. Phys. A-Math. & Gen. **22**, 2847 (1989).

[184] H. P. Laba and V. M. Tkachuk, "Quantum-mechanical analogy and supersymmetry of electromagnetic wave modes in planar waveguides", Phys. Rev. **A 89**, 033826 (2014).

[185] L. D. Landau and E. M. Lifshitz, *Quantum Mechanics*, NY: Pergamon Press (1977).

[186] R. E. Langer, "On the connection formulas and solutions of the wave equation", Phys. Rev. **51**, 669 (1937).

[187] R. A. Leacock and M. J. Padgett, "Hamilton-Jacobi/action-angle quantum mechanics", Phys. Rev. **D 28**, 2491 (1983).

[188] N. N. Lebedev, *Special Functions and Their Applications*, Dover, NY, (1972).

[189] H.-W. Lee and M. O. Scully, "Wigner phase-space description of a Morse oscillator", J. Chem. Phys. **77**, 4604 (1982).

[190] C. J. Lee, "Supersymmetry of a relativistic electron in a uniform magnetic field", Phys. Rev. **A 50**, 2053 (1994).

[191] C. Lee, "Equivalence of logarithmic perturbation theory and expansion of the superpotential in supersymmetric quantum mechanics", Phys. Lett. A **267**, 101–108 (2000).

[192] G. Lévai, P. Roy, "Conditionally exactly solvable potentials and supersymmetric transformations", Phys. Lett. A **264**, 117–123 (1999).

[193] L. Liao and Y.-C. Huang, "Path integral quantization corresponding to Faddeev-Jackiw canonical quantization', Phys. Rev. D **75**, 025025 (2007).

[194] Q. K. K. Liu, "Investigation of the partner potentials from supersymmetric quantum mechanics by bremsstrahlung", Nucl. Phys. A **550**, 263–280 (1992).

[195] K. Mahdi, Y. Kasri, Y. Grandati, and A. Bérard, "SWKB and proper quantization conditions for translationally shape invariant potentials", arXiv:1602.03295

[196] J. Mañez, B. Zumino, "WKB method, SUSY quantum mechanics and the index theorem", Nucl. Phys. B **270**, 651–686 (1986).

[197] A. Messiah, *Quantum Mechanics*, v.1, Paris: Dunod Publ. (1964).

[198] B. Midya, "Quasi-Hermitian Hamiltonians associated with exceptional orthogonal polynomials", Phys. Lett. A **376**, 2851–2854 (2012).

[199] W. Miller, Jr. *Lie Theory and Special Functions (Mathematics in Science and Engineering)*; Academic Press: New York, NY, USA, 1968.

[200] J. E. Moyal, "Quantum Mechanics as a Statistical Theory", Proc. Camb. Phil. Soc. **45**, 99 (1949).

[201] Y. Murayama, "Supersymmetric Wentzel-Kramers-Brilluoin (SWKB) method", Phys. Lett. A **136**, 455–457 (1989).

[202] A. Nanayakkara and T. Mathanaranjan, "Equivalent Hermitian Hamiltonians for some non-Hermitian Hamiltonians", Phys. Rev. A **86**, 022106 (2012).

[203] F. J. Narcowich and R. F. O'Connell, "Necessary and sufficient conditions for a phase-space function to be a Wigner distribution", Phys. Rev. A **34**, 1 (1986).

[204] F. J. Narcowich, "Conditions for the convolution of two Wigner distributions to be itself a Wigner distribution", J. Math. Phys. **29**, 2036 (1988).

[205] G. A. Natanzon, "Study of the one-dimensional Schroedinger equation generated from the hypergeometric equation", Leningrad University Publication [Vestnik Leningradskogo Universiteta], **10**, 22 (1971.); [English Translation by Haret Rosu is available at arXiv:physics/9907032v1].

[206] J. von Neumann, "Die Eindeutigkeit der Schrödingerschen Operatoren", Math. Ann. **104**, 570 (1931).

[207] Y. Nogami and F. M. Toyama, "Supersymmetry aspects of the Dirac equation in one dimension with a Lorentz scalar potential", Phys. Rev. A **47**, 1708 (1993).

[208] S. Odake and R. Sasaki, "Infinitely many shape invariant discrete quantum mechanical systems and new exceptional orthogonal polynomials related to the Wilson and Askey-Wilson polynomials", Phys. Lett. B **682**, 130–136 (2009).

[209] S. Odake and R. Sasaki, "Another set of infinitely many exceptional (X_ℓ) laguerre polynomials", Phys. Lett. B **684**, 173–176 (2010).

[210] S. Odake and R. Sasaki, "Exactly Solvable Quantum Mechanics and Infinite Families of Multi-indexed Orthogonal Polynomials", Phys. Lett. **B 702**, 164–170 (2011).

[211] S. Odake and R. Sasaki, "Extensions of solvable potentials with finitely many discrete eigenstates", J. Phys. **A 46**, 235205 (2013).

[212] V. K. Oikonomou, "F-theory Yukawa couplings and supersymmetric quantum mechanics", Nucl. Phys. **B 856**, 1–25 (2012).

[213] V. K. Oikonomou, "A relation between Z3-graded symmetry and shape invariant supersymmetric systems", Jour. Phys A: Math. and Gen. **47**, 435304 (2014). doi:10.1088/1751-8113/47/43/435304

[214] Y. Ou, Z. Cao, Q. Shen, "Exact energy eigenvalues for spherically symmetrical three-dimensional potential", Phys. Lett. **A 318**, 36–39 (2003).

[215] S. Paban, S. Sethi, and M. Stern, "Constraints from extended supersymmetry in quantum mechanics", Nucl. Phys. **B 534**, 137–154 (1998).

[216] A. Pagnamenta and U. Sukhatme, "Non-divergent semiclassical wave functions in supersymmetric quantum mechanics", Phys. Lett. **A 151**, 7–11 (1990).

[217] P. K. Panigrahi and U. P. Sukhatme, "Singular superpotentials in supersymmetric quantum mechanics", Phys. Lett. **A 178**, 251–257 (1993).

[218] R. C. Paschoal, J. A. Helayël-Neto, and L. P. G. de Assis, "Planar supersymmetric quantum mechanics of a charged particle in an external electromagnetic field", Phys. Lett. **A 349**, 67–74 (2006).

[219] C. Pedder, J. Sonner, and D. Tong, "Geometric phase in supersymmetric quantum mechanics", Phys. Rev. **D 77**, 025009 (2008).

[220] J. S. Petrović, V. Milanović, and Z. Ikonić, "Bound states in continuum of complex potentials generated by supersymmetric quantum mechanics", Phys. Lett. **A 300**, 595–602 (2002).

[221] B. Piette and L. Vinet, "Spectrum-generating algebras for the supersymmetric Morse and Pöschl-Teller Hamiltonians", Phys. Lett. **A 125**, 380–384 (1987).

[222] A. R. Plastino, A. Rigo, M. Casas, F. Garcias, and A. Plastino, "Supersymmetric approach to quantum systems with position-dependent effective mass", Phys. Rev. **A 60**, 4318 (1999).

[223] M. S. Plyushchay and L.-M. Nieto, "Self-isospectrality, mirror symmetry, and exotic nonlinear supersymmetry", Phys. Rev. **D 82**, 065022 (2010).

[224] M. Porrati and A. Rozenberg, "Bound states at threshold in super symmetric quantum mechanics", Nucl. Phys. **B 515**, 184–202 (1998).

[225] S. Post, L. Vinet and A. Zhedanov, "Supersymmetric quantum mechanics with reflections", Jour. Phys A: Math. and Theor. **44**, 435301 (2011). doi:10.1088/1751-8113/44/43/435301

[226] A. M. Pupasov, B. F. Samsonov, and J.-M. Sparenberg, "Exactly solvable coupled-channel potential models of atom-atom magnetic Feshbach resonances from supersymmetric quantum mechanics", Phys. Rev. **A 77**, 012724 (2008).

[227] D. L. Pursey, "Isometric operators, isospectral Hamiltonians, and supersymmetric quantum mechanics", Phys. Rev. **D 33**, 2267 (1986).

[228] C. Quesne, "Exceptional orthogonal polynomials, exactly solvable potentials and supersymmetry", J. Phys. **A 41**, 392001:1–392001:6 (2008).

[229] C. Quesne, "Solvable rational potentials and exceptional orthogonal polynomials in supersymmetric quantum mechanics", Sigma **5**, 084:1–084:24 (2009).

[230] C. Quesne, "Novel Enlarged Shape Invariance Property and Exactly Solvable Rational Extensions of the Rosen-Morse II and Eckart Potentials", SIGMA **8**, 080 (2012). arXiv:1208.6165

[231] C. Quesne, "Revisiting (quasi-)exactly solvable rational extensions of the Morse potential", Int. J. Mod. Phys. **A 27**, 1250073 (2012), arXiv:1203.1812

[232] K. Raghunathan, M. Seetharaman and S. S. Vasan, "On The Exactness of the SUSY Semiclassical Quantization Rule", Phys. Lett. **B 188**, 351–352 (1987).

[233] A. Ramos, "On the new translational shape-invariant potentials", Jour. Phys A: Math. and Gen. **44**, 342001 (2011).

[234] S. Sree Ranjani, P. K. Panigrahi, A. K. Kapoor, A. Khare, and A. Gangopadhyaya, "Exceptional orthogonal polynomials, QHJ formalism and SWKB quantization condition", Jour. of Phys. **A 45**, 055210 (2012).

[235] S. Sree Ranjani, R. Sandhya, and A. K. Kapoor, "Shape Invariant Rational Extensions and Potentials Related to Exceptional Polynomials", arXiv:1503.01394.

[236] C. Rasinariu, J. Dykla, A. Gangopadhyaya, and J. V. Mallow, "Exactly solvable systems and the Quantum Hamilton-Jacobi formalism", Phys. Lett. **A 338**, 197–202 (2005).

[237] C. Rasinariu, J. V. Mallow and A. Gangopadhyaya, "Exactly solvable problems of quantum mechanics and their spectrum generating algebras: A review", Central Euro. Jour. of Phys. **5**, 111 (2007).

[238] C. Rasinariu, "Shape invariance in phase space," Fortschritte der Physik **61**, 4–19 (2013).

[239] D. Ridikas, J. S. Vaagen, and J. M. Bang, "Phase equivalent potentials for one-neutron halo systems", Nucl. Phys. **A 609**, 21–37 (1996).

[240] M. Rocek, "Representation theory of the nonlinear SU (2) algebra", Phys. Lett. **B 255** (1991) 554–557.

[241] R. de Lima Rodrigues and X. Sigaud, "The quantum mechanics SUSY algebra: an introductory review", Monograph CBPF-MO-03-01 (December/2001).

[242] E. S. Rodrigues, A. F. de Lima, and R. de Lima Rodrigues; "Dirac Equation with vector and scalar potentials via supersymmetry in quantum mechanics", arXiv:1301.6148.

[243] H. Rosu and J. R. Guzman, "Gegenbauer polynomials and supersymmetric quantum mechanics", Nuovo Cimento della Societa Italiana di Fisica **B 112**, 941 (1997).

[244] H. Rosu and J. R. Guzman, "Gegenbauer Polynomials and Supersymmetric Quantum Mechanics", Nuovo Cimento della Societa Italiana di Fisica **B 112**, 941–942 (1997).

276   *Supersymmetric Quantum Mechanics: An Introduction*

[245] P. Roy and R. Roychoudhury, "Question of degenerate states in supersymmetric quantum mechanics", Phys. Rev. **D 32**, 1597 (1985).

[246] P. Roy, B. Roy, and R. Roychoudhury, "Partial algebraization of spectral problems and supersymmetry", Phys. Lett. **A 139**, 427–430 (1989).

[247] B. F. Samsonov, V. V. Shamshutdinova, A. V. Osipov, "Equivalent Hermitian operator from supersymmetric quantum mechanics", Phys. Lett. **A 374**, 1962–1965 (2010).

[248] E. Schrödinger, "Further studies on solving eigenvalue problems", Proc. Roy, Irish Acad. **A 46**, 183–206 (1941).

[249] E. Schrödinger, "A Method of Determining Quantum-Mechanical Eigenvalues a Eigenfunctions", Proc. R. Irish Acad. **A 46**, 9 (1940).

[250] A. Schulze-Halberg, J. M. Rivas, J. J. P. Gil, J. García-Ravelo, and P. Roy, "Exact solutions of the Fokker-Planck equation from an nth order supersymmetric quantum mechanics approach", Phys. Lett. **A 373**, 1610–1615 (2009).

[251] M. A. Shifman, A. V. Smilga, and A. I. Vainshtein, "On the Hilbert space of supersymmetric quantum systems", Nucl. Phys. **B 299**, 79–90 (1988).

[252] T. Sil, A. Mukherjee, R. Dutt, and Y. P. Varshni, "Supersymmetric WKB approach to tunneling through a one-dimensional barrier", Phys. Lett. **A 184**, 209–214 (1994).

[253] A. N. Sissakian, V. M. Ter-Antonyan, G. S. Pogosyan, and I. V. Lutsenko, "Supersymmetry of a one-dimensional hydrogen atom", Phys. Lett. **A 143**, 247–249 (1990).

[254] P. Solomonson and J. W. Van Holten, "Fermionic coordinates and supersymmetry in quantum mechanics", Nucl. Phys. **B 196**, 509–531 (1982).

[255] V. A. Soroka, "Supersymmetry and the odd Poisson bracket", Nucl. Phys. **B 101**, Proceedings Supplements, 26–42 (2001).

[256] V. P. Spiridonov, "Exactly solvable potentials and quantum algebras", Phys. Rev. Lett. **69**, 298 (1992).

[257] A. A. Stahlhofen, "Completely transparent potentials for the Schrödinger equation", Phys. Rev. **A 51**, 934 (1995).

[258] M. Stone, "Supersymmetry and the quantum mechanics of spin", Nucl. Phys. **B 314**, 557–586 (1989).

[259] M. Stone, "Supersymmetry and the quantum mechanics of spin", Nucl. Phys. **B 314**, 557–586 (1989).

[260] A. Sukumar, "Supersymmetry and the Dirac equation for a central Coulomb field", Jour. Phys. A-Math. & Gen. **18**, L697 (1985).

[261] U. Sukhatme, C. Rasinariu, and A. Khare, "Cyclic shape invariant potentials," Phys. Lett. **A 234**, 401 (1997).

[262] T. Tanaka, "N-fold supersymmetry and quasi-solvability associated with X-2-laguerre polynomials", J. Math. Phys. **51**, 032101:1–032101:20 (2010).

[263] R. D. Tangerman and J. A. Tjon, "Exact supersymmetry in the nonrelativistic hydrogen atom", Phys. Rev. **A 48**, 1089 (1993).

[264] V. I. Tatarskii, "The Wigner representation of quantum mechanics", Sov. Phys. Usp. **26**, 311 (1983).

[265] J. Tosiek and M. Przanowski, "Weyl-Wigner-Moyal Formalism. I. Operator Ordering", Acta Phys. Pol. **26**, 1703 (1995).

[266] T. Uchino and I. Tsutsui, "Supersymmetric quantum mechanics with a point singularity", Nucl. Phys **B 662**, 447–460 (2003).

[267] Y. P. Varshni, "Relative convergences of the WKB and SWKB approximations", Jour. of Phys. A: Math and General, **25**, 5761 (1992).

[268] P. Di Vecchia and F. Ravndal, "Supersymmetric Dirac particles", Phys. Lett. **A 73**, 371–373 (1979).

[269] J. Vey, "Déformation du crochet de poisson sur une variété symplectique", Comment. Math. Helv. **50**, 421 (1975).

[270] H. Weyl, "Quantenmechanik und Gruppentheorie", Z. Phys. **46**, 1 (1927).

[271] E. P. Wigner, "On the Quantum Correction For Thermodynamic Equilibrium", Phys. Rev. **40**, 749 (1932).

[272] E. P. Wigner, *Perspectives in Quantum Theory*, Dover, NY, (1979).

[273] M. de Wilde and P. Lecomte, "Existence of star-products and of formal deformations of the Poisson Lie algebra of arbitrary symplectic manifolds", Lett. Math. Phys. **7**, 487 (1983).

[274] B. W. Williams, "Exact solutions of a Schrödinger equation based on the Lambert function", Phys. Lett. **A 334**, 117–122 (2005).

[275] E. Witten, "Dynamical breaking of supersymmetry", Nucl. Phys. **B 185**, 513–554 (1981).

[276] E. Witten, "Constraints on supersymmetry breaking", Nucl. Phys. **B 202**, 253 (1982).

[277] J. Wu, Y. Alhassid, and F. Gürsey: "Group theory approach to scattering. IV. Solvable potentials associated with SO(2,2)", Ann. Phys. **196**, 163 (1989).

[278] J. Wu and Y. Alhassid, "The Potential Group Approach and Hypergeometric Differential Equations", Phys. Rev. **A 31** (1990) 557.

[279] R. K. Yadav, A. Khare, and B. P. Mandal, "The scattering amplitude for rationally extended shape invariant Eckart potentials", Phys. Lett. **A 379**, 67–70 (2015).

[280] C. Yin, Z. Cao, and Q. Shen, "Why SWKB approximation is exact for all SIPs", Ann. of Phys. **325**, 528 (2010).

[281] C. K. Zachos, D. B. Fairlie, and T. L. Curtright, *Quantum Mechanics in Phase Space*, World Scientific, (2005).

[282] G.-D. Zhang, J.-Y. Liu, L.-H. Zhang, W. Zhou, and C.-S. Jia, "Modified Rosen-Morse potential-energy model for diatomic molecules", Phys. Rev. **A 86**, 062510 (2012).

Index

Printed in the United States
By Bookmasters